The Receptors

Volume 31

Series Editor

Giuseppe di Giovanni
Department of Physiology & Biochemistry
Faculty of Medicine and Surgery
University of Malta
Malta, Italy

The Receptors book Series, founded in the 1980's, is a broad-based and well-respected series on all aspects of receptor neurophysiology. The series presents published volumes that comprehensively review neural receptors for a specific hormone or neurotransmitter by invited leading specialists. Particular attention is paid to in-depth studies of receptors' role in health and neuropathological processes. Recent volumes in the series cover chemical, physical, modeling, biological, pharmacological, anatomical aspects and drug discovery regarding different receptors. All books in this series have, with a rigorous editing, a strong reference value and provide essential up-to-date resources for neuroscience researchers, lecturers, students and pharmaceutical research.

More information about this series at http://www.springer.com/series/7668

Richard Teke Ngomba • Giuseppe Di Giovanni
Giuseppe Battaglia • Ferdinando Nicoletti
Editors

mGLU Receptors

Humana Press

Editors
Richard Teke Ngomba
Neuropharmacology Unit I.N.M. Neuromed
I.N.M. Neuromed
Pozzilli, Isernia, Italy

Giuseppe Battaglia
Neuropharmacology Unit
I.N.M. Neuromed
Pozzilli, Isernia, Italy

Giuseppe Di Giovanni
Department of Physiology and Biochemistry
University of Malta
Malta, Italy

Ferdinando Nicoletti
Dipartimento di Fisiologia Umana
Università di Roma
Roma, Italy

The Receptors
ISBN 978-3-319-85842-5 ISBN 978-3-319-56170-7 (eBook)
DOI 10.1007/978-3-319-56170-7

Softcover reprint of the hardcover 1st edition 2017

Printed on acid-free paper

This Humana Press imprint is published by Springer Nature
The registered company is Springer International Publishing AG
The registered company address is: Gewerbestrasse 11, 6330 Cham, Switzerland

Preface

Metabotropic glutamate (mGlu) receptors have been discovered in the mid-1980s by a French group of scientists (Sladeczek et al., Nature, 1985), who include some of the current leaders in the field. Since then, the field has grown exponentially, and now subtype-selective ligands of mGlu receptors [orthosteric agonists and antagonists, positive and negative allosteric modulators (PAMs and NAMs), and agonists/PAMs] are under development for the treatment of neurological and psychiatric disorders. The present book is the follow-up of the 8th International Meeting on Metabotropic Glutamate Receptors (Taormina, Italy, 2014) and incorporates chapters from some of the authorities in the mGlu receptor field.

The chapter by Philippe Rondard, Xavier Rovira, Cyril Goudet, and Jean-Philippe Pin is the state of the art of mechanisms regulating the structural and functional dynamics of mGlu receptors and their relevance to mGlu receptor pharmacology. This group of scientists has highly contributed to our current knowledge of physical interactions (homo- and heterodimerization) and allosteric modulation of mGlu receptors. Recent findings obtained by the authors and their collaborators lay the groundwork for the development of light-regulated ligands of mGlu receptors (i.e., drugs that can be either activated or inactivated by light). These molecules represent new valuable tools for the study of the role played by individual mGlu receptor subtypes in physiology and pathology with a high spatial and temporal resolution. Some of these drugs have recently appeared in the literature and hold promise for the treatment of pain and anxiety.

The chapter by Hardy Hagena and Denise Manahan-Vaughan is an excellent synopsis of the role played by mGlu receptors in mechanisms of hippocampal synaptic plasticity underlying information processing and long-term memory. This is a theme of great relevance from a therapeutic standpoint considering that some mGlu receptor ligands (e.g., mGlu5 receptor PAMs and mGlu2 receptor NAMs) are under development as cognition enhancers. In vivo studies on synaptic plasticity performed in Denise's lab are milestones in the mGlu receptor field.

The chapter by Zhengping Jia and Graham Collingridge focuses on mechanisms underlying mGlu receptor-dependent long-term depression (LTD) of excitatory synaptic transmission, a particular form of activity-dependent synaptic plasticity that has attracted the interest of scientists working on fragile X and other forms of monogenic autism. Graham is an absolute authority in the field of synaptic plasticity. The authors discuss the role played by the GluA2 subunit of AMPA receptors in mGlu receptor-dependent LTD proposing a molecular model that links functional and structural plasticity through molecular events mediating actin remodeling.

Three chapters by (i) Paolo Gubellini, Yoland Smith, and Marianne Amalric; (ii) Gunasingh Masilamoni and Yoland Smith; and (iii) Nicolas Morin and Therese di Paolo focus on the role played by mGlu5 receptors in the pathophysiology of Parkinson's disease (PD) and L-DOPA-induced dyskinesias (LIDs). These chapters highlight the importance of translation research in the mGlu field describing how data obtained in preclinical models (e.g., 6-hydroxydopamine-treated rats and 1-methyl-4-phenyl-1,2,3,6-tetrahydropyridine-treated mice and monkeys) laid the groundwork for clinical studies with mGlu5 receptor NAMs in patients with PD and LIDs. Of note, mGlu5 receptor NAMs not only produce symptomatic benefit in PD and LIDs but also exert neuroprotective effects in "parkinsonian" mice and monkeys, suggesting that these drugs cater the potential to behave as disease modifiers.

The chapter by Javier Gonzalez-Maeso is a nice synopsis of the epigenetic and functional mechanisms regulating the cross talk between mGlu2 and 5-HT_{2A} receptors. This mechanism, which has been described in detail in some seminal papers by Javier and his collaborators, is of great relevance to the pathophysiology and treatment of schizophrenia.

The chapter by Francesco Ferraguti focuses on mGlu receptors in the amygdala, a complex brain structure that plays a key role in fear memory and anxiety. Francesco is one of the best neuroanatomists and pharmacologists in Europe, and he is highly contributing to our current knowledge of the complex neuronal circuits linking the input and output nuclei of the amygdaloid complex.

The chapter by Tom Salt and Carolina Copeland examines the role played by mGlu receptors in the regulation of synaptic transmission in the thalamus. Tom Salt's lab is pioneer in the study of thalamic function in response to sensory inputs.

The chapter by Gilles van Luijtelaar, Valerio D'Amore, Ines Santolini, and Richard Ngomba focuses on mGlu5 receptors as a new candidate drug target for the treatment of absence epilepsy. Absence seizures are characterized by spike-and-wave discharges at the EEG, which are generated by an abnormal oscillatory activity within a cortico-thalamic-cortical circuit. A significant percentage of patients with absence epilepsy is refractory to current medication. mGlu5 receptor PAMs hold promise as new drugs for the treatment of absence epilepsy and may act in the thalamus by restraining GABAergic transmission.

The chapter by Francesca Guida, Enza Palazzo, L. Longo, Ida Marabese, Vito de Novellis, and Sabatino Maione examines the role played by mGlu receptors in the pain pathways focusing on supraspinal mechanisms. Supraspinal mechanisms are involved in the top-down regulation of pain transmission and mediate the affective and cognitive aspects of pain, being a linking bridge between chronic pain and affective disorders. Dino Maione's group is leader in the study of mGlu receptors and pain regulation.

The chapter by Andrew Lawrence and Christina Perry focuses on mGlu receptors as candidate drug targets for the treatment of drug addiction. Several lines of evidence indicate that mGlu5 receptor NAMs inhibit both drug taking and drug seeking. Here, the authors comment on mGlu5 receptors and drug addiction from a different angle. Moving from the evidence that mGlu5 receptors contribute to mechanisms of activity-dependent synaptic plasticity, Andrew and Christina suggest that mGlu5 receptor PAMs may serve as useful add-on treatment to behavioral therapy in addiction.

Finally, the chapter by Suzy Chen is a nice synopsis of what we currently know about mGlu receptors and cancer and focuses on the link between mGlu1 receptors and the pathophysiology of malignant melanomas. Melanoma is one of the most aggressive tumors originating from melanocytes, which are cells present in the skin, uvea, and leptomeninges and originate from the neural crest. Although the current use of BRAF and MEK inhibitors and immunotherapies has extended the progression-free survival and overall survival of patients, the treatment of metastatic melanomas is still suboptimal. Suzy Chen and her collaborators have demonstrated that ectopic expression of mGlu1 receptors in melanocytes is sufficient to generate melanomas in mice and that human melanoma samples and melanoma cell lines express mGlu1 receptors. Riluzole, a drug that lowers the concentrations of ambient glutamate, limits the growth of melanomas and shows radio-sensitizing activity in the treatment of brain metastasis of melanoma. This paves the way to the clinical use of drugs that restrain the activation of mGlu1 receptors as adjunctive treatment in patients with melanoma.

In conclusion, this is an excellent book that is easy to read and critically reviews some of the most relevant aspects related to the physiology and pharmacology of mGlu receptors.

Isernia, Italy — Richard Teke Ngomba, Pharm.D., Ph.D.
Msida, Malta — Giuseppe Di Giovanni, Ph.D.
Isernia, Italy — Giuseppe Battaglia, M.D., Ph.D.
Roma, Italy — Ferdinando Nicoletti, M.D., Ph.D.

Contents

Chapter 1
mGlu5 Signaling: A Target for Addiction Therapeutics?

Christina J. Perry, M. Foster Olive, and Andrew J. Lawrence

Abstract The role of metabotropic glutamate 5 (mGlu5) receptors in substance abuse disorders has been a focus of research for over a decade. In animal models, mGlu5 antagonists not only decrease drug taking but also drug seeking. It follows that mGlu5 antagonists are promising potential pharmacotherapeutic agents for the treatment of substance abuse. More recently, however, evidence has emerged that such compounds may in fact interfere with cognitive behavioral strategies for treatment of such disorders. mGlu5 receptors are linked to N-methyl-D-aspartate (NMDA) receptors via scaffold proteins and consequently are critical for NMDA receptor-dependent neural plasticity, giving them a prominent role in learning and memory. This is important because these processes are critical for rehabilitation treatment during recovery from substance abuse disorders. Therefore, although an antagonist or negative allosteric modulator (NAM) for mGlu5 may serve to decrease the reinforcing value of drugs such as cocaine or methamphetamine, it may also interfere with the process of behavioral change during treatment. Conversely, mGlu5 stimulation may actually serve to enhance this process. This chapter will follow the line of evidence supporting the idea that compounds that enhance mGlu5 receptor function may serve as useful adjuncts to behavioral therapy for substance abuse. We will also discuss the effects of chronic drug use on mGlu5 expression and function. We propose that mGlu5 PAMs in fact show promise as short-term adjuncts to behavioral therapy and could improve the long-term prognosis of such strategies.

Keywords Metabotropic glutamate 5 • Addiction • Cognition • Learning • Memory • Reinforcement • Treatment • Negative allosteric modulator • Positive allosteric modulator

C.J. Perry, Ph.D. (✉) • A.J. Lawrence, Ph.D., FBPhS
Behavioural Neuroscience Division, Florey Institute of Neuroscience and Mental Health, 30 Royal Parade, Parkville, VIC 3052, Australia
e-mail: Christina.perry@florey.edu.au; Andrew.lawrence@florey.edu.au

M.F. Olive, Ph.D.
Behavioral Neuroscience Area, Department of Psychology, Arizona State University, Tempe, AZ, USA
e-mail: foster.olive@asu.edu

R.T. Ngomba et al. (eds.), *mGLU Receptors*, The Receptors 31,
DOI 10.1007/978-3-319-56170-7_1

Abbreviations

CDPPB	3-cyano-N-(1,3-diphenyl-1H-pyrazol-5-yl)benzamide
CET	cue exposure therapy
LTD	long-term depression
LTP	long-term potentiation
mGlu5	metabotropic glutamate 5
MPEP	2-Methyl-6-(phenylethynyl)pyridine
MSN	medium spiny neuron
MTEP	3-((2-Methyl-4-thiazolyl)ethynyl)pyridine
NAM	negative allosteric modulator
NMDA	N-Methyl-D-aspartate
PAM	positive allosteric modulator

1.1 Introduction: Pharmacotherapy and Allosteric Modulators

In recent years there has been increasing interest in the potential for allosteric modulators of the metabotropic glutamate 5 (mGlu5) receptor to be used in the treatment of a number of different disorders. Targeting metabotropic rather than ionotropic receptors is regarded as a safer strategy because it avoids unwanted side effects that arise from suppression of fast glutamatergic excitatory transmission (Carroll 2008). Likewise, allosteric modulators have increased receptor selectivity and fewer contraindications than agonists or antagonists that bind to the orthosteric binding site (Carroll 2008; Nickols and Conn 2014). Such favorable characteristics result from these compounds having no intrinsic agonist or antagonist ability but rather an influence over receptor activity only when the endogenous agonist itself is present (Gregory et al. 2011). This, however, is not to say that they are free of side effects. mGlu5 negative allosteric modulators (NAM) at high doses can induce psychotomimetic effects and cognitive impairments in animal models (Campbell et al. 2004) and in clinical trials (Pecknold et al. 1982; Friedmann et al. 1980), while positive allosteric modulators (PAMs) at high doses can induce seizures (Nickols and Conn 2014). Nevertheless, such improvements in drug development have allowed for mGlu5 NAMs to be tested at the clinical trial phase and beyond for a number of different disorders, including anxiety, depression, Parkinson's disease, and Fragile X syndrome (Gregory et al. 2011), while PAMs might be useful for treating schizophrenia (Nickols and Conn 2014; Gregory et al. 2011).

Substance abuse is an area of mental health where there is a pressing need for novel and effective medication (Kim and Lawrence 2014; Douaihy et al. 2013), and increasingly it seems that mGlu5 allosteric modulators are promising candidates (Gregory et al. 2011; Olive 2010; Bird and Lawrence 2009). However, there is conflicting evidence as to how best to apply these treatments to achieve long-term resistance to relapse. On the one hand, mGlu5 NAMs may reduce the reinforcing properties of drugs of abuse (Watterson et al. 2013; Keck et al. 2013; Kumaresan et al. 2009) and, as a consequence, the severity of an acute relapse episode. On the other, an mGlu5 PAM has the potential to act as a cognitive aid (Olive 2010; Homayoun and Moghaddam 2010), facilitating behavioral therapy and creating longer-term cognitive resistance to relapse. The purpose of this chapter is to review the evidence for both of these possibilities and evaluate the most effective strategy (in theory) for treating substance abuse using mGlu5 allosteric modulators.

1.2 Negative Allosteric Modulators Reduce Drug Reward

It has been reported that mice lacking the mGlu5 receptor show decreased sensitivity to the rewarding and locomotor-stimulating properties of cocaine and morphine (Chiamulera et al. 2001; Veeneman et al. 2011), although these findings have not been replicated with regard to cocaine (Bird et al. 2014). Nevertheless, NAMs reduce behavioral sensitization to cocaine (McGeehan and Olive 2003; Scheggi et al. 2007; Brown et al. 2012; Martinez-Rivera et al. 2013), alcohol (Kotlinska et al. 2006), and amphetamine (McGeehan et al. 2004). mGlu5 NAMs increase the threshold for intracranial self-stimulation (ICSS), which indicates a decrease in brain reward function (Kenny et al. 2005; Cleva et al. 2012). Likewise, mGlu5 NAMs reduce perseveration for drug reinforcement in a progressive ratio test (Paterson and Markou 2005). Together, these findings suggest a role for mGlu5 receptors in the reinforcing and motivational properties of addictive drugs.

A medication that can reduce the motivation to seek drugs is certainly a desirable candidate for substance abuse treatment for drug addiction. Indeed, in preclinical models of relapse, mGlu5 NAMs reduce drug-primed reinstatement of drug seeking for cocaine (Kumaresan et al. 2009; Wang et al. 2013; Schmidt et al. 2014), ethanol (Backstrom et al. 2004), or methamphetamine (Watterson et al. 2013). In addition, reinstatement triggered by drug-associated cues (Wang et al. 2013; Backstrom and Hyytia 2006; Martin-Fardon et al. 2009) or contexts (Knackstedt et al. 2014) is also reduced following administration of an mGlu5 NAM. These findings are promising, because they indicate that NAMs might reduce the severity of an acute relapse episode, and also decrease craving and motivation to seek drug in recovering addicts.

There are, however, some important impediments to application of mGlu5 NAMs for treating substance abuse. First, there is evidence that tolerance to these compounds can develop quite quickly (Cleva et al. 2012). This is critical, because effective medication with a NAM would involve a long-term prescription to avoid potential negative ramifications of exposure to drugs or drug-associated cues and future time points undefined. Therefore, it is quite possible that the protection afforded by the medication would lessen over time. Second, mGlu5 receptor expression changes following increased drug exposure. Given that much of the data discussed above derives from preclinical models where there is only limited exposure to drug, it may be that they are not directly applicable to the case of substance abuse clients, who by definition have had extended exposure to the drug in question. Indeed, we will describe how changes to mGlu5 receptor availability have been documented in cocaine and nicotine addicts undergoing withdrawal (Martinez et al. 2014; Milella et al. 2014; Hulka et al. 2014). Finally, mGlu5 NAMs have negative impact on cognitive function (Campbell et al. 2004), in particular to memory formation (Naie and Manahan-Vaughan 2004; Rodrigues et al. 2002; Simonyi et al. 2010; Homayoun et al. 2004), spatial learning (Manahan-Vaughan and Braunewell 2005; Christoffersen et al. 2008; Petersen et al. 2002), and inhibitory learning (Kim et al. 2014; Xu et al. 2009). This last point is a critical consideration because there is ample evidence that cognitive capacity is positively correlated with treatment outcome for substance abuse disorders (e.g. Aharonovich et al. 2006). Decreased mGlu5 function is associated with extinction deficits (Bird et al. 2014; Kim et al. 2014; Chesworth et al. 2013), and extinction forms an integral part of many behavioral strategies targeting addiction (Conklin and Tiffany 2002). Therefore there is a very good chance that, although mGlu5 NAMs may offer acute benefits in the form of reducing the reinforcing properties of addictive drugs, they may simultaneously create long-term cognitive impairments that interfere with behavioral therapy, hence ultimately increasing the likelihood of future relapse. These issues will be addressed in the following sections.

1.3 mGlu5 Receptors and Synaptic Plasticity

mGlu5 receptors are widely distributed throughout the mammalian brain (Shigemoto and Mizuno 2000), with highest levels of expression in forebrain regions such as the cerebral cortex, dorsal and ventral striatum, olfactory bulb and tubercle, lateral septum, and hippocampus (Shigemoto et al. 1993; Romano et al. 1995). Within these regions, mGlu5 receptors are predominantly expressed on postsynaptic elements, particularly the perisynaptic annulus of dendritic spines (Shigemoto and Mizuno 2000), although some investigators have reported localization of mGlu5 receptors to presynaptic terminals and glia (Mitrano and Smith 2007; Anwyl 1999).

mGlu5 receptors are integral to both the initiation and maintenance of synaptic plasticity in the form of long-term potentiation (LTP) or depression (LTD) of synaptic efficacy. This is achieved via a variety of subcellular signaling mechanisms including facilitation of N-methyl-D-aspartate (NMDA) receptor function, interactions with postsynaptic scaffolding proteins, release of calcium into the cytosol from the endoplasmic reticulum, and resulting activation of downstream effector proteins such as MAP and ERK kinases which in turn activate transcription factors to modulate gene expression and initiation of phospholipid signaling (Anwyl 2009; Bellone et al. 2008; Gladding et al. 2009). Perhaps the most widely studied and understood mechanism for mGlu5-mediated enhancement of LTP is via positive coupling of these receptors to postsynaptic NMDA receptor function (Alagarsamy et al. 2001; Attucci et al. 2001; Awad et al. 2000; Benquet et al. 2002; Doherty et al. 1997; Kotecha and MacDonald 2003; Pisani et al. 2001; Ugolini et al. 1999), which is largely mediated by mGlu5-induced activation of PKC and phosphorylation of (among other substrates) specific subunits of the NMDA (Lu et al. 1999). In addition to these biochemical interactions, there is also evidence for interactions between mGlu5 and NMDA receptors that are either direct or indirect via various scaffolding proteins including PSD-95, Shank, and the Homer family of proteins (Perroy et al. 2008; Hermans and Challiss 2001). mGlu5 receptors also regulate the synthesis of retrograde signaling molecules such as endocannabinoids, which stimulate presynaptic CB_1 receptors to initiate LTD (Bellone et al. 2008).

1.4 mGlu5 Receptor Expression Changes Across Development of Addiction

Many of the animal models used to investigate the role of mGlu5 in drug seeking and drug taking involve animals with only limited drug exposure. Models such as the extinction-reinstatement paradigm are useful for studying relapse because they encompass well the types of situations that trigger relapse in human addiction (Bossert et al. 2013). They do not, however, reflect the types of compulsive behaviors that are characteristic of long-term substance abuse. In fact, it is frequently not viable to study true addiction using animal models, because only a small percentage of those animals that are exposed to drugs will go onto develop addiction as judged by diagnostic criteria (Piazza and Deroche-Gamonet 2013), meaning that the number of experimental animals to model such behaviors would be greatly inflated.

Studying the behavior of nondependent animals provides important insights into mechanisms that lead to the development of addiction and addictive-like behavior (Bossert et al. 2013). However, it is important to remember that neural chemistry changes with increasing drug exposure, and this should be factored in when evaluating potential medications for use with substance abuse. We described previously how preclinical models have shown that mGlu5 NAMs reduce the reinforcing properties of drugs and drug-associated cues. It is also clear that chronic drug exposure causes glutamate receptor redistribution and that this affects the responsiveness of drug-seeking behavior to compounds that modulate mGlu5 receptor activity (McCutcheon et al. 2011).

These differences are apparent in models of escalated drug use, such as the long-access paradigm where laboratory animals are provided with extended opportunity to self-administer drug each day (Ahmed 2012). For example, the mGlu5 NAM MTEP decreased motivation to seek cocaine only after short but not after escalated self-administration (Hao et al. 2010). Furthermore, it was shown that an mGlu2/3 receptor agonist was more effective than MTEP in reducing ethanol seeking in dependent rats when compared with nondependent rats (Sidhpura et al. 2010). This suggests that while mGlu5 receptors mediate the incentive properties of addictive drugs during early use, following extended and escalated drug use, drug seeking may become less dependent on this receptor. In other words, treatment with an mGlu5 NAM may be ineffective against the type of behavior that will likely be in play in addicts.

Withdrawal from substance abuse also influences mGlu5 expression. In the prefrontal cortex (PFC), mGlu5 expression was reduced in rats following escalated cocaine self-administration followed by withdrawal from the drug. This difference appeared after rats were exposed to cocaine-associated cues and was correlated with the intensification of cue-induced drug seeking that occurs in cocaine-experienced animals following a period of abstinence (Ben-Shahar et al. 2013). After extended withdrawal from escalated cocaine self-administration, there is also a switch from mGlu5- to mGlu1-regulated synaptic rectification in accumbal medium spiny neurons (MSNs) (McCutcheon et al. 2011).

Changes in mGlu5 sensitivity after withdrawal are also apparent in human drug users. For example, fMRI revealed a decrease in mGlu5 availability in key cortical and limbic regions in abstinent cocaine users (Martinez et al. 2014; Milella et al. 2014), and in abstinent smokers (Hulka et al. 2014), which magnified with increasing length of abstinence (Milella et al. 2014). Thus, it seems that, although mGlu5 NAMs reduce drug seeking and relapse in animal models where there is only limited drug exposure, it is likely that they may be less efficient following escalated drug taking and withdrawal, a pattern that is typical in addicts seeking treatment. This would greatly decrease their value as medications.

1.5 Cognitive Capacity Is Positively Correlated with Treatment Outcome for Substance Abuse Disorders

Current behavioral treatments for addiction show poor prognosis, with roughly 60% relapsing in the first year after treatment (Conklin and Tiffany 2002; McLellan et al. 2000), and there is a strong correlation between cognitive function and treatment outcome (Aharonovich et al. 2006; Fox et al. 2009; Turner et al. 2009). This relationship most likely derives from the fact that there is an important cognitive component to treatment. Standard behavioral treatment for drug addiction consists of single or group psychotherapy (Douaihy et al. 2013). This involves assimilation of new information; hence, cognitive capacity is an important determinant of treatment outcome. In addition, many behavioral treatments involve learning new responses to drug-associated cues. This may take the form of extinction learning, such as in cue exposure therapy, where repeated presentation of previously drug-associated stimuli in the absence of further drug reward leads to a decrease in cue reactivity (i.e. craving), in abstinent substance abuse clients (Conklin and Tiffany 2002). Another, more novel approach is to provide reinforcement for abstinence behavior (Higgins et al. 2012; Silverman et al. 2012). Although these approaches are quite distinct in the way they seek to change behavior, all share the common feature of integrating new learning to implement more adaptive behavioral patterns.

Given that decline in cognitive function is in fact a well-documented consequence of long-term exposure to drugs of abuse (Fox et al. 2009; Crews and Boettiger 2009), overcoming the cognitive dysfunction that arises as a direct consequence of addiction should be a priority in therapeutic strategies to help prevent relapse. In the next section, we will describe how mGlu5 NAMs are problematic in this regard, because they actually have the tendency to produce cognitive deficits (Campbell et al. 2004). On the other hand, mGlu5 PAMs such as CDPPB have been shown to reverse cognitive deficits produced by phencyclidine (PCP) (Horio et al. 2013) and after adolescent alcohol exposure (Gass et al. 2014).

1.6 Amnestic Effects of mGlu5 Inactivation

Given the established role of mGlu5 in synaptic plasticity under normal physiological conditions, it is not surprising that genetic or systemic pharmacological blockade of mGlu5 receptors attenuates the expression of LTP and LTD in regions such as the hippocampus (dentate gyrus and CA1 region), amygdala, and dorsal and ventral striatum, with concomitant deficits in various forms of learning and memory (Homayoun and Moghaddam 2010; Naie and Manahan-Vaughan 2004; Rodrigues et al. 2002; Homayoun et al. 2004; Manahan-Vaughan and Braunewell 2005; Christoffersen et al. 2008; Xu et al. 2009; Alagarsamy et al. 2001; Attucci et al. 2001; Awad et al. 2000; Benquet et al. 2002; Doherty et al. 1997; Kotecha and MacDonald 2003; Pisani et al. 2001; Ugolini et al. 1999; Bikbaev et al. 2008; Lu et al. 1997). Studies utilizing site-specific microinjections of mGlu5 NAMs including 2-methyl-6-(phenylethynyl)-pyridine (MPEP) and 3-[2-methyl-1,3-thiazol-4-yl)ethynyl]pyridine (MTEP) have confirmed a critical role for these receptors in spatial navigation, inhibitory avoidance, and instrumental and/or habit learning (Rodrigues et al. 2002; Simonyi et al. 2010; Packard et al. 2001; Jacob et al. 2009). Deficits in acquisition, but not expression of a spatial learning task, were also observed following targeted knockdown of mGlu5 receptors in the mouse dorsal hippocampus (Tan et al. 2015). However, it should be noted that other laboratories have not found spatial learning memory impairments following administration of mGlu5 antagonists (Petersen et al. 2002; Semenova and Markou 2007), and some studies have indicated that mGlu5 antagonists selectively impair reference but not working memory (Naie and Manahan-Vaughan 2004; Gravius et al. 2008).

1.7 Facilitatory Effects of mGlu5 Activation on Learning, Memory, and Extinction Processes

In the context of behavior modification, extinction can be defined as the targeted reduction of specific maladaptive responses or behaviors, including pathological fear, anxiety, or avoidance, as well as excessive seeking of rewarding or reinforcing stimuli such as drugs of abuse. Extinction is a form of new and active inhibitory learning (Bouton 2000) and thus engages the neural mechanisms responsible for synaptic plasticity and learning and memory, including mGlu5 receptors. This is supported by findings that genetic deletion or pharmacological inhibition of mGlu5 receptors results in decrements in the extinction of conditioned fear (Handford et al. 2014) as well as cocaine and methamphetamine seeking (Kim et al. 2014; Chesworth et al. 2013).

In agreement with this, it has recently been established that PAMs selective for mGlu5, such as 3-cyano-N-(1,3-diphenyl-1H-pyrazol-5-yl)benzamide (CDPPB), 4-nitro-N-(1,3-diphenyl-1H-pyrazol-5-yl)benzamide (VU-29), and (S)-(4-fluoro-phenyl)-(3-[3-(4-fluorophenyl)-[1,2,4]-oxadiazol-5-yl]piperidin-1-yl)methanone (ADX47273) facilitate the induction of various markers of synaptic plasticity including long-term synaptic potentiation (Kroker et al. 2011; Ayala et al. 2009) and increased phosphorylation of the NR1 and NR2B subunits of the NMDA receptor, the GluR1 subunit of the AMPA receptor, and activate α-calmodulin-dependent kinase II, CREB, and ERK (Uslaner et al. 2009; Liu et al. 2008). Many of these effects can be blocked by NMDA antagonists, confirming that increased NMDA receptor functioning is necessary for mGlu5 PAM-induced synaptic potentiation (Ayala et al. 2009). However, interestingly, it has recently been reported that newer mGlu5 PAMs with biased agonist properties can produce pro-cognitive effects that are independent of indirect NMDA receptor activation (Rook et al. 2015).

Consistent with the notion of mGlu5 PAMs as enhancers of synaptic plasticity, these ligands have been shown to improve spatial memory in normal animals (Ayala et al. 2009; Balschun et al. 2006; Fowler et al. 2013) and reverse pharmacologically induced impairments in object recognition (Uslaner et al. 2009; Reichel et al. 2011), conditioned avoidance (Schlumberger et al. 2010; Spear et al. 2011), reversal learning (Xu et al. 2013; LaCrosse et al. 2015; Darrah et al. 2008), and five-choice serial reaction time tests (Liu et al. 2008). Many of these pro-cognitive effects of mGlu5 PAMs appear to be mediated by increased prefrontal cortical functioning (Homayoun and Moghaddam 2006, 2010; Gass et al. 2014; Stefani and Moghaddam 2010).

In the context of drug addiction, mGlu5 PAMs facilitate the extinction of a cocaine-induced conditioned place preference (Gass and Olive 2009) and reduce extinction responding following intravenous cocaine and methamphetamine and oral alcohol self-administration (Gass et al. 2014; Cleva et al. 2011; Kufahl et al. 2012). However, CDPPB failed to facilitate extinction of methamphetamine seeking when extinction sessions were conducted in a context different to the methamphetamine self-administration context (Widholm et al. 2011). Importantly, while one study reported that mGlu5 PAMs produce excitotoxicity at high doses (Parmentier-Batteur et al. 2014), others have shown that repeated mGlu5 PAM administration at more moderate doses does not produce evidence of neurotoxicity (Gass and Olive 2009) and is devoid of effects on brain reward function (Cleva et al. 2012). Thus, mGlu5 receptor PAMs may be a novel class of compounds by which to facilitate the extinction of drug stimuli or drug-context associations.

1.8 Conclusion

A drug that reduces the severity of a relapse episode presents an enticing therapeutic potential for addiction treatment. This is made all the more so in the case of an mGlu5 NAM because these drugs are already in clinical trials or approved for use with other disorders. However, mGlu5 receptors are integral to both the initiation and maintenance of synaptic plasticity (Alagarsamy et al. 2001; Attucci et al. 2001;

Awad et al. 2000; Benquet et al. 2002; Doherty et al. 1997; Kotecha and MacDonald 2003; Pisani et al. 2001; Ugolini et al. 1999), and this relationship affords these receptors an important role in cognitive capacity, learning, and memory. They consequently play a critical role in assimilating information and updating memories following new experiences and changes to the environment (Qi et al. 2013). Behavioral therapy for substance abuse is focused around learning new and more adaptive responses to replace drug-seeking behaviors. Whether this involves behavioral extinction or some other type of training, the need to assimilate new information is a central tenet to this process. In this regard, mGlu5 NAMs represent a "double-edged sword" – simultaneously reducing drug-seeking responses acutely while interfering with cognitive load involved in behavioral therapy. On the other hand, the pro-cognitive properties of PAMs facilitate change in animal models of behavioral therapy. Given that tolerance to the reward-reducing properties of a NAM can develop following repeat administration, and further that they are less effective in animals with more extensive drug experience, we propose that short-term application of a PAM in conjunction with behavioral therapy should in theory at least be a more effective way of treating substance abuse in the long term than a medication that has acute effects on the reinforcing effects of drug (i.e., NAM).

Acknowledgments AJL is a Principal Research Fellow of the National Health and Medical Research Council (NHMRC) of Australia, and CJP is supported by NHMRC project grant number 1022201. These authors acknowledge support from the Victorian Government's Operational Infrastructure Support Program. MFO wishes to acknowledge the support of the US National Institute of Health grants DA024355, DA025606, DA037741, and AA013852.

The authors have no other relevant affiliations or financial involvement with any organization or entity with a financial interest in or financial conflict with the subject matter or materials discussed in the manuscript apart from those disclosed.

References

Aharonovich E, Hasin DS, Brooks AC, Liu X, Bisaga A, Nunes EV (2006) Cognitive deficits predict low treatment retention in cocaine dependent patients. Drug Alcohol Depend 81(3):313–322

Ahmed SH (2012) The science of making drug-addicted animals. Neuroscience 211:107–125

Alagarsamy S, Sorensen SD, Conn PJ (2001) Coordinate regulation of metabotropic glutamate receptors. Curr Opin Neurobiol 11(3):357–362

Anwyl R (1999) Metabotropic glutamate receptors: electrophysiological properties and role in plasticity. Brain Res Rev 29(1):83–120

Anwyl R (2009) Metabotropic glutamate receptor-dependent long-term potentiation. Neuropharmacology 56(4):735–740

Attucci S, Carla V, Mannaioni G, Moroni F (2001) Activation of type 5 metabotropic glutamate receptors enhances NMDA responses in mice cortical wedges. Br J Pharmacol 132(4):799–806

Awad H, Hubert GW, Smith Y, Levey AI, Conn PJ (2000) Activation of metabotropic glutamate receptor 5 has direct excitatory effects and potentiates NMDA receptor currents in neurons of the subthalamic nucleus. J Neurosci 20(21):7871–7879

Ayala JE, Chen Y, Banko JL, Sheffler DJ, Williams R, Telk AN et al (2009) mGluR5 positive allosteric modulators facilitate both hippocampal LTP and LTD and enhance spatial learning. Neuropsychopharmacology 34:2057–2071

Backstrom P, Hyytia P (2006) Ionotropic and metabotropic glutamate receptor antagonism attenuates cue-induced cocaine seeking. Neuropsychopharmacology 31(4):778–786

Backstrom P, Bachteler D, Koch S, Hyytia P, Spanagel R (2004) mGluR5 antagonist MPEP reduces ethanol-seeking and relapse behavior. Neuropsychopharmacology 29(5):921–928

Balschun D, Zuschratter W, Wetzel W (2006) Allosteric enhancement of metabotropic glutamate receptor 5 function promotes spatial memory. Neuroscience 142(3):691–702

Bellone C, Luscher C, Mameli M (2008) Mechanisms of synaptic depression triggered by metabotropic glutamate receptors. Cell Mol Life Sci 65(18):2913–2923

Benquet P, Gee CE, Gerber U (2002) Two distinct signaling pathways upregulate NMDA receptor responses via two distinct metabotropic glutamate receptor subtypes. J Neurosci 22(22):9679–9686

Ben-Shahar O, Sacramento AD, Miller BW, Webb SM, Wroten MG, Silva HE et al (2013) Deficits in ventromedial prefrontal cortex group 1 metabotropic glutamate receptor function mediate resistance to extinction during protracted withdrawal from an extensive history of cocaine self-administration. J Neurosci 33(2):495–506a

Bikbaev A, Neyman S, Ngomba RT, Conn J, Nicoletti F, Manahan-Vaughan D (2008) MGluR5 mediates the interaction between late-LTP, network activity, and learning. PLoS One 3(5):e2155

Bird MK, Lawrence AJ (2009) The promiscuous mGlu5 receptor--a range of partners for therapeutic possibilities? Trends Pharmacol Sci 30(12):617–623

Bird MK, Lohmann P, West B, Brown RM, Kirchhoff J, Raymond CR et al (2014) The mGlu5 receptor regulates extinction of cocaine-driven behaviours. Drug Alcohol Depend 137:83–89

Bossert JM, Marchant NJ, Calu DJ, Shaham Y (2013) The reinstatement model of drug relapse: recent neurobiological findings, emerging research topics, and translational research. Psychopharmacology 229(3):453–476

Bouton ME (2000) A learning theory perspective on lapse, relapse, and the maintenance of behavior change. Health Psychol 19(1 Suppl):57–63

Brown RM, Duncan JR, Stagnitti MR, Ledent C, Lawrence AJ (2012) mGlu5 and adenosine A2A receptor interactions regulate the conditioned effects of cocaine. Int J Neuropsychopharmacol 15(7):995–1001

Campbell UC, Lalwani K, Hernandez L, Kinney GG, Conn PJ, Bristow LJ (2004) The mGluR5 antagonist 2-methyl-6-(phenylethynyl)-pyridine (MPEP) potentiates PCP-induced cognitive deficits in rats. Psychopharmacology 175(3):310–318

Carroll FI (2008) Antagonists at metabotropic glutamate receptor subtype 5: structure activity relationships and therapeutic potential for addiction. Ann N Y Acad Sci 1141:221–232

Chesworth R, Brown RM, Kim JH, Lawrence AJ (2013) The metabotropic glutamate 5 receptor modulates extinction and reinstatement of methamphetamine-seeking in mice. PLoS One 8(7):e68371

Chiamulera C, Epping-Jordan MP, Zocchi A, Marcon C, Cottiny C, Tacconi S et al (2001) Reinforcing and locomotor stimulant effects of cocaine are absent in mGluR5 null mutant mice. Nat Neurosci 4(9):873–874

Christoffersen GR, Simonyi A, Schachtman TR, Clausen B, Clement D, Bjerre VK et al (2008) MGlu5 antagonism impairs exploration and memory of spatial and non-spatial stimuli in rats. Behav Brain Res 191(2):235–245

Cleva RM, Hicks MP, Gass JT, Wischerath KC, Plasters ET, Widholm JJ et al (2011) mGluR5 positive allosteric modulation enhances extinction learning following cocaine self-administration. Behav Neurosci 125:10–19

Cleva RM, Watterson LR, Johnson MA, Olive MF (2012) Differential modulation of thresholds for intracranial self-stimulation by mGlu5 positive and negative allosteric modulators: implications for effects on drug self-administration. Front Pharmacol 2(93):1–7

Conklin CA, Tiffany ST (2002) Applying extinction research and theory to cue-exposure addiction treatments. Addiction 97(2):155–167

Crews FT, Boettiger CA (2009) Impulsivity, frontal lobes and risk for addiction. Pharmacol Biochem Behav 93(3):237–247

Darrah JM, Stefani MR, Moghaddam B (2008) Interaction of N-methyl-D-aspartate and group 5 metabotropic glutamate receptors on behavioral flexibility using a novel operant set-shift paradigm. Behav Pharmacol 19(3):225–234

Doherty AJ, Palmer MJ, Henley JM, Collingridge GL, Jane DE (1997) (RS)-2-chloro-5-hydroxyphenylglycine (CHPG) activates mGlu5, but not mGlu1, receptors expressed in CHO cells and potentiates NMDA responses in the hippocampus. Neuropharmacology 36(2):265–267

Douaihy AB, Kelly TM, Sullivan C (2013) Medications for substance use disorders. Social work in public health 28(3–4):264–278

Fowler SW, Walker JM, Klakotskaia D, Will MJ, Serfozo P, Simonyi A et al (2013) Effects of a metabotropic glutamate receptor 5 positive allosteric modulator, CDPPB, on spatial learning task performance in rodents. Neurobiol Learn Mem 99:25–31

Fox HC, Jackson ED, Sinha R (2009) Elevated cortisol and learning and memory deficits in cocaine dependent individuals: relationship to relapse outcomes. Psychoneuroendocrinology 34(8):1198–1207

Friedmann CTH, Davis LJ, Ciccone PE, Rubin RT (1980) Phase-Ii double-blind controlled-study of a new anxiolytic, fenobam (Mcn-3377) Vs placebo. Curr Ther Res Clin E 27(2):144–151

Gass JT, Olive MF (2009) Positive allosteric modulation of mGluR5 receptors facilitates extinction of a cocaine contextual memory. Biol Psychiatry 65:717–720

Gass JT, Trantham-Davidson H, Kassab AS, Glen WB Jr, Olive MF, Chandler LJ (2014) Enhancement of extinction learning attenuates ethanol-seeking behavior and alters plasticity in the prefrontal cortex. J Neurosci 34(22):7562–7574

Gladding CM, Fitzjohn SM, Molnar E (2009) Metabotropic glutamate receptor-mediated long-term depression: molecular mechanisms. Pharmacol Rev 61(4):395–412

Gravius A, Dekundy A, Nagel J, More L, Pietraszek M, Danysz W (2008) Investigation on tolerance development to subchronic blockade of mGluR5 in models of learning, anxiety, and levodopa-induced dyskinesia in rats. J Neural Transm 115(12):1609–1619

Gregory KJ, Dong EN, Meiler J, Conn PJ (2011) Allosteric modulation of metabotropic glutamate receptors: structural insights and therapeutic potential. Neuropharmacology 60(1):66–81

Handford CE, Tan S, Lawrence AJ, Kim JH (2014) The effect of the mGlu5 negative allosteric modulator MTEP and NMDA receptor partial agonist D-cycloserine on Pavlovian conditioned fear. Int J Neuropsychopharmacol 17(9):1521–1532

Hao Y, Martin-Fardon R, Weiss F (2010) Behavioral and functional evidence of metabotropic glutamate receptor 2/3 and metabotropic glutamate receptor 5 dysregulation in cocaine-escalated rats: factor in the transition to dependence. Biol Psychiatry 68(3):240–248

Hermans E, Challiss RA (2001) Structural, signalling and regulatory properties of the group I metabotropic glutamate receptors: prototypic family C G-protein-coupled receptors. Biochem J 359(Pt 3):465–484

Higgins ST, Silverman K, Sigmon SC, Naito NA (2012) Incentives and health: an introduction. Prev Med 55(Suppl):S2–S6

Homayoun H, Moghaddam B (2006) Bursting of prefrontal cortex neurons in awake rats is regulated by metabotropic glutamate 5 (mGlu5) receptors: rate-dependent influence and interaction with NMDA receptors. Cereb Cortex 16(1):93–105

Homayoun H, Moghaddam B (2010) Group 5 metabotropic glutamate receptors: role in modulating cortical activity and relevance to cognition. Eur J Pharmacol 639(1–3):33–39

Homayoun H, Stefani MR, Adams BW, Tamagan GD, Moghaddam B (2004) Functional interaction between NMDA and mGlu5 receptors: effects on working memory, instrumental learning, motor behaviors, and dopamine release. Neuropsychopharmacology 29(7):1259–1269

Horio M, Fujita Y, Hashimoto K (2013) Therapeutic effects of metabotropic glutamate receptor 5 positive allosteric modulator CDPPB on phencyclidine-induced cognitive deficits in mice. Fundam Clin Pharmacol 27(5):483–488

Hulka LM, Treyer V, Scheidegger M, Preller KH, Vonmoos M, Baumgartner MR et al (2014) Smoking but not cocaine use is associated with lower cerebral metabotropic glutamate receptor 5 density in humans. Mol Psychiatry 19(5):625–632

Jacob W, Gravius A, Pietraszek M, Nagel J, Belozertseva I, Shekunova E et al (2009) The anxiolytic and analgesic properties of fenobam, a potent mGlu5 receptor antagonist, in relation to the impairment of learning. Neuropharmacology 57:97–108

Keck TM, Yang HJ, Bi GH, Huang Y, Zhang HY, Srivastava R et al (2013) Fenobam sulfate inhibits cocaine-taking and cocaine-seeking behavior in rats: implications for addiction treatment in humans. Psychopharmacology 229(2):253–265

Kenny PJ, Boutrel B, Gasparini F, Koob GF, Markou A (2005) Metabotropic glutamate 5 receptor blockade may attenuate cocaine self-administration by decreasing brain reward function in rats. Psychopharmacology 179(1):247–254

Kim JH, Lawrence AJ (2014) Drugs currently in phase II clinical trials for cocaine addiction. Expert Opin Investig Drugs 23(8):1105–1122

Kim JH, Perry C, Luikinga S, Zbukvic I, Brown RM, Lawrence AJ (2014) Extinction of a cocaine-taking context that protects against drug-primed reinstatement is dependent on the metabotropic glutamate 5 receptor. Addict Biol 20(3):482–489

Knackstedt LA, Trantham-Davidson HL, Schwendt M (2014) The role of ventral and dorsal striatum mGluR5 in relapse to cocaine-seeking and extinction learning. Addict Biol 19(1):87–101

Kotecha SA, MacDonald JF (2003) Signaling molecules and receptor transduction cascades that regulate NMDA receptor-mediated synaptic transmission. Int Rev Neurobiol 54:51–106

Kotlinska J, Bochenski M, Danysz W (2006) N-methyl-D-aspartate and group I metabotropic glutamate receptors are involved in the expression of ethanol-induced sensitization in mice. Behav Pharmacol 17(1):1–8

Kroker KS, Rast G, Rosenbrock H (2011) Differential effect of the mGlu5 receptor positive allosteric modulator ADX-47273 on early and late hippocampal LTP. Neuropharmacology 61(4):707–714

Kufahl PR, Hood LE, Nemirovsky NE, Barabas P, Halstengard C, Villa A et al (2012) Positive allosteric modulation of mGluR5 accelerates extinction learning but not relearning following methamphetamine self-administration. Front Pharmacol 3:194

Kumaresan V, Yuan M, Yee J, Famous KR, Anderson SM, Schmidt HD et al (2009) Metabotropic glutamate receptor 5 (mGluR5) antagonists attenuate cocaine priming- and cue-induced reinstatement of cocaine seeking. Behav Brain Res 202(2):238–244

LaCrosse AL, Burrows BT, Angulo RM, Conrad PR, Himes SM, Mathews N et al (2015) mGluR5 positive allosteric modulation and its effects on MK-801 induced set-shifting impairments in a rat operant delayed matching/non-matching-to-sample task. Psychopharmacology 232(1):251–258

Liu F, Grauer S, Kelley C, Navarra R, Graf R, Zhang G et al (2008) ADX47273: a novel metabotropic glutamate receptor 5 selective positive allosteric modulator with preclinical antipsychotic-like and pro-cognitive activities. J Pharmacol Exp Ther 327(3):827–839

Lu YM, Jia Z, Janus C, Henderson JT, Gerlai R, Wojtowicz JM et al (1997) Mice lacking metabotropic glutamate receptor 5 show impaired learning and reduced CA1 long-term potentiation (LTP) but normal CA3 LTP. J Neurosci 17(13):5196–5205

Lu WY, Xiong ZG, Lei S, Orser BA, Dudek E, Browning MD et al (1999) G-protein-coupled receptors act via protein kinase C and Src to regulate NMDA receptors. Nat Neurosci 2(4):331–338

Manahan-Vaughan D, Braunewell KH (2005) The metabotropic glutamate receptor, mGluR5, is a key determinant of good and bad spatial learning performance and hippocampal synaptic plasticity. Cereb Cortex 15(11):1703–1713

Martinez D, Slifstein M, Nabulsi N, Grassetti A, Urban NB, Perez A et al (2014) Imaging glutamate homeostasis in cocaine addiction with the metabotropic glutamate receptor 5 positron emission tomography radiotracer [(11)C]ABP688 and magnetic resonance spectroscopy. Biol Psychiatry 75(2):165–171

Martinez-Rivera A, Rodriguez-Borrero E, Matias-Aleman M, Montalvo-Acevedo A, Guerrero-Figuereo K, Febo-Rodriguez LJ et al (2013) Metabotropic glutamate receptor 5 within nucleus accumbens shell modulates environment-elicited cocaine conditioning expression. Pharmacol Biochem Behav 110:154–160

Martin-Fardon R, Baptista MA, Dayas CV, Weiss F (2009) Dissociation of the effects of MTEP [3-[(2-methyl-1,3-thiazol-4-yl)ethynyl]piperidine] on conditioned reinstatement and reinforcement: comparison between cocaine and a conventional reinforcer. J Pharmacol Exp Ther 329(3):1084–1090

McCutcheon JE, Loweth JA, Ford KA, Marinelli M, Wolf ME, Tseng KY (2011) Group I mGluR activation reverses cocaine-induced accumulation of calcium-permeable AMPA receptors in nucleus accumbens synapses via a protein kinase C-dependent mechanism. J Neurosci 31:14536–14541

McGeehan AJ, Olive MF (2003) The mGluR5 antagonist MPEP reduces the conditioned rewarding effects of cocaine but not other drugs of abuse. Synapse 47(3):240–242

McGeehan AJ, Janak PH, Olive MF (2004) Effect of the mGluR5 antagonist 6-methyl-2-(phenylethynyl) pyridine (MPEP) on the acute locomotor stimulant properties of cocaine, D-amphetamine, and the dopamine reuptake inhibitor GBR12909 in mice. Psychopharmacology 174(2):266–273

McLellan AT, Lewis DC, O'Brien CP, Kleber HD (2000) Drug dependence, a chronic medical illness: implications for treatment, insurance, and outcomes evaluation. JAMA 284(13):1689–1695
Milella MS, Marengo L, Larcher K, Fotros A, Dagher A, Rosa-Neto P et al (2014) Limbic system mGluR5 availability in cocaine dependent subjects: a high-resolution PET [(11)C]ABP688 study. NeuroImage 98:195–202
Mitrano DA, Smith Y (2007) Comparative analysis of the subcellular and subsynaptic localization of mGluR1a and mGluR5 metabotropic glutamate receptors in the shell and core of the nucleus accumbens in rat and monkey. J Comp Neurol 500(4):788–806
Naie K, Manahan-Vaughan D (2004) Regulation by metabotropic glutamate receptor 5 of LTP in the dentate gyrus of freely moving rats: relevance for learning and memory formation. Cereb Cortex 14(2):189–198
Nickols HH, Conn PJ (2014) Development of allosteric modulators of GPCRs for treatment of CNS disorders. Neurobiol Dis 61:55–71
Olive MF (2010) Cognitive effects of group I metabotropic glutamate receptor ligands in the context of drug addiction. Eur J Pharmacol 639(1–3):47–58
Packard MG, Vecchioli SF, Schroeder JP, Gasbarri A (2001) Task-dependent role for dorsal striatum metabotropic glutamate receptors in memory. Learn Mem 8(2):96–103. Epub 2001/03/29
Parmentier-Batteur S, Hutson PH, Menzel K, Uslaner JM, Mattson BA, O'Brien JA et al (2014) Mechanism based neurotoxicity of mGlu5 positive allosteric modulators – development challenges for a promising novel antipsychotic target. Neuropharmacology 82:161–173
Paterson NE, Markou A (2005) The metabotropic glutamate receptor 5 antagonist MPEP decreased break points for nicotine, cocaine and food in rats. Psychopharmacology 179(1):255–261
Pecknold JC, McClure DJ, Appeltauer L, Wrzesinski L, Allan T (1982) Treatment of anxiety using fenobam (a nonbenzodiazepine) in a double-blind standard (diazepam) placebo-controlled study. J Clin Psychopharmacol 2(2):129–133
Perroy J, Raynaud F, Homburger V, Rousset MC, Telley L, Bockaert J et al (2008) Direct interaction enables cross-talk between ionotropic and group I metabotropic glutamate receptors. J Biol Chem 283(11):6799–6805
Petersen S, Bomme C, Baastrup C, Kemp A, Christoffersen GRJ (2002) Differential effects of mGluR1 and mGluR5 antagonism on spatial learning in rats. Pharmacol Biochem Behav 73:381–389
Piazza PV, Deroche-Gamonet V (2013) A multistep general theory of transition to addiction. Psychopharmacology 229(3):387–413
Pisani A, Gubellini P, Bonsi P, Conquet F, Picconi B, Centonze D et al (2001) Metabotropic glutamate receptor 5 mediates the potentiation of N-methyl-D-aspartate responses in medium spiny striatal neurons. Neuroscience 106(3):579–587
Qi Y, Hu NW, Rowan MJ (2013) Switching off LTP: mGlu and NMDA receptor-dependent novelty exploration-induced depotentiation in the rat hippocampus. Cereb Cortex 23(4):932–939
Reichel CM, Schwendt M, McGinty JF, Olive MF, See RE (2011) Loss of object recognition memory produced by extended access to methamphetamine self-administration is reversed by positive allosteric modulation of metabotropic glutamate receptor 5. Neuropsychopharmacology 36:782–792
Rodrigues SM, Bauer EP, Farb CR, Schafe GE, LeDoux JE (2002) The group I metabotropic glutamate receptor mGluR5 is required for fear memory formation and long-term potentiation in the lateral amygdala. J Neurosci 22(12):5219–5229
Romano C, Sesma MA, McDonald CT, O'Malley K, van den Pol AN, Olney JW (1995) Distribution of metabotropic glutamate receptor mGluR5 immunoreactivity in rat brain. J Comp Neurol 355(3):455–469
Rook JM, Xiang Z, Lv X, Ghoshal A, Dickerson JW, Bridges TM et al (2015) Biased mGlu5-positive allosteric modulators provide in vivo efficacy without potentiating mGlu5 modulation of NMDAR currents. Neuron 86(4):1029–1040
Scheggi S, Raone A, De Montis MG, Tagliamonte A, Gambarana C (2007) Behavioral expression of cocaine sensitization in rats is accompanied by a distinct pattern of modifications in the PKA/DARPP-32 signaling pathway. J Neurochem 103(3):1168–1183
Schlumberger C, Pietraszek M, Gravius A, Danysz W (2010) Effects of a positive allosteric modulator of mGluR5 ADX47273 on conditioned avoidance response and PCP-induced hyperlocomotion in the rat as models for schizophrenia. Pharmacol Biochem Behav 95(1):23–30

Schmidt HD, Kimmey BA, Arreola AC, Pierce RC (2014) Group I metabotropic glutamate receptor-mediated activation of PKC gamma in the nucleus accumbens core promotes the reinstatement of cocaine seeking. Addict Biol 20(2):285–296

Semenova S, Markou A (2007) The effects of the mGluR5 antagonist MPEP and the mGluR2/3 antagonist LY341495 on rats' performance in the 5-choice serial reaction time task. Neuropharmacology 52(3):863–872

Shigemoto R, Mizuno N (2000) Metabotropic glutamate receptors – immunocytochemical and in situ hybridization analysis. In: Ottersen OP, Storm-Mathisen J (eds) Handbook of chemical neuroanatomy: metabotropic glutamate receptors: immunocytochemical and in situ hybridization analyses. Elsevier, London, pp 63–98

Shigemoto R, Nomura S, Ohishi H, Sugihara H, Nakanishi S, Mizuno N (1993) Immunohistochemical localization of a metabotropic glutamate receptor, mGluR5, in the rat brain. Neurosci Lett 163(1):53–57

Sidhpura N, Weiss F, Martin-Fardon R (2010) Effects of the mGlu2/3 agonist LY379268 and the mGlu5 antagonist MTEP on ethanol seeking and reinforcement are differentially altered in rats with a history of ethanol dependence. Biol Psychiatry 67(9):804–811

Silverman K, DeFulio A, Sigurdsson SO (2012) Maintenance of reinforcement to address the chronic nature of drug addiction. Prev Med 55(Suppl):S46–S53

Simonyi A, Schachtman TR, Christoffersen GR (2010) Metabotropic glutamate receptor subtype 5 antagonism in learning and memory. Eur J Pharmacol 639(1–3):17–25

Spear N, Gadient RA, Wilkins DE, Do M, Smith JS, Zeller KL et al (2011) Preclinical profile of a novel metabotropic glutamate receptor 5 positive allosteric modulator. Eur J Pharmacol 659(2–3):146–154

Stefani MR, Moghaddam B (2010) Activation of type 5 metabotropic glutamate receptors attenuates deficits in cognitive flexibility induced by NMDA receptor blockade. Eur J Pharmacol 639(1–3):26–32

Tan SZ, Ganella DE, Dick AL, Duncan JR, Ong-Palsson E, Bathgate RA et al (2015) Spatial Learning Requires mGlu5 Signalling in the Dorsal Hippocampus. Neurochem Res 40(6):1303–1310

Turner TH, LaRowe S, Horner MD, Herron J, Malcolm R (2009) Measures of cognitive functioning as predictors of treatment outcome for cocaine dependence. J Subst Abus Treat 37(4):328–334

Ugolini A, Corsi M, Bordi F (1999) Potentiation of NMDA and AMPA responses by the specific mGluR5 agonist CHPG in spinal cord motoneurons. Neuropharmacology 38(10):1569–1576

Uslaner JM, Parmentier-Batteur S, Flick RB, Surles NO, Lam JS, McNaughton CH et al (2009) Dose-dependent effect of CDPPB, the mGluR5 positive allosteric modulator, on recognition memory is associated with GluR1 and CREB phosphorylation in the prefrontal cortex and hippocampus. Neuropharmacology 57:531–538

Veeneman MM, Boleij H, Broekhoven MH, Snoeren EM, Guitart Masip M, Cousijn J et al (2011) Dissociable roles of mGlu5 and dopamine receptors in the rewarding and sensitizing properties of morphine and cocaine. Psychopharmacology 214(4):863–876

Wang X, Moussawi K, Knackstedt L, Shen H, Kalivas PW (2013) Role of mGluR5 neurotransmission in reinstated cocaine-seeking. Addict Biol 18(1):40–49

Watterson LR, Kufahl PR, Nemirovsky NE, Sewalia K, Hood LE, Olive MF (2013) Attenuation of reinstatement of methamphetamine-, sucrose-, and food-seeking behavior in rats by fenobam, a metabotropic glutamate receptor 5 negative allosteric modulator. Psychopharmacology 225(1):151–159

Widholm JJ, Gass JT, Cleva RM, Olive MF (2011) The mGluR5 positive allosteric modulator CDPPB does not alter extinction or contextual renewal of methamphetamine-seeking behavior in rats. J Addict Res Ther S1:004

Xu J, Zhu Y, Contractor A, Heinemann SF (2009) mGluR5 has a critical role in inhibitory learning. J Neurosci 29(12):3676–3684

Xu J, Zhu Y, Kraniotis S, He Q, Marshall JJ, Nomura T et al (2013) Potentiating mGluR5 function with a positive allosteric modulator enhances adaptive learning. Learn Mem 20(8):438–445

Chapter 2
Supraspinal Metabotropic Glutamate Receptors: An Endogenous Substrate for Alleviating Chronic Pain and Related Affective Disorders

Francesca Guida, Enza Palazzo, Livio Luongo, Ida Marabese, Vito de Novellis, Sabatino Maione, and Francesco Rossi

Abstract Metabotropic glutamate receptors (mGluRs) are key players in modulating excitatory transmission and important regulators of synaptic plasticity. mGluRs are G-protein-coupled receptors (GPCRs) that have been subdivided into three groups (mGluR1–mGluR8) based on sequence homology, intracellular pathways, and pharmacological profile. mGluRs are widely localized all along the nociceptive neuroaxis, including brain circuits controlling pain often overlapping those controlling affective/cognitive behaviors which prove deeply altered in several neurological disorders including chronic pain.

This chapter summarizes current outcomes related to the supraspinal mGluRs in chronic pain states. Due to their wide expression within the pain descending system, a particular highlighting will be given to the pharmacological manipulation of mGluRs in PAG-RVM pathway, a key circuitry of the pain descending system. The current development of novel subtype-selective mGluR positive and negative allosteric modulators will allow a more stringent assessment of each mGluR subtype role in controlling chronic pain and pain-related affective cognitive behavior.

Keywords Metabotropic glutamate receptors • Chronic pain • Antinociceptive descending pathway • Positive and negative allosteric modulators • Pain-related affective and cognitive disorders

---

F. Guida, Ph.D. • E. Palazzo, Ph.D. • L. Luongo, Ph.D. • I. Marabese, Ph.D.
V. de Novellis, Ph.D. • S. Maione, Ph.D. (✉) • F. Rossi, Ph.D.
Department of Experimental Medicine, Section of Pharmacology L. Donatelli, Università degli Studi della Campania "Luigi Vanvitelli", 80138 Naples, Italy
e-mail: franc.guida@gmail.com; enza.palazzo@unicampania.it; livio.luongo@gmail.com; ida.marabese@unicampania.it; vito.denovellis@unicampania.it; sabatino.maione@unicampania.it; francesco.rossi@unina2.it

R.T. Ngomba et al. (eds.), *mGLU Receptors*, The Receptors 31,
DOI 10.1007/978-3-319-56170-7_2

Abbreviations

BLA	basolateral amygdala
CeA	central nucleus of the amygdala
CNS	central nervous system
CPCCOEt	7-(hydroxyimino)cyclopropa[b]chromen-1a-carboxylate ethyl ester
(S)-3,4-DCPG	(S)-3,4-dicarboxyphenylglycine
GPCRs	G-protein coupled receptors
iGluR	ionotropic glutamate receptor
IL	infra-limbic
mGluR	metabotropic glutamate receptor
MPEP	2-methyl-6-(phenylethynyl)pyridine
mPFC	medial prefrontal cortex
NAAG	N-acetylaspartylglutamate
NMDA	N-methyl-D-aspartate
NTS	nucleus tractus solitarius
PAG	periaqueductal gray
PL	pre-limbic
PLC	phospholipase C
RVM	rostral ventromedial medulla
TRPV1	transient receptor potential vanilloid 1

2.1 Introduction

As the most abundant excitatory neurotransmitter in the central nervous system (CNS), glutamate plays a pivotal role in main physiological brain functions. Glutamate exerts its effects through the activation of ligand-gated ionotropic glutamate receptors (iGluRs) and metabotropic glutamate receptors (mGluRs). iGluRs are ion channel receptors, divided into N-methyl-D-aspartate (NMDA), α-amino-3-hydroxy-5-methylisoxazole-4-propionic acid (AMPA), and kainate (KA) receptors, which mediate fast responses, associated with long-lasting modifications in synaptic transmission (Bleakman and Lodge 1998; Yamakura and Shimoji 1999), while mGluRs are G-protein-coupled receptors (GPCRs) subdivided into three groups (mGluR1–mGluR8) based on sequence homology, intracellular pathways, and pharmacological profile. Group I mGluRs, consisting of $mGluR_1$ and $mGluR_5$, are postsynaptically located and positively coupled to phospholipase C, and their activation leads to the intracellular calcium mobilization. Group II mGluRs, consisting of mGluR2 and mGluR3, and group III mGluRs, consisting of mGluR4, mGluR6, mGluR7, and mGluR8, are mainly presynaptic and coupled to the inhibition of adenylyl cyclase activity and to other signaling pathway, including the activation of mitogen-activated protein kinase cascade (Tian et al. 2010). mGluRs, by modulating ion channel activity and neurotransmitter release, play a modulatory role on

CNS synaptic excitability (Maione et al. 1998a; Cartmell and Schoepp 2000). Thus, signaling via these receptors is slower and longer-lasting driving to a fine-tuning of glutamate transmission.

Hyperexcitability of glutamatergic system is the main event occurring in central sensitization associated to chronic pain. Apart from classical symptoms, such as allodynia (pain experience following a not-painful stimulus) and hyperalgesia (increased pain perception from a painful stimulus), chronic pain presents comorbidity with affective and cognitive impairments. mGluRs are suitable pharmacological substrate for producing a fine modulation of glutamate transmission, whose hyperactivity or altered functioning is often associated with neurodegenerative, neurological, and psychiatric diseases. The effects of mGluRs stimulation in supraspinal areas of the pain pathway are summarized in Fig. 2.1. By this subject, mGluRs

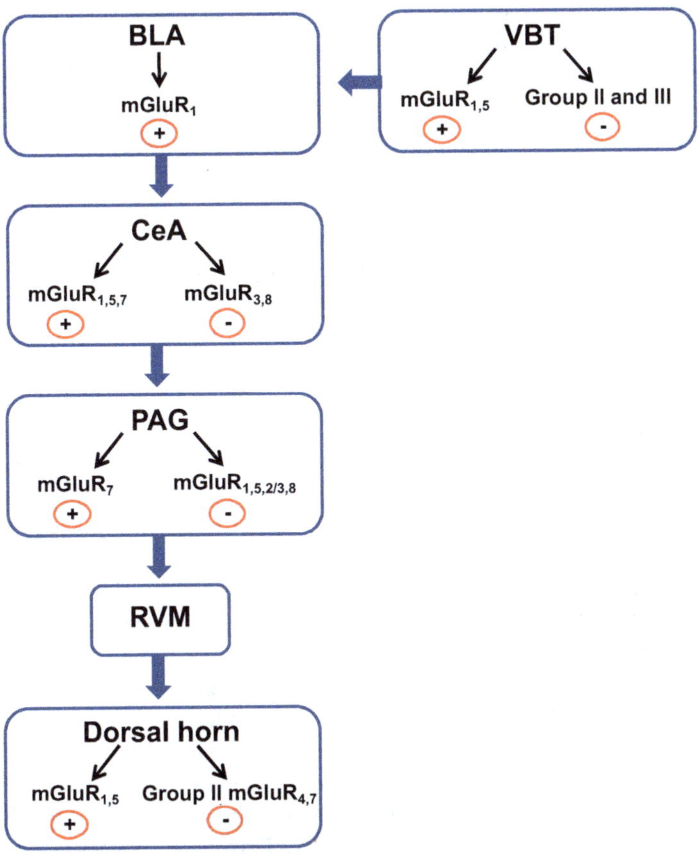

Fig. 2.1 Scheme indicating the effect of mGluRs in supraspinal areas of the pain pathway. + indicates a facilitatory effect on pain transmission, while – indicates an inhibitory effect on pain transmission. *VBT* ventrobasal thalamus

pharmacological manipulation may represent an optimal strategy to treat chronic pain, including untreatable forms such as neuropathic pain, and associated affective and cognitive impairments.

2.2 mGluRs and Pain Processing

Increasing evidence supports a crucial role of mGluRs in nociceptive transmission, given their extensive expression all along the nociceptive neuroaxis, such as the peripheral sensory terminals, dorsal root ganglia, spinal cord dorsal horn, rostral ventromedial medulla (RVM), periaqueductal gray (PAG), thalamus, amygdala, and cortex. The large variety and different synaptic distribution of mGluR subtypes determine their diverse role in pain transmission. Group I mGluRs are mainly expressed in the postsynaptic membrane, whereas group II and III mGluRs are localized presynaptically where they serve as auto- or hetero-receptors (Ohishi et al. 1995). Generally, stimulation of group I mGlu receptors facilitates pain (Hama 2003; Chiechio and Nicoletti 2012; Palazzo et al. 2014a, b). Conversely, activation of presynaptic mGluRs on glutamatergic terminals leads to a decrease in glutamate release (Attwell et al. 1995; Battaglia et al. 1997) correcting hyperactivity of glutamatergic system associated with chronic pain (Gerber et al. 2000). Within the pain descending pathway, high glutamate levels are associated with antinociception throughout a facilitation of the descending pain system functioning (Behbehani and Field 1979). In fact, the activation of group I or groups II and III mGluRs exerts antinociceptive or pronociceptive opposite effects (Marabese et al. 2005, 2007a, b) depending on the mGluR subtype signaling and its location on glutamatergic or GABAergic terminals.

Over the last decade, accumulating evidence has strengthened the evidence that dysfunction of the glutamate system is linked with the pathophysiological mechanisms responsible for chronic pain development (Neugebauer 2001). Neuropathic or inflammatory injury triggers structural and functional changes in the peripheral or central sensory circuits, resulting in altered nociceptive signal processes, such as spontaneous pain, allodynia, and hyperalgesia (Neugebauer et al. 2009; Goudet et al. 2008; Byrnes et al. 2009). Under neuropathic pain conditions, enhanced glutamate release and overactivation of glutamate receptors have been observed in cortical areas involved in pain-related responses, including the anterior cingulate (ACC), insular, and prelimbic-infralimbic (PL-IL) cortex (Giordano et al. 2012; Hung et al. 2014). Furthermore, chronic pain interferes with specific limbic brain areas affecting neuropsychological processes which are glutamate-dependent, such as cognition, memory, and decision-making. mGluRs are supraspinally expressed within limbic system thus controlling negative affective disorders associated with chronic pain (Ferraguti and Shigemoto 2006; Latremoliere and Woolf 2009).

2.3 Group I mGluRs

The difficulty to develop selective agonists or antagonists for specific mGluR subtype arises from the sequence homology of mGluRs within the ligand binding site, which presents a hydrophilic moiety which makes the compound too hydrophilic to penetrate the blood-brain barrier for brain exposure. The identification, in the last years, of novel compounds acting as PAMs or NAMs, has guaranteed selectivity and lipophilicity since allosteric binding sites proved less conserved among the other mGluRs and do not require hydrophilic molecules. Activation of group I mGluRs generally facilitates nociception (Bhave et al. 2001) and mediates the development of inflammatory hyperalgesia (Walker et al. 2000; Palazzo et al. 2004). The role of supraspinal group I mGluRs in controlling pain responses has been investigated in the ventrobasal thalamus, periaqueductal gray, rostral ventromedial medulla, and amygdala.

Thalamus Ventrobasal thalamus is a crucial relay point processing the somatosensory information which from the spinal cord reach the cerebral cortex. At this level, mGluR1, mGluR5, and NMDA receptor stimulation enhances neuronal responses to nociceptive stimuli. Conversely, NMDA receptor, mGluR1, or mGluR5 blockade reduced the neuronal responses (Salt and Binns 2000).

Amygdala and Cortex Basolateral (BLA) and central nucleus (CeA) of the amygdala are deeply involved in the emotional consequences of pain, such as pain-related anxiety and depression-like behaviors. While the electrical manipulation of CeA activity does not modify the spontaneous nociceptive behaviors (Carrasquillo and Gereau 2008; Veinate et al. 2013), chronic pain leads to increased synaptic transmission in the CeA, associated with the induction or maintenance of hypersensitivity observed in different pain models.

The activation of group I mGluRs in the amygdala is generally associated with pronociceptive effects (Neugebauer et al. 2003a, b; Kolber et al. 2010). Conversely, the blockade of mGluR1 inhibits pain stimuli-induced audible and ultrasonic vocalizations (Han and Neugebauer 2005) and decreases excitatory postsynaptic currents in neurons within the CeA in arthritic rats (Neugebauer et al. 2003a; Ren and Neugebauer 2010). It has been recently shown that group I mGluRs are involved in the plastic changes that develop in the BLA and medial prefrontal cortex (mPFC) circuitry in chronic pain states (Ji and Neugebauer 2011; Guida et al. 2015). Later on, Luongo et al. (2013) have showed that intra-BLA microinjection of (S)-3,5-dihydroxyphenylglycine (DHPG), a group I mGluR agonist, reverted neuronal phenotypical changes occurring in the mPFC, under chronic pain condition. The 7-(hydroxyimino)cyclopropa[b]chromen-1a-carboxylate ethyl ester, CPCOOEt, a selective mGluR1 antagonist, but not 2-methyl-6-(phenylethynyl)-pyridine, MPEP, a selective mGluR5 antagonist, prevented alteration in mPFC under chronic pain condition, suggesting that mGluR1, but not mGluR5, plays a role in the modulation of BLA-mPFC circuitry, which is thought to be a key substrate for affective/cognitive impairments associated with chronic pain.

Group I mGluRs may control inflammation and chronic pain state also through a functional cross talk with transient receptor potential vanilloid type 1, TRPV1 (Hu et al. 2002; Palazzo et al. 2002), a sort of a polymodal pain transducer, activated by chemical and thermal stimuli which open a cationic influx driving to action potentials. mGluR5 activation sensitizes TRPV1 through the PLC activation and protein kinase A pathway (Hu et al. 2002). Reciprocally, brain TRPV1 stimulation induces glutamate release through depolarization and intracellular Ca^{2+} mobilization (Hu et al. 2002; Kim et al. 2009). The spared nerve injury (SNI) of the sciatic nerve, a model of neuropathic pain in rodents (Decosterd and Woolf 2000), increased extracellular level of glutamate in the PL/IL portion of the mPFC. Blockade of mGluR5 by MPEP microinjection in the PL/IL worsened mechanical allodynia in the SNI mice, possibly by braking the enhancement of the antinociceptive descending pathway played by glutamate outflow, an effect which requires mGluR5 receptor stimulation (Giordano et al. 2012).

Periaqueductal Gray (PAG)/Rostral Ventromedial Medulla (RVM) The presence of mGluRs within the PAG has a strategic importance for modulating pain processing, since it is a key station in the pain descending pathway. Indeed, its electrical or chemical stimulation produces deep analgesia in rats (Reynolds 1969; Behbehani 1995). The PAG-mediated analgesia is correlated with the modulation of neuron firing activity within the nucleus raphe magnus, the adjacent reticular formation, and the nucleus gigantocellularis, which constitute the RVM (Behbehani and Fields 1979). RVM neurons, in turn, project to the spinal cord dorsal horn (Fields and Basbaum 1999), thus inhibiting the nociceptive neuronal response to noxious stimuli. Stimulation of group I mGluRs at the PAG level triggers glutamate-induced analgesia. It has been shown that selective activation of mGluR1 in the PAG prevents nocifensive behavior in formalin-induced persistent pain (Maione et al. 2000). mGluR5, at supraspinal level, contributes to hypersensitivity following peripheral neuropathy. Indeed, shown that MPEP, the selective mGluR5 antagonist, dose-dependently reduces the neuropathy-induced cold hypersensitivity, when microinjected directly into the RVM (Urban et al. 2003). Conversely, groups II and III mGluR stimulation facilitated nociceptive responses by depressing PAG-RVM activity within the antinociceptive descending pathway (Maione et al. 1998b). The mechanism through which mGluRs mediate analgesia at the PAG level remains not fully understood. Activation of group I mGluRs produces antinociception via presynaptic inhibition of GABA release (Maione et al.1998a, b, 2000), which in turn enhances pain descending system activity. Moreover, several studies have shown that activation of group I mGluR promotes the production of endocannabinoids, which in turn diffuse out of postsynaptic neurons in a retrograde way to presynaptic terminals. Endocannabinoids bind to cannabinoid type 1 receptor (CB1R) and reduce neurotransmitter release. A functional cross talk between CB1R and group I mGluRs in the PAG-RVM circuitry has also been shown in healthy (Palazzo et al. 2001) and in neuropathic pain condition induced by chronic constriction injury (CCI) of the sciatic nerve. In this study, the effects of intra-ventrolateral (VL) PAG microinjection of WIN 55,212-2, a nonselective cannabinoid receptor agonist, evaluated as behavioral analgesia and electrophysiological RVM neuronal responses, were prevented by MPEP, a selective mGluR5 antagonist (Palazzo et al. 2012), supporting a role for mGluR5 in cannabinoid-induced analgesia at PAG level in neuropathic pain conditions.

2.4 Group II mGluRs

Group II agents have shown being involved in a number of animal models of psychiatric, neurological, and neurodegenerative disorders, associated with glutamate overexcitation. Group II mGluRs are located on the preterminal regions of primary afferent nerve endings and in supraspinal pain-regulatory centers such as the thalamus, amygdala, and PAG. They negatively regulate glutamate and other neurotransmitter release and inhibit TRPV1 and sodium channel activities (Yang and Gereau 2002, 2003; Carlton et al. 2009). mGluR2 agonists produce analgesia in models of inflammatory and neuropathic pain, but their use is limited by the development of tolerance, usually occurring after 4–5 days of repeated doses (Jones et al. 2005). Their supraspinal involvement in pain processing has been investigated in the amygdala, PAG, and ventrobasal and reticular thalamus.

Amygdala In contrast to group I, group II mGluR stimulation inhibits nociceptive processing in CeA neurons. In anesthetized rats, the increased function and endogenous activation of group II mGluRs reduces the ongoing, as well as, evoked cell activity, in the knee-joint kaolin-/carrageenan-induced arthritis (Li and Neugebauer 2006). Further and more recent studies have indicated that mGluR2 is more important in controlling pain transmission than mGluR3 (Zammataro et al. 2011). In the amygdala, mGluR2 and mGluR3 stimulation inhibits GABA release at the BLA-CeA synapse leading to an increased activity of GABAergic interneurons in the CeA which in turn reduces the output of the medial central amygdala to the brain regions mediating the autonomic, motor, and endocrine features of anxiety (Linden et al. 2005, 2006). By this subject group II mGluRs in the amygdala are well positioned to inhibit both pain responses and the pain-related anxiety behavior.

Ventrobasal Thalamus and Reticular Thalamus Group II mGluRs in the ventrobasal thalamus inhibit GABAergic interneuron responses evoked by sensory stimulation in vivo (Salt and Eaton 1995; Salt and Turner 1998). Conversely, their blockade in the reticular thalamic nucleus produced antinociception in monoarthritic rats, probably by disinhibiting local GABAergic inhibitory neurons (Neto and Castro-Lopes 2000; Neto et al. 2000).

PAG-RVM Circuitry The effects of supraspinal group II mGluRs on pain responses have been recently carried out exploiting the compound ZJ43, a peptidase inhibitor degrading N-acetylaspartylglutamate (NAAG) which acts as a selective mGluR3 agonist. The microinjection of ZJ43 into the PAG and RVM reduced both phases of the nocifensive reaction induced by the peripheral injection of formalin into the hind paw. The effect of ZJ43 was antagonized by LY341495, a group II mGluR antagonist. However, ZJ43 did not affect the thermal withdrawal response in the hot plate test in healthy animals when microinjected into the same brain sites (Yamada et al. 2012). This study is consistent with previous studies showing that the intra-PAG microinjection of a group II mGluR agonist, L-CCG-I, or the intracerebroventricular administration of NAAG peptidase inhibitors reduced nocifensive behavior in the formalin test (Maione et al. 2000; Yamamoto et al. 2000) and with those which conversely show that intra-PAG L-CCG-I decreased the thermal threshold in the hot plate test, possibly by depressing the activity of the pain descending system at PAG level (Maione et al. 1998b). These studies suggest that group II mGluR stimulation within the PAG produces analgesia in chronic pain conditions, but proved ineffective (or even facilitated pain responses) in acute pain conditions (Maione et al. 1998b; Palazzo et al. 2001).

2.5 Group III mGluRs

Group III is the largest mGluR group which remains the less investigated due to the lack of selective ligands. Fortunately, selective compounds have been developed recently thus permitting a clarification of group III mGluR roles in the CNS diseases (Acher and Goudet 2015). Apart from mGluR6 which mediates responses of ON bipolar cells at retinal synapses (Vardi et al. 2000), mGluR4, mGluR7, and mGluR8 are extensively distributed throughout the CNS. Group III mGluR activation in the presynaptic active zone of transmitter release leads to glutamate and GABA release inhibition (Niswender and Conn 2010). Low release of excitatory transmitters results in the presynaptic inhibition of transmission. On the other hand, the decrease of release of GABA may indirectly have excitatory effects on neuronal excitability (disinhibition). Thus, the overall effect of group III mGluR activity may be a balance between pain facilitation and inhibition. Moreover, glutamate shows high affinity for mGluR4 and mGluR8 and very low affinity for mGluR7 (Schoepp 2001) suggesting that mGluR7 may be recruited under pathological conditions characterized by excessive glutamate transmission activity. mGluR4 proved to be highly expressed in the cerebellum where it inhibits glutamate release at the synapses between granule cell axons and Purkinje cells. mGluR4 expression has been also found in the thalamus, basal ganglia, and olfactory bulb. mGluR7 is the most widely distributed among the presynaptic mGluRs. High expression of mGluR7 has been found in locus coeruleus, hippocampus, septum, frontal cortex, amygdala, cingulate cortex, bed nucleus of the stria terminalis, PAG, nucleus accumbens, thalamus, hypothalamus, and pituitary gland (Kinzie et al. 1995; Kinoshita et al. 1998). The mGluR8 is also widely distributed in brain areas such as the amygdala, pontine gray, lateral septum, lateral reticular nucleus of the thalamus, olfactory bulb, piriform cortex, cerebral cortex, hippocampus, hypothalamus, and cerebellum (Saugstat et al. 1997; Ferraguti et al. 2006). Notwithstanding the wide distribution information regarding group III mGluR, the role and contribution to pain modulation has only just started to emerge owing to the limitations of the ligands used until recently. The majority of the drugs which have been applied until the last years are nonselective among the group III mGluR subtypes and display scarce solubility or metabolite instability. The recent progress in the selective ligand development has driven the investigations toward a better understanding of their modulatory function and their therapeutic potential in neurological and psychiatric disorders.

Thalamus The thalamus shows a relevant expression for group III mGluRs (Neto and Castro-Lopes 2000). In the corticothalamic synapses, broad-spectrum group III mGluR agonists inhibited excitatory postsynaptic potentials, showing that glutamate released at this level depresses cortical control of thalamic function (Turner and Salt 1999).

Amygdala L-AP4 is a broad-spectrum group III mGluR agonist which does not discriminate between mGluR4 and mGluR8 and shows scarce affinity for mGluR7. In the CeA, L-AP4 inhibits neuron sensitization in chronic pain conditions induced by the knee-joint injection of kaolin/carrageen which develops in arthritic pain. The inhibition of neuron plasticity by L-AP4 within the CeA involves a presynaptic mechanism. L-AP4 slightly changed CeA neuronal responses under normal conditions (Han et al. 2004). The N,N′-bis(diphenylmethyl)-1,2-ethanediamine dihydrochloride, AMN082, has been introduced as the first selective PAMs toward mGluR7 (Mitsukawa et al. 2005). In the CeA, AMN082 perfusion throughout reverse microdialysis facilitated nociception, whereas the selective activation of mGluR8 with (S)-3,4-DCPG (Thomas et al. 2001) produced the opposite effect driving to antinociception (Palazzo et al.). mGluR7 and mGluR8 in the CeA also changed pain-related affective responses conversely. AMN082 increased audible and ultrasonic vocalizations, which reflect supraspinally organized nocifensive and affective responses to aversive stimuli (Borszcz and Leaton 2003) and reduced the open-arm preference, a positive indicator of anxiolytic-like effect, in the elevated plus maze in healthy rats, whereas it was devoid of activity in the model of arthritic inflammatory pain. In contrast, (S)-3,4-DCPG had no effect in normal animals, but inhibited audible and ultrasonic vocalizations and enhanced the open-arm access in the arthritic animals. It seems like under normal conditions, mGluR7, but not mGluR8, facilitates pain and anxiety responses, whereas mGluR8 can inhibit pain response and affective behavior in chronic pain conditions such as arthritic pain (Palazzo et al.). The effect of mGluR8 activation at CeA level in chronic pain condition only has also investigated in inflammatory pain conditions induced by the intraplantar injection of carrageenan (Palazzo et al.). Intra-CeA (S)-3,4-DCPG decreased GABA and simultaneously increased serotonin and glutamate release in carrageenan-given rats, while it did not change neurotransmitter levels in control animals. Intra-CeA (S)-3,4-DCPG also increased the thermoceptive threshold and modify the activity of RVM pain-responding neurons: the ON cells which are activated and the OFF cells which are inhibited by nociceptive stimuli (Fields et al. 1983; Heinricher et al. 1989). Consistently with behavioral analgesia, intra-CeA (S)-3,4-DCPG inhibited the ON cell activity and enhanced the OFF cell activity in rats treated with carrageenan. (S)-3,4-DCPG was instead ineffective on pain responses and ON and OFF cell activity in control rats. Thus, it seems like a pain-related plasticity in the CeA is required for mGluR8 stimulation being effective. An increase in mGluR8 expression on vesicular GABA transporter (vGAT)-positive profiles has been found in the CAA of carrageenan-given rats thus providing a functional target for inhibiting pain through the disinhibition of the descending facilitatory system (Maione et al. 2000; Berrino et al. 2001).

PAG/RVM Being PAG a crucial area within the antinociceptive descending pathway, whose stimulation induces analgesia (Reynolds et al. 1969; Behbehani 1995), mGluR expression at this level (Catania et al. 1994; Leyva et al. 1995) has a strategic importance. First evidence that group III mGluRs at PAG level modulate pain threshold came from a study in which L-SOP, a phosphate analog of L-AP4 with the same group III mGluR agonist property, microinjected into the PAG facilitated pain response under normal conditions (Maione et al. 1998b) and in the late phase of the formalin pain test, possibly by inhibiting descending pain inhibition (disinhibition) or by activating the descending facilitatory system (Maione et al. 2000; Berrino et al. 2001). Later on, by using more selective compounds, it has been shown that the individual roles of mGluR7 and mGluR8 in the regulation of pain threshold within the PAG were not univocal as in the CeA. The intra-PAG administration of (S)-3,4-DCPG alleviated pain responses, an effect which proved to depend from the time of administration with respect to the peripheral injection of formalin and carrageenan. Interestingly, the intra-PAG microinjection of MSOP, a group III receptor antagonist, antagonized the analgesic response of systemically administered (S)-3,4-DCPG suggesting that the analgesic effect of systemic (S)-3,4-DCPG requires mGuR8 within the PAG (Marabese et al. 2007a). The analgesic effect of intra-PAG (S)-3,4-DCPG was also associated with simultaneous changes in the RVM activity (Marabese et al. 2007b). Indeed, it increased the ongoing and tail flick-related activity of OFF cells and reduced that of ON cells. Conversely, the mGluR7 stimulation in the PAG by AMN082 increased the activity of ON cells and reduced that of OFF cells and consistently decreased tail flick latency (Marabese et al. 2007b). mGluR7 and mGluR8 stimulation within the PAG produced opposite effect on GABA and glutamate release too. Indeed, intra-PAG (S)-3,4-DCPG increased glutamate and decreased the GABA levels, whereas AMN082 reduced the glutamate concentration and increased GABA release, conversely (Marabese et al. 2007b). The role of mGluR7 at PAG level has been recently deepened by microinjecting into the PAG the novel selective negative allosteric modulator for mGluR7, the 6-(4-methoxyphenyl)-5-methyl-3-pyridin-4-ylisoxazolo[4,5-c]pyridin-4(5H)-one, MMPIP (Suzuki et al. 2007; Nakamura et al. 2010). Intra-VL PAG MMPIP inhibited pain responses in formalin and neuropathic pain models remaining ineffective in healthy rats. MMPIP also modulated RVM cell activity by increasing the activity of the antinociceptive OFF cells and decreased that of the pronociceptive ON cells, consistently with antinociception only in neuropathic rats and proved to be ineffective in healthy controls (Palazzo et al. 2012). The context-dependent MMPIP effect was later on confirmed in a recent study where systemic MMPIP has proved to revert pain and pain-related affective and cognitive responses in the SNI model of neuropathic while remaining ineffective in control mice (Palazzo et al. 2015).

Nucleus Tractus Solitarius (NTS) Interestingly and accordingly to the opposite effects on pain modulation played by mGluR7 and mGluR8 stimulation in the PAG and amygdala, opposite roles of these receptors have also been reported in the NTS. However, the function played by mGluR7 (pain facilitation) and mGluR8 (pain inhibition) on cardiac somatic reflex induced by pericardial capsaicin was exactly the opposite to those played in the amygdala and PAG. Indeed, mGluR7 has an inhibitory, whereas mGluR8 has a facilitatory role in modulating cardiac nociception in the NTS (Liu et al. 2012). The effects of intra-NTS stimulation of mGluR7 and mGluR8 were antagonized by intra-NTS MSOP, a selective group III mGluR antagonist. The reason of the discrepancy among mGluR7 and mGluR8 stimulation effect within the NTS and PAG/amygdala may originate from the fact that NTS has a pronociceptive action and the stimulation of mGluR8 on GABAergic terminals by disinhibiting NTS-descending activity increases nociceptive information at the spinal cord level. The stimulation of mGluR7 on glutamatergic terminals would have the opposite effect (Liu et al. 2012). It has been also recently shown that group III mGluR blockade in NTS prevents sensitization of neural firing response to intratracheal application of capsaicin in both naïve- and allergen-sensitized rats (Spaziano et al. 2015).

Dorsal Striatum (DS) The role of the mGluR8 in modulating pain responses in neuropathic pain conditions has been recently investigated also in the dorsal striatum (DS). mGluR8 stimulation did not modify the nociceptive behavior and RVM neural activity in control rats but increased the mechanical withdrawal threshold and the activity of the OFF cells while decreasing that of the ON cells in the SNI model of neuropathic pain. The effect of mGluR8 manipulation (and of other group III mGluRs) being analgesic in chronic pain conditions only is consistent with previous reports (2011b, 2014a,b, 2015). AZ12216052, a mGluR8 PAM, has shown an effect similar to that of (S)-3,4-DCPG. An increase in tail flick latency and in the OFF cell activity and a decrease in the ON cell activity have been observed in SNI rats only. However, the effect of AZ12216052 was less potent than that of (S)-3,4-DCPG. Differently from mGluR8, the mGluR4 stimulation by VU0155041, a selective mGluR4 PAM, was ineffective in changing pain responses in both shams and SNI rats thus confirming the specificity of (S)-3,4-DCPG effect was due by mGluR8 stimulation (Thomas et al. 2001). mGluR8 protein expression was increased in the DS and associated with vesicular glutamate transporter vGAT-positive profiles in SNI rats suggesting that disinhibition of the DS inhibits pain transmission throughout the descending pain pathway activation at RVM level in chronic pain condition (Rossi et al. 2014).

2.6 Conclusion

Altogether the potential and the variety of the effects which may be obtained by mGluR manipulation on CNS functioning are unlimited. A large number of neurological and psychiatric diseases are associated with glutamate transmission dysregulation; thus, its modulation by mGluR fine-tuning appears strategic. The mGluR-wide expression in nervous tissue of the pain pathway suggests that multiple avenues exist to exploit mGluRs for designing pain-killers. In particular, pain-controlling supraspinal sites often overlap those controlling affective and cognitive behaviors which are seriously compromised in chronic pain suffering subjects. Moreover, their expression in supraspinal structures such as those belonging to the endogenous antinociceptive descending pathway, including PAG and RVM, whose task is to counteract pain, offers an endogenous substrate for evoking analgesia. Taken as a whole, the mGluR ligand effects on pain and pain-related affective/cognitive responses depend not only from their selectivity for a particular mGluR subtype and the transduction cascades to which they are coupled but also from the neuron type (glutamate or GABA) and the brain site which can be pronociceptive or antinociceptive on which they are located. The recent increasing development of NAMs and PAMs which have shown selectivity toward a specific mGluR subtype offers a large variety of pharmacological tools for investigating mGluR potential on psychiatric and neurological diseases including chronic pain and its emotional and cognitive consequences. Moreover, these allosteric modulators present better lipophilicity and selectivity than their orthosteric counterparts. While preliminary studies investigating the effect of mGluR8 PAMs and mGluR7 NAMs on chronic pain and related affective cognitive consequences appear encouraging, further studies will be needed to clarify the precise contribution of each mGluR subtype and its therapeutic potential.

References

Acher F, Goudet C (2015) Therapeutic potential of group III metabotropic glutamate receptor ligands in pain. Curr Opin Pharmacol 20:64–72

Attwell PJE, Kaura S, Sigala G, Bradford HF, Croucher MJ, Jane DE et al (1995) Blockade of both epileptogenesis and glutamate release by (1S,3S)-ACPD, a presynaptic glutamate receptor agonist. Brain Res 698(1–2):155–162

Battaglia G, Monn JA, Schoepp DD (1997) In vivo inhibition of veratridine-evoked release of striatal excitatory amino acids by the group II metabotropic glutamate receptor agonist LY354740 in rats. Neurosci Lett 229(3):161–164

Behbehani MM (1995) Functional characteristics of the midbrain periaqueductal gray. Prog Neurobiol 46(6):575–605

Behbehani MM, Fields HL (1979) Evidence that an excitatory connection between the periaqueductal gray and nucleus raphe magnus mediates stimulation produced analgesia. Brain Res 170(1):85–93

Berrino L, Oliva P, Rossi F, Palazzo E, Nobili B, Maione S (2001) Interaction between metabotropic and NMDA glutamate receptors in the periaqueductal grey pain modulatory system. Naunyn Schmiedebergs Arch Pharmacol 364(5):437–443

Bhave G, Karim F, Carlton SM, Gereau RW (2001) Peripheral group I metabotropic glutamate receptors modulate nociception in mice. Nat Neurosci 4(4):417–423

Bleakman D, Lodge D (1998) Neuropharmacology of AMPA and kainate receptors. Neuropharmacol 37(10–11):1187–1204

Borszcz GS, Leaton RN (2003) The effect of amygdala lesions on conditional and unconditional vocalizations in rats. Neurobiol Learn Mem 79(3):212–225

Byrnes KR, Stoica B, Loane DJ, Riccio A, Davis MI, Faden AI (2009) Metabotropic glutamate receptor 5 activation inhibits microglial associated inflammation and neurotoxicity. Glia. 57(5):550–560

Carlton SM, Du J, Zhou S (2009) Group II metabotropic glutamate receptor activation on peripheral nociceptors modulates TRPV1 function. Brain Res 1248:86–95

Carrasquillo Y, Gereau RW (2008) Hemispheric lateralization of a molecular signal for pain modulation in the amygdala. Mol Pain 4:24

Cartmell J, Schoepp DD (2000) Regulation of neurotransmitter release by metabotropic glutamate receptors. J Neurochem 75(3):889–907

Catania MV, De SH, Penney JB, Young AB (1994) Metabotropic glutamate receptor heterogeneity in rat brain. Mol Pharmacol 45(4):626–636

Chiechio S, Nicoletti F (2012) Metabotropic glutamate receptors and the control of chronic pain. Curr Opin Pharmacol. 12(1):28–34

Decosterd I, Woolf CJ (2000) Spared nerve injury: an animal model of persistent peripheral neuropathic pain. Pain. 87(2):149–158

Ferraguti F, Shigemoto R (2006) Metabotropic glutamate receptors. Cell Tissue Res 326(2):483–504

Fields HL, Basbaum AI (1999) Central nervous system mechanisms of pain modulation. In: Wall PD, Melzack R (eds) Textbook of pain. Churchill Livingstone, New York, pp 243–257

Fields HL, Bry J, Hentall I, Zorman G (1983) The activity of neurons in the rostral medulla of the rat during withdrawal from noxious heat. J Neurosci 3(12):2545–2552

Gerber G, Zhong J, Youn D, Randic M (2000) Group II and group III metabotropic glutamate receptor agonists depress synaptic transmission in the rat spinal cord dorsal horn. Neurosci. 100(2):393–406

Giordano C, Cristino L, Luongo L, Siniscalco D, Petrosino S, Piscitelli F et al (2012) TRPV1-dependent and-independent alterations in the limbic cortex of neuropathic mice: impact on glial caspases and pain perception. Cereb Cortex. 22(11):2495–2518

Goudet C, Chapuy E, Alloui A, Acher F, Pin JP, Eschalier A (2008) Group III metabotropic glutamate receptors inhibit hyperalgesia in animal models of inflammation and neuropathic pain. Pain. 137(1):112–124

Guida F, Luongo L, Marmo F, Romano R, Iannotta M, Napolitano F, Belardo C, Marabese I, D'Aniello A, Bellini G, Rossi F, Piscitelli F, Lattanzi R, de Bartolomeis A, Usiello A, Di Marzo V, de Novellis V, Maione S (2015) Palmitoylethanolamide reduces pain-related behaviors and restores glutamatergic synapses homeostasis in the medial prefrontal cortex of neuropathic mice. Mol Brain 8:47

Hama AT (2003) Acute activation of the spinal cord metabotropic glutamate subtype-5 receptor leads to cold hypersensitivity in the rat. Neuropharmacology 44(4):423–430

Han JS, Neugebauer V (2005) mGluR1 and mGluR5 antagonists in the amygdala inhibit different components of audible and ultrasonic vocalizations in a model of arthritic pain. Pain 113(1–2):211–222

Han JS, Bird GC, Neugebauer V (2004) Enhanced group III mGluR-mediated inhibition of pain-related synaptic plasticity in the amygdala. Neuropharmacology 46(7):918–926

Heinricher MM, Barbaro NM, Fields HL (1989) Putative nociceptive modulating neurons in the rostral ventromedial medulla of the rat: firing of on- and off-cells is related to nociceptive responsiveness. Somatosens Mot Res 6(4):427–439

Hu H-J, Bhave G, Gereau RW (2002) Prostaglandin and protein kinase A-dependent modulation of vanilloid receptor function by metabotropic glutamate receptor 5: potential mechanism for thermal hyperalgesia. J Neurosci. 22(17):7444–7452

Hung KL, Wang SJ, Wang YC, Chiang TR, Wang CC (2014) Upregulation of presynaptic proteins and protein kinases associated with enhanced glutamate release from axonal terminals (synaptosomes) of the medial prefrontal cortex in rats with neuropathic pain. Pain. 155(2):377–387

Ji G, Neugebauer V (2010) Reactive oxygen species are involved in group I mGluR-mediated facilitation of nociceptive processing in amygdala neurons. J Neurophysiol 104(1):218–229

Ji G, Neugebauer V (2011) Pain-related deactivation of medial prefrontal cortical neurons involves mGluR1 and GABAA receptors. J Neurophysiol 106(5):2642–2652

Jones CK, Eberle EL, Peters SC, Monn JA, Shannon HE (2005) Analgesic effects of the selective group II (mGlu2/3) metabotropic glutamate receptor agonists LY379268 and LY389795 in persistent and inflammatory pain models after acute and repeated dosing. Neuropharmacology 49(Suppl):206–218

Kim H, Cui L, Kim J, Kim SJ (2009) Transient receptor potential vanilloid type 1 receptor regulates glutamatergic synaptic inputs to the spinothalamic tract neurons of the spinal cord deep dorsal horn. Neuroscience 160(2):508–516

Kinoshita A, Shigemoto R, Ohishi H, van der Putten H, Mizuno N (1998) Immunohistochemical localization of metabotropic glutamate receptors, mGluR7a and mGluR7b, in the central nervous system of the adult rat and mouse: a light and electron microscopic study. J Comp Neurol 393(3):332–352

Kinzie JM, Saugstad JA, Westbrook GL, Segerson TP (1995) Distribution of metabotropic glutamate receptor 7 messenger RNA in the developing and adult rat brain. Neuroscience 69(1):167–176

Kolber BJ, Montana MC, Carrasquillo Y, Xu J, Heinemann SF, Muglia LJ et al (2010) Activation of metabotropic glutamate receptor 5 in the amygdala modulates pain-like behavior. J Neurosci 30(24):8203–8213

Latremoliere A, Woolf CJ (2009) Central sensitization: a generator of pain hypersensitivity by central neural plasticity. J Pain 10(9):895–926

Leyva J, Maione S, Pallotta M, Berrino L, Rossi F (1995) Metabotropic and ionotropic glutamate receptors mediate opposite effects on periaqueductal gray matter. Eur J Pharmacol 285(2):123–126

Li W, Neugebauer V (2006) Differential changes of group II and group III mGluR function in central amygdala neurons in a model of arthritic pain. J Neurophysiol 96(4):1803–1815

Linden AM, Shannon H, Baez M, Yu JL, Koester A, Schoepp DD (2005) Anxiolytic-like activity of the mGLU2/3 receptor agonist LY354740 in the elevated plus maze test is disrupted in metabotropic glutamate receptor 2 and 3 knock-out mice. Psychopharmacology 179(1):284–291

Linden A-M, Baez M, Bergeron M, Schoepp DD (2006) Effects of mGlu2 or mGlu3 receptor deletions on mGlu2/3 receptor agonist (LY354740)-induced brain c-Fos expression: specific roles for mGlu2 in the amygdala and subcortical nuclei, and mGlu3 in the hippocampus. Neuropharmacology 51(2):213–228

Liu XH, Han M, Zhu JX, Sun N, Tang JS, Huo FQ et al (2012) Metabotropic glutamate subtype 7 and 8 receptors oppositely modulate cardiac nociception in the rat nucleus tractus solitarius. Neuroscience [Internet] 220:322–329

Luongo L, De Novellis V, Gatta L, Palazzo E, Vita D, Guida F et al (2013) Role of metabotropic glutamate receptor 1 in the basolateral amygdala-driven prefrontal cortical deactivation in inflammatory pain in the rat. Neuropharmacology 66:317–329

Maione S, Palazzo E, de Novellis V et al (1998a) Metabotropic glutamate receptors modulate serotonin release in the rat periaqueductal gray matter. Naunyn Schmiedebergs Arch Pharmacol 358(4):411–417

Maione S, Marabese I, Leyva J, Palazzo E, De Novellis V, Rossi F (1998b) Characterisation of mGluRs which modulate nociception in the PAG of the mouse. Neuropharmacology 37(12):1475–1483

Maione S, Oliva P, Marabese I, Palazzo E, Rossi F, Berrino L et al (2000) Periaqueductal gray matter metabotropic glutamate receptors modulate formalin-induced nociception. Pain 85(1–2):183–189

Marabese I, de Novellis V, Palazzo E, Mariani L, Siniscalco D, Rodella L et al (2005) Differential roles of mGlu8 receptors in the regulation of glutamate and gamma-aminobutyric acid release at periaqueductal grey level. Neuropharmacology 1:157–166

Marabese I, de Novellis V, Palazzo E, Scafuro MA, Vita D, Rossi F et al (2007a) Effects of (S)-3,4-DCPG, an mGlu8 receptor agonist, on inflammatory and neuropathic pain in mice. Neuropharmacology 52(2):253–262

Marabese I, Rossi F, Palazzo E, de Novellis V, Starowicz K, Cristino L et al (2007b) Periaqueductal gray metabotropic glutamate receptor subtype 7 and 8 mediate opposite effects on amino acid release, rostral ventromedial medulla cell activities, and thermal nociception. J Neurophysiol 98(1):43–53

Mitsukawa K, Yamamoto R, Ofner S, Nozulak J, Pescott O, Lukic S, Stoehr N, Mombereau C, Kuhn R, KH MA, van der Putten H, Cryan JF, Flor PJ (2005) A selective metabotropic glutamate receptor 7 agonist: activation of receptor signaling via an allosteric site modulates stress parameters in vivo. Proc Natl Acad Sci USA 102:18712–18717

Nakamura M, Kurihara H, Suzuki G, Mitsuya M, Ohkubo M, Ohta H (2010) Isoxazolopyridone derivatives as allosteric metabotropic glutamate receptor 7 antagonists. Bioorganic Med Chem Lett. 20(2):726–729

Neto FL, Castro-Lopes JM (2000) Antinociceptive effect of a group II metabotropic glutamate receptor antagonist in the thalamus of monoarthritic rats. Neurosci Lett. 296(1):25–28

Neto FL, Schadrack J, Berthele A, Zieglgänsberger W, Tölle TR, Castro-Lopes JM (2000) Differential distribution of metabotropic glutamate receptor subtype mRNAs in the thalamus of the rat. Brain Res 854(1–2):93–105

Neugebauer V (2001) Metabotropic glutamate receptors: novel targets for pain relief. Expert Rev Neurother 1(2):207–224

Neugebauer V, Li W, Bird GC, Bhave G, Gereau RW (2003a) Synaptic plasticity in the amygdala in a model of arthritic pain: differential roles of metabotropic glutamate receptors 1 and 5. J Neurosci 23(1):52–63

Neugebauer V, Galhardo V, Maione S, Mackey SC (2009) Forebrain pain mechanisms. Brain Res Rev 60(1):226–242

Niswender CM, Conn PJ (2010) Metabotropic glutamate receptors: physiology, pharmacology, and disease. Annu Rev Pharmacol Toxicol 50:295–322

Ohishi H, Akazawa C, Shigemoto R, Nakanishi S, Mizuno N (1995) Distributions of the mRNAs for L-2-amino-4- phosphonobutyrate-sensitive metabotropic glutamate receptors, mGluR4 and mGluR7, in the rat brain. J Comp Neurol 360(4):555–570

Palazzo E, Marabese I, De Novellis V, Oliva P, Rossi F, Berrino L et al (2001) Metabotropic and NMDA glutamate receptors participate in the cannabinoid-induced antinociception. Neuropharmacology 40(3):319–326

Palazzo E, De Novellis V, Marabese I, Cuomo D, Rossi F, Berrino L et al (2002) Interaction between vanilloid and glutamate receptors in the central modulation of nociception. Eur J Pharmacol 439(1–3):69–75

Palazzo E, Genovese R, Mariani L, Siniscalco D, Marabese I, De Novellis V et al (2004) Metabotropic glutamate receptor 5 and dorsal raphe serotonin release in inflammatory pain in rat. Eur J Pharmacol 492(2–3):169–176

Palazzo E, Marabese I, Soukupova M, Luongo L, Boccella S, Giordano C et al (2011b) Metabotropic glutamate receptor subtype 8 in the amygdala modulates thermal threshold, neurotransmitter release, and rostral ventromedial medulla cell activity in inflammatory pain. J Neurosci 31(12):4687–4697

Palazzo E, Luongo L, Bellini G, Guida F et al (2012) Changes in Cannabinoid Receptor Subtype 1 Activity and Interaction with Metabotropic Glutamate Subtype 5 Receptors in the Periaqueductal Gray-Rostral Ventromedial Medulla Pathway in a Rodent Neuropathic Pain Model. CNS Neurol Disord Drug Targets 11(2)

Palazzo E, Marabese I, de Novellis V, Rossi F, Maione S (2014a) Supraspinal metabotropic glutamate receptors: a target for pain relief and beyond. Eur J Neurosci. 39(3):444–454

Palazzo E, de Novellis V, Rossi F, Maione S (2014b) Supraspinal metabotropic glutamate receptor subtype 8: a switch to turn off pain. Amino Acids 46(6):1441–1448

Palazzo E, Romano R, Luongo L, Boccella S, De Gregorio D, Giordano ME et al (2015) MMPIP, an mGluR7-selective negative allosteric modulator, alleviates pain and normalizes affective and cognitive behavior in neuropathic mice. Pain 156(6):1060–1073

Ren W, Neugebauer V (2010) Pain-related increase of excitatory transmission and decrease of inhibitory transmission in the central nucleus of the amygdala are mediated by mGluR1. Mol Pain 6:93

Reynolds DV (1969) Surgery in the rat during electrical analgesia induced by focal brain stimulation. Science 164(878):444–445

Rossi F, Marabese I, De Chiaro M, Boccella S, Luongo L, Guida F et al (2014) Dorsal striatum metabotropic glutamate receptor 8 affects nocifensive responses and rostral ventromedial medulla cell activity in neuropathic pain conditions. J Neurophysiol 111(11):2196–2209

Salt TE, Binns KE (2000) Contributions of mGlu1 and mGlu5 receptors to interactions with N-methyl-D-aspartate receptor-mediated responses and nociceptive sensory responses of rat thalamic neurons. Neuroscience 100(2):375–380

Salt TE, Eaton SA (1995) Modulation of sensory neurone excitatory and inhibitory responses in the ventrobasal thalamus by activation of metabotropic excitatory amino acid receptors. Neuropharmacology 34(8):1043–1051

Salt TE, Turner JP (1998) Modulation of sensory inhibition in the ventrobasal thalamus via activation of group II metabotropic glutamate receptors by 2R,4R-aminopyrrolidine-2,4-dicarboxylate. Exp Brain Res 121(2):181–185

Saugstad JA, Kinzie JM, Shinohara MM, Segerson TP, Westbrook GL (1997) Cloning and expression of rat metabotropic glutamate receptor 8 reveals a distinct pharmacological profile. Mol Pharmacol 51(1):119–125

Saugstad JA, Marino MJ, Folk JA, Hepler JR, Conn PJ (1998) RGS4 inhibits signaling by group I metabotropic glutamate receptors. J Neurosci 18(3):905–913

Schoepp DD (2001) Unveiling the functions of presynaptic metabotropic glutamate receptors in the central nervous system. J Pharmacol Exp Ther 299(1):12–20

Spaziano G, Luongo L, Guida F, Petrosino S, Matteis M, Palazzo E et al (2015) Exposure to allergen causes changes in NTS neural activities after intratracheal capsaicin application, in endocannabinoid levels and in the glia morphology of NTS. Biomed Res Int 2015:980983

Suzuki G, Tsukamoto N, Fushiki H, Kawagishi A, Nakamura M, Kurihara H et al (2007) In vitro pharmacological characterization of novel isoxazolopyridone derivatives as allosteric metabotropic glutamate receptor 7 antagonists. J Pharmacol Exp Ther 323(1):147–156

Thomas NK, Wright RA, Howson PA, Kingston AE, Schoepp DD, Jane DE (2001) (S)-3,4-DCPG, a potent and selective mGlu8a receptor agonist, activates metabotropic glutamate receptors on primary afferent terminals in the neonatal rat spinal cord. Neuropharmacology 40(3):311–318

Tian Y, Liu Y, Chen X, Kang Q, Zhang J, Shi Q et al (2010) AMN082 promotes the proliferation and differentiation of neural progenitor cells with influence on phosphorylation of MAPK signaling pathways. Neurochem Int 57(1):8–15

Turner JP, Salt TE (1999) Group III metabotropic glutamate receptors control corticothalamic synaptic transmission in the rat thalamus in vitro. JPhysiol 519:481–491

Urban MO, Hama AT, Bradbury M, Anderson J, Varney MA, Bristow L (2003) Role of metabotropic glutamate receptor subtype 5 (mGluR5) in the maintenance of cold hypersensitivity following a peripheral mononeuropathy in the rat. Neuropharmacology 44(8):983–993

Vardi N, Duvoisin R, Wu G, Sterling P (2000) Localization of mGluR6 to dendrites of ON bipolar cells in primate retina. J Comp Neurol 423(3):402–412

Veinante P, Yalcin I, Barrot M (2013) The amygdala between sensation and affect: a role in pain. J Mol Psychiat 1(1):9

Walker K, Reeve A, Bowes M, Winter J, Wotherspoon G, Davis A et al (2000) mGlu5 receptors and nociceptive function II. mGlu5 receptors functionally expressed on peripheral sensory neurones mediate inflammatory hyperalgesia. Neuropharmacology 40(1):10–19

Yamada T, Zuo D, Yamamoto T, Olszewski RT, Bzdega T, Moffett JR, Neale JH (2012) Peptidase inhibition in the periaqueductal gray and rostral ventromedial medulla reduces flinching in the formalin model inflammation. Mol Pain 12:8–67

Yamakura T, Shimoji K (1999) Subunit- and site-specific pharmacology of the NMDA receptor channel. Prog Neurobiol 59(3):279–298

Yamamoto T, Kozikowski A, Zhou J, Neale JH (2000) Intracerebroventricular administration of N-acetylaspartylglutamate (NAAG) peptidase inhibitors is analgesic in inflammatory pain. Mol Pain 81:4–31

Yang D, Gereau RW (2002) Peripheral group II metabotropic glutamate receptors (mGluR2/3) regulate prostaglandin E2-mediated sensitization of capsaicin responses and thermal nociception. J Neurosci. 22(15):6388–6393

Yang D, Gereau RW (2003) Peripheral group II metabotropic glutamate receptors mediate endogenous anti-allodynia in inflammation. Pain:411–417

Zammataro M, Chiechio S, Montana MC, Traficante A, Copani A, Nicoletti F et al (2011) mGlu2 metabotropic glutamate receptors restrain inflammatory pain and mediate the analgesic activity of dual mGlu2/mGlu3 receptor agonists. Mol Pain 7:6

Chapter 3
Metabotropic Glutamate Receptors and Parkinson's Disease: Basic and Preclinical Neuroscience

Paolo Gubellini, Liliya Iskhakova, Yoland Smith, and Marianne Amalric

Abstract Since the late 1990s, data from various laboratories have provided strong evidence that the three groups of mGlu receptors are widely and specifically expressed in the basal ganglia (BG), where they modulate neuronal excitability as well as synaptic transmission and plasticity. Therefore, targeting specific mGlu receptor subtypes by means of selective drugs could be a possible strategy for restoring normal synaptic function and neuron activity in the BG in Parkinson's disease (PD). In this context, the spectacular development of subtype-selective mGlu receptor agonists, antagonists, and allosteric modulators has provided scientists with a wide range of neuropharmacological tools that have largely supported this hypothesis. This review provides data showing that drugs acting on mGlu receptors can alleviate PD motor symptoms and reduce levodopa-induced dyskinesia in animal models of PD and recent clinical trials that confirm these findings.

P. Gubellini, Ph.D. (✉)
Aix-Marseille University, CNRS, UMR7288 Institut de Biologie du Développement de Marseille (IBDM), Case 907 Parc Scientifique de Luminy, 163 Avenue de Luminy, 13288 Marseille cedex 9, France
e-mail: paolo.gubellini@univ-amu.fr

L. Iskhakova
Department of Medical Neurobiology, Hebrew University of Jerusalem-Hadassah Medical School, P.O. Box 12272, Jerusalem, Israel, 91120

Yerkes National Primate Research Center, Emory University, Atlanta, GA 30329, USA

Y. Smith, Ph.D.
Yerkes National Primate Research Center, Emory University, Atlanta, GA 30329, USA

Department of Neurology, Emory University School of Medicine, Atlanta, GA 30322, USA

NIH UDALL Center of Excellence for Parkinson's Disease Research, Emory University, Atlanta, GA 30322, USA
e-mail: ysmit01@emory.edu

M. Amalric, Ph.D.
Aix-Marseille University, CNRS, UMR7291, FR3C 3512, Laboratoire de Neurosciences Cognitives, 3 place Victor-Hugo, 13331 Marseille cedex 3, France
e-mail: marianne.amalric@univ-amu.fr

R.T. Ngomba et al. (eds.), *mGLU Receptors*, The Receptors 31,
DOI 10.1007/978-3-319-56170-7_3

Keywords Basal ganglia • Globus pallidus • Dyskinesia • mGlu • Monkey • Rodent • Striatum • Substantia nigra • Subthalamic nucleus • Synaptic transmission

3.1 Parkinson's Disease

Parkinson's disease (PD) is a chronic neurodegenerative disorder primarily due to the progressive loss of *substantia nigra pars compacta* (SNc) neurons releasing dopamine (DA) and characterized by the "classical" motor symptoms: resting tremor, bradykinesia, akinesia, rigidity, and postural instability (Olanow 2007; Jankovic 2008; Schapira and Jenner 2011). These DAergic neurons project to the basal ganglia (BG), a highly organized network of brain nuclei implicated in movement control and motor learning, as well as limbic and cognitive functions, and receive massive projections from the cortex and thalamus (Fig. 3.1a). The BG nuclei include the caudate-putamen (or striatum in non-primates), which is the main input station and is composed by >95% of projection medium spiny neurons (MSNs) using γ-aminobutyric acid (GABA) as neurotransmitter and by cholinergic and GABAergic interneurons; the external *globus pallidus* (GPe or GP in non-primates); the so-called output structures, i.e., the *substantia nigra pars reticulata* (SNr) and the internal *globus pallidus* (GPi or entopeduncular nucleus, EP, in non-primates) that project to the thalamus and brain stem; and the subthalamic nucleus (STN), which is the second input station and the only intrinsic glutamatergic structure of the BG, the others being mainly GABAergic. Two distinct efferent pathways [although such segregation is not complete (Calabresi et al. 2014)] originate from the striatum: the "direct" pathway, consisting in striatal GABAergic projection neurons expressing mainly DA D1-like receptors and projecting directly to the BG output structures (SNr/GPi), and the "indirect" pathway, whose neurons express D2-like receptors and project to the SNr/GPi indirectly, via two intermediate stations—the GPe and the STN (Fig. 3.1a) (Levy et al. 1997; DeLong and Wichmann 2007). In PD (Fig. 3.1b), the loss of SNc DAergic neurons leads to two opposite effects in the BG: hypoactivity of direct-pathway neurons and, conversely, hyperactivity of the indirect pathway, resulting in increased inhibition of the GPe which, in turn, disinhibits the STN (Blandini et al. 2000). The overall effect is an increased glutamatergic and a decreased GABAergic tone at the level of the SNr/GPi, which become hyperactive leading to excessive thalamic and brain stem inhibition and to the PD motor symptoms (Obeso et al. 2000; DeLong and Wichmann 2007; Wichmann and Dostrovsky 2011). Moreover, an increased corticostriatal glutamatergic input and an increased striatal cholinergic tone are thought to contribute to such imbalance in the cortico-BG circuit (Chase and Oh 2000; Chase et al. 2003; Bamford et al. 2004; Yin and Lovinger 2006; Pisani et al. 2007; Obeso et al. 2008).

To date, available therapies for PD are only symptomatic. The gold standard treatment is the administration of the DA precursor L-3,4-dihydroxyphenylalanine

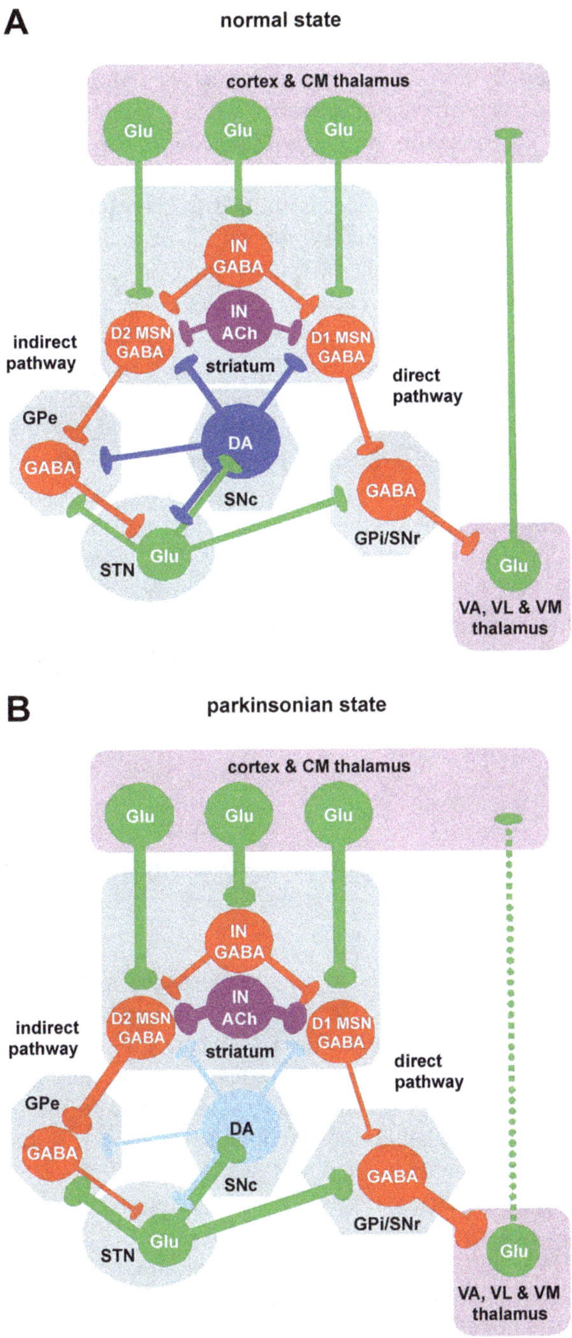

Fig. 3.1 Simplified scheme representing the basal ganglia circuitry in normal (**a**) and in parkinsonian state (**b**). The "hyperdirect" pathway (from cortex to STN), the thalamostriatal afferences, the connection between SNc and GPi/SNr, as well as the cortical outputs to the brain stem and spinal cord are omitted. The strenght of synaptic pathways is represented by their thickness (the thicker the stronger and vice versa), and the neurotransmitters utilized are indicated by different colors in fibers and neuronal cell bodies (*ACh* acetylcholine, *Glu* glutamate, *GABA* γ-aminobutyric acid, *DA* dopamine).Fig. 3.1 (continued) The degeneration of the SNc and dopaminergic fibers in parkinsonian condition is represented in **b** (the dotted line represents altered thalamic output to the cortex). Abbreviations: *CM* centromedian, *GPe/i* globus pallidus pars externa/interna, *IN* interneuron, *MSN* medium spiny neuron, *SNc/r* substantia nigra pars compacta/reticulata, *STN* subthalamic nucleus, *VA* ventral anterior, *VL* ventral lateral, *VM* ventral medial

(L-DOPA), DA agonists, and/or DA degradation blockers. However, the relief provided by such therapy is impaired in the long term by loss of efficacy and the appearance of dose-dependent side effects, notably L-DOPA-induced dyskinesia (LID) and motor fluctuations, probably due to the pulsatile, non-physiological stimulation of DA receptors by these treatments (Chase 1998; Ahlskog and Muenter 2001; Olanow et al. 2009; Bastide et al. 2015). Since the mid-1990s, deep brain stimulation has become a successful and frequently practiced neurosurgical approach for alleviating PD motor symptoms, in particular high-frequency stimulation of the STN (Benabid et al. 2001; Gubellini et al. 2009; Benabid et al. 2009). However, its clinical use is proceeding without a complete understanding of its action mechanisms, and its long-term consequences on non-motor symptoms are still under study (Rosa et al. 2012; Karas et al. 2013). Furthermore, some patients show a progressive worsening of both motor and non-motor symptoms that are likely due to the progression of the disease, although long-term changes induced by chronic STN stimulation cannot be excluded (Fasano et al. 2012; Rizzone et al. 2014). Finally, several other treatments are currently under investigation, including pharmacologically targeting other neurotransmitter systems, cell transplantation, gene therapy, and neuroprotective treatments (Smith et al. 2012; Brichta et al. 2013; Schapira et al. 2014; Buttery and Barker 2014). Among the new therapeutic strategies for PD, probably the best option would be an alternative non-DAergic treatment that could relieve PD symptoms by restoring the excitation/inhibition balance in the BG and, at the same time, avoid the development of side effects, notably LID (Smith et al. 2012; Brichta et al. 2013). In this context, the metabotropic glutamate (mGlu) receptors constitute interesting targets.

3.2 Rationale for Targeting mGlu Receptors for Parkinson's Disease

The three classes of mGlu receptors, i.e., group I (mGlu1 and mGlu5), group II (mGlu2 and mGlu3), and group III (mGlu4, mGlu7, and mGlu8), are specifically expressed in the BG (Fig. 3.2), where they regulate neuron excitability, synaptic transmission and synaptic plasticity, in particular in the striatum (mGlu6, another group III mGlu receptor, is expressed only in the retina). Briefly, group I mGlu receptors are largely postsynaptic, and their activation increases neuronal excitability and response to glutamate, while groups II and III are mainly located presynaptically, where they act as auto- and hetero-receptors and their stimulation leads to decreased neurotransmitter release (Testa et al. 1994, 1995, 1998; Conn and Pin 1997; Tallaksen-Greene et al. 1998; Bradley et al. 1999b; Pin and Acher 2002; Pisani et al. 2003; Gubellini et al. 2004; Conn et al. 2005; Bonsi et al. 2007). Given such ubiquitous but rather specific expression on neurons and synapses of the BG, and due to their differential effects, mGlu receptors are interesting tools to modulate

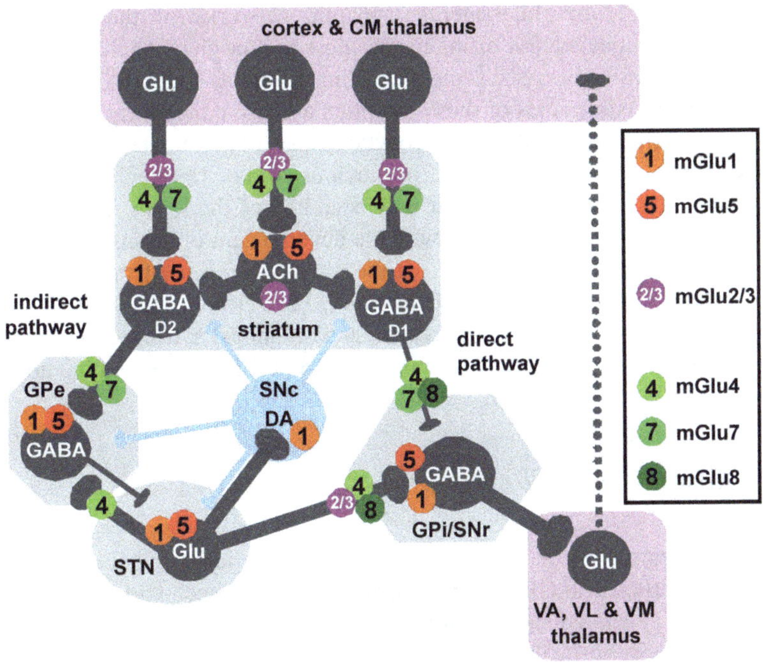

Fig. 3.2 Pre- and postsynaptic localization of the different mGlu receptor subtypes in the basal ganglia (for abbreviations, see Fig. 3.1). The circuitry represented here is in parkinsonian condition, in order to highlight the over- and hypoactive synapses and neuronal subtypes targeted by selectively acting on specific mGlu receptor subtypes (See Fig. 3.1 for abbreviations)

neuronal excitability and synaptic strength in these brain structures. Therefore, they represent potential pharmacological targets that could allow restoring physiological activity in BG circuitry in PD state by, for example, inhibiting a given synapse or neuronal type or targeting a specific BG nucleus.

3.3 Localization and Function of mGlu Receptors in the Basal Ganglia

The cellular localization of mGlu receptors in the BG has been examined using in situ hybridization and immunohistochemical approaches. Further electron microscopic studies with high-resolution pre-embedding immunogold techniques provided detailed maps of the subcellular and subsynaptic distribution of these receptors throughout the BG circuitry. These studies also highlighted the complex and intriguing pattern of localization of the various mGlu receptor subtypes suggesting that these receptors are located to subserve pre- and postsynaptic regulation of glutamatergic and non-glutamatergic synaptic networks of the BG circuitry. Furthermore, the localization of mGlu receptors at non-glutamatergic synapses as well as extrasynaptically raised some interesting questions about the sources of activation and the possible roles of these receptors at these nonconventional sites. The importance of glutamate spillover, astrocytic release of glutamate, and protein-protein interactions has been discussed in previous studies [for review, see (Galvan et al. 2006)]. These advances in our understanding of the localization of specific mGlu receptor subtypes, combined with the development of specific mGlu-related compounds, have helped dissecting out the modulatory role of these receptors at various synapses throughout the BG circuitry. Furthermore, these studies set the foundation for the development of preclinical work aimed at testing the therapeutic efficacy of drugs acting on mGlu receptors in various brain disorders including PD.

3.3.1 Group I mGlu Receptors

Virtually all neurons of the BG nuclei express to a variable degree group I mGlu receptors, in particular mGlu5 (Fig. 3.2), and because of such widespread distribution, they have generated significant interest as potential targets for drug therapy in PD (Paquet and Smith 2003; Conn et al. 2005).

In the dorsal and ventral striatum, both mGlu1a (a splice variant of mGlu1) and mGlu5 receptors are strongly expressed in both populations of projection neurons, cholinergic interneurons, and subtypes of GABAergic interneurons (Testa et al. 1994; Tallaksen-Greene et al. 1998; Pisani et al. 2001a). Co-localization studies in the nucleus accumbens (or ventral striatum) revealed that ~60% of dendritic spines co-express mGlu1a and mGlu5 receptors and that a significant amount of the postsynaptic labeling for both group I subtypes is expressed in D1 receptor-positive and receptor-negative profiles, suggesting that both direct and indirect pathway neurons express group I mGlu receptors (Mitrano and Smith 2007).

At the subcellular level, the two receptor subtypes are mainly found postsynaptically in dendritic spines and shafts, although presynaptic expression of mGlu1a receptor (but not mGlu5) in glutamatergic and DAergic terminals has also been reported in the caudate nucleus and putamen of rhesus monkeys (Paquet and Smith 2003), but not in the nucleus accumbens where presynaptic mGlu1a and mGlu5 labeling is confined to rare unmyelinated axons (Mitrano and Smith 2007). In general, the bulk of postsynaptic labeling for both receptor subtypes in the dorsal striatum and nucleus accumbens is located at non-synaptic sites along the plasma membrane of dendrites and spines, but a significant amount of labeling is also specifically found at the edges of the postsynaptic densities (i.e., perisynaptic) of asymmetric glutamatergic synapses from the cortex, thalamus, and amygdala (Paquet and Smith 2003; Mitrano and Smith 2007), a pattern reminiscent of the localization of group I mGlu receptors at glutamatergic synapses throughout the CNS (Baude et al. 1993; Nusser et al. 1994; Galvan et al. 2006) (Fig. 3.3a). In the monkey dorsal striatum, mGlu5 receptor (but not mGlu1a) is also expressed perisynaptically at symmetric synapses formed by DAergic terminals, suggesting that it could be involved in mediating some of the physiological effects of nigrostriatal afferents in the primate striatum [see discussion in (Paquet and Smith 2003)] (Fig. 3.3a). In light of evidence that subsets of nigral DAergic terminals co-express and release glutamate in the striatum (Chuhma et al. 2009; Hnasko et al. 2010; Trudeau et al. 2014), the potential role of mGlu5 receptor at nigrostriatal synapses warrants further investigation. In the monkey dorsal striatum, a significant amount of labeling for both receptor subtypes is also found in the main body of putative GABAergic symmetric synapses formed by terminals that ultrastructurally resemble those from local axon collaterals of striatal projection neurons (Paquet and Smith 2003) (Fig. 3.3a). In addition to this heterogeneous pattern of plasma membrane expression of both group I mGlu receptor subtypes, a significant amount of immunoreactivity (mainly mGlu5) is expressed in various intracellular compartments of neurons in the dorsal and ventral striatum (Paquet and Smith 2003; Mitrano and Smith 2007). Whether this labeling is accounted for by receptors being trafficked to and from the plasma membrane, or by a pool of functional intracellular mGlu5 receptors, still remains to be established.

At the presynaptic level, mGlu1a receptor is expressed predominantly in subsets of thalamic glutamatergic terminals from the caudal intralaminar nuclei and nigrostriatal DAergic boutons in the dorsal striatum (Paquet and Smith 2003), while in the nucleus accumbens it is particularly abundant in unmyelinated axons (Mitrano and Smith 2007). The functional significance of this presynaptic labeling remains to be established (Battaglia et al. 2001; Paquet and Smith 2003). Altogether, these studies demonstrate that group I mGlu receptors display a highly heterogeneous pattern of cellular, subcellular, and subsynaptic distribution in the rodent and primate striatum, through which they can regulate both glutamatergic and non-glutamatergic transmission.

From a neurophysiological point of view, the pharmacological stimulation of mGlu5 receptor has been shown to enhance N-methyl-D-aspartate (NMDA) receptor-mediated glutamatergic signaling in striatal MSNs (Awad et al. 2000; Pisani et al. 2001b), while in the GP this effect is achieved by stimulating mGlu1 receptor (Sun et al. 2012). Moreover, long-term synaptic plasticity (long-term depression (LTD) and long-term potentiation (LTP)) in the striatum, known to underlie motor control and learning, also depends on group I mGlu receptor stimulation (Gubellini et al. 2001, 2003; Sung et al. 2001; Lovinger 2010; Bagetta et al. 2010). Furthermore, a form of corticostriatal LTD is achieved by the stimulation of group I mGlu receptors via the synthesis of nitric oxide and the release of endocannabinoids (Sergeeva et al. 2007). Given the evidence that corticostriatal LTD and LTP are lost in animal models of PD (Picconi et al. 2012), a possible strategy for PD treatment should be to rescue synaptic plasticity and/or restore physiological synaptic transmission by acting on group I mGlu receptors. In particular, inhibition of mGlu5 receptor could reduce the increased excitability of striatal MSNs and cholinergic interneurons, and thus the subsequent overinhibition of the GPe, therefore counteracting the hyperactive glutamatergic inputs that the striatum and GPe receive from the cortex and the STN, respectively, in PD condition (Figs. 3.1b and 3.2).

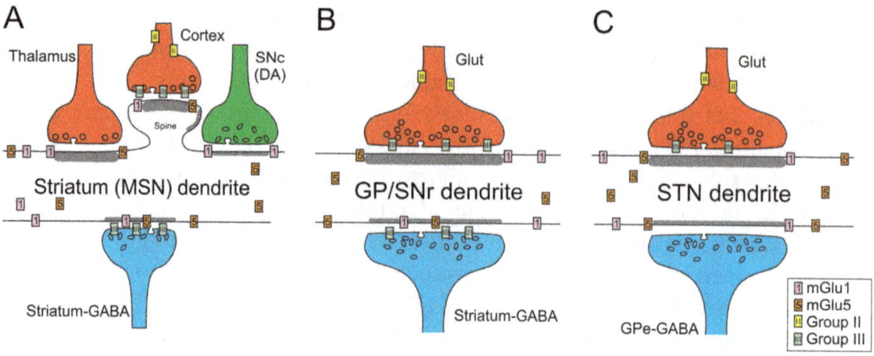

Fig. 3.3 Schematics showing the subcellular and subsynaptic distribution of groups I, II, and III mGlu receptors in relation to synapses formed by glutamatergic (*red*), GABAergic (*blue*), and dopaminergic (*green*) terminals in the striatum (**a**), GP/SNr (**b**), and STN (**c**). The legend of the different receptor subtypes is shown at the *bottom right* of the figure (See Fig. 3.1 for abbreviations)

In the GPe and GPi, both mGlu1a and mGlu5 receptors are also co-expressed in most neurons (Poisik et al. 2003). Overall, the pattern of subcellular postsynaptic labeling for both receptor subtypes is similar to that described in the striatum, i.e., strong extrasynaptic dendritic labeling of GPe and GPi neurons with perisynaptic expression at the edges of asymmetric glutamatergic synapses most likely originating from the STN (Hanson and Smith 1999; Smith et al. 2001, 2000). In line with data from the dorsal striatum, prominent mGlu1a and mGlu5 receptor labeling is expressed in the main body of putative striatopallidal GABAergic synapses on dendrites of GPe and GPi neurons (Hanson and Smith 1999). Similar findings were also obtained for striatonigral GABAergic synapses in the rat SNr (Hubert et al. 2001) (Fig. 3.3b). Although the role and mechanisms of activation of postsynaptic group I mGlu receptors at GABAergic striatofugal synapses remain unknown, it is noteworthy that these observations have recently been confirmed and extended to other brain regions using different imaging methods (Mansouri et al. 2015).

In the GP, the pharmacological stimulation of mGlu1 receptor has been shown to increase neuronal spontaneous firing rate and induce a long-lasting excitation in both rat (Kaneda et al. 2007) and monkey (Kaneda et al. 2005), suggesting that this receptor may play an important role in the control of pallidal neuron activity. Although mGlu1a and mGlu5 receptors display a high degree of co-localization in the GP and other BG nuclei, little is known about the functional significance of this co-expression (Valenti et al. 2002). However, data from the rat GP indicate that mGlu1 and mGlu5 receptors functionally interact at the level of single pallidal neurons so that mGlu5 activation regulates the desensitization of mGlu1 (Poisik et al. 2003). Supporting such possible interaction, it has been shown that mGlu1 receptor stimulation or mGlu5 receptor blockade produces similar excitatory effects on pallidal neurons (Sun et al. 2012). Interestingly, in reserpine-treated rats, these effects are inverted (Poisik et al. 2007). This novel mechanism of interaction expands the possibility by which group I mGlu receptors can regulate neuronal activity in the pallidal complex (Valenti et al. 2002), especially in parkinsonian conditions in which the expression and excitatory action of mGlu1 in the pallidal complex are downregulated (Kaneda et al. 2005).

In the STN, both group I mGlu receptor subtypes also frequently co-localize in cell bodies and dendritic profiles of single neurons (Testa et al. 1998; Awad et al. 2000; Marino et al. 2002; Kuwajima et al. 2004, 2007). At the subcellular and subsynaptic level, mGlu1a and mGlu5 receptors display a preferential perisynaptic localization at the edges of glutamatergic and GABAergic synapses, some of which being established by putative pallido-subthalamic terminals (Awad et al. 2000; Kuwajima et al. 2004) (Fig. 3.3c). In contrast to other BG nuclei, a significant amount of mGlu1a receptor immunoreactivity displays glial expression in the monkey STN (Kuwajima et al. 2004, 2007). Despite significant co-expression of both group I mGlu receptor subtypes in single STN neurons, it appears that mGlu5 is the key mediator of slow depolarization, direct excitation, increased burst-firing activity, and potentiation of NMDA receptor-mediated currents in STN neurons, while activation of both receptor subtypes regulates the levels of intracellular calcium (Awad et al. 2000; Valenti et al. 2002; Marino et al. 2002). However, other studies showed that stimulation of group I mGlu receptors (most likely of mGlu1) generates a hyperpolarizing K-ATP current by a nitric oxide- and protein kinase G-dependent process that is mediated by the release of calcium from intracellular stores (Shen and Johnson 2013) and depresses glutamatergic transmission (Awad-Granko and Conn 2001), suggesting that these receptors might also exert crucial inhibitory influences on the activity and synaptic transmission of STN neurons.

In both the SNc and SNr, the pattern of plasma membrane-bound subsynaptic labeling of the two receptor subtypes is the same as in the GP, with the exception that the relative prevalence of intracellular mGlu5 labeling is much higher in dendrites of SNr/SNc neurons than in GPe/GPi neurons. In contrast, most mGlu1a labeling is located on the dendritic plasma membrane in both structures (Hanson and Smith 1999; Hubert et al. 2001). Strong synaptic labeling in the core of putative striatonigral GABAergic synapses and perisynaptic expression at glutamatergic synapses is the most typical pattern of mGlu1a and mGlu5 labeling in the SNr (Hubert et al. 2001) (Fig. 3.3b). As in the GP, glial and presynaptic expression of group I mGlu receptors is almost nonexistent in the SNr of adult rats and monkeys (Hanson and Smith 1999; Hubert et al. 2001). However, strong enrichment of both group I mGlu receptor subtypes in astrocytic processes and axon terminals has been reported in young rats (up to at least P18), suggesting a critical role of these receptors in early brain development (Hubert and Smith 2004).

In both the SNc and SNr, the pharmacological stimulation of group I mGlu receptors triggers an inward current (Shen and Johnson 2013). In the SNc, in particular, this results in increased internal sodium and calcium concentration in DAergic neurons (Marino et al. 2002; Katayama et al. 2003), altering their firing from regular to bursty, suggesting that these receptors can modulate the activity as well as the vulnerability of these cells (Guatteo et al. 1999; Prisco et al. 2002). In the SNr, the activation of both mGlu1a and mGlu5 receptors depolarizes GABAergic neurons and inhibits glutamatergic transmission (Wittmann et al. 2001; Marino et al. 2001; Marino et al. 2002).

3.3.2 Group II mGlu Receptors

The two group II mGlu receptor subtypes (mGlu2 and mGlu3) are expressed to variable extent throughout the BG (Testa et al. 1994; Petralia et al. 1996; Tamaru et al. 2001; Samadi et al. 2008a), the striatum being the most enriched nucleus (Fig. 3.2). Overall, striatal mGlu2 receptor immunoreactivity is mainly expressed in glutamatergic axons and terminals and in cholinergic interneurons (Testa et al. 1994; Pisani et al. 2002), while mGlu3 labeling is found in glia and at low levels in various subtypes of neuronal cell bodies, except large cholinergic cells (Testa et al. 1994; Petralia et al. 1996). Albeit both being located presynaptically in glutamatergic terminals, mGlu2 and mGlu4 receptors (the latter belonging to group III; see below) display a strikingly different pattern of subcellular localization, mGlu2 being expressed away from the release site of glutamate, while mGlu4 is densely aggregated in the presynaptic grid of active zones at glutamatergic synapses (Petralia et al. 1996; Bradley et al. 1999a; Bogenpohl et al. 2013) (Fig. 3.3), suggesting that these two populations of presynaptic mGlu autoreceptors rely on different sources and mechanisms of activation (Galvan et al. 2006). In line with these anatomical data, group II mGlu receptor activation exerts powerful inhibitory control over glutamatergic corticostriatal transmission and acetylcholine release in rodents (Lovinger and McCool 1995; Marti et al. 2001; Kahn et al. 2001; Kupferschmidt and Lovinger 2015).

Other BG nuclei display low to moderate levels of neuronal group II mGlu receptor mRNA (Testa et al. 1994) and protein immunoreactivity (Petralia et al. 1996; Testa et al. 1998; Tamaru et al. 2001). Electrophysiological studies have shown that activation of group II mGlu receptors reduces glutamatergic transmission in the STN (Shen and Johnson 2003), the GP (Poisik et al. 2005) and the SN (Bradley et al. 2000; Wittmann et al. 2002; Katayama et al. 2003; Johnson et al. 2011). Thus, group II mGlu receptor agonists should help in restoring BG circuitry in parkinsonian state by reducing the output of glutamatergic corticostriatal and subthalamofugal synapses and by inhibiting cholinergic neurons in the striatum (Figs. 3.1b and 3.2).

3.3.3 Group III mGlu Receptors

The two main group III mGlu receptors, mGlu4 and mGlu7, are expressed throughout the BG circuitry (Fig. 3.2), although evidence for low level of mGlu8 expression has also been suggested (Kinoshita et al. 1998; Kosinski et al. 1999; Bradley et al. 1999a, b; Corti et al. 2002; Messenger et al. 2002; Conn et al. 2005; Bogenpohl et al. 2013; Johnson et al. 2013). In the striatum, mGlu4a and mGlu7a receptors are mainly expressed presynaptically in glutamatergic corticostriatal terminals (Kosinski et al. 1999; Bradley et al. 1999a, b; Corti et al. 2002; Bogenpohl et al. 2013) (Fig. 3.3a), which is in line with the powerful inhibitory effects of mGlu4 receptor activation on corticostriatal transmission in the rodent striatum (Cuomo et al. 2009; Bennouar et al. 2013; Gubellini et al. 2014). The effects of mGlu7 receptor stimulation have been mainly characterized in the nucleus accumbens, where it lowers extracellular GABA and increases extracellular glutamate levels (Li et al. 2008). Recent findings indicate that mGlu4-containing cortical boutons preferentially target spines of indirect-pathway projection neurons (Iskhakova and Smith 2015), suggesting that mGlu4 receptor-related drugs may have a differential modulatory effects upon the activity of the two main populations of striatofugal neurons. In addition to their role as autoreceptors on glutamatergic terminals, striatal mGlu4 receptors are also expressed presynaptically in putative GABAergic terminals, most likely from intrinsic axon collaterals of striatofugal GABAergic neurons (Fig. 3.3a). Interestingly, the main postsynaptic targets of these intrastriatal mGlu4-containing terminals are dendrites of cholinergic interneurons (Kuramoto et al. 2007). Together with recent findings showing that axon collaterals of direct- and indirect-pathway neurons terminate massively upon striatal cholinergic interneurons (Gonzales et al. 2013; Gonzales and Smith 2015) and that mGlu4 receptor activation inhibits intrastriatal GABAergic transmission (Cuomo et al. 2009), these findings confirm that mGlu4 receptor can regulate intrastriatal GABAergic microcircuits and suggest that it could also influence striatal cholinergic activity. In both glutamatergic and GABAergic terminals, mGlu4 and mGlu7 receptors display a specific pattern of subsynaptic localization, being mainly aggregated at the presynaptic active zones of these terminals (Kinoshita et al. 1998; Kosinski et al. 1999; Bradley et al. 1999a, b; Corti et al. 2002; Bogenpohl et al. 2013). Detailed information about the localization and synaptic effects of mGlu8 receptor modulation in the BG is lacking. Thus, as for group III mGlu receptors, mGlu4 activation is viewed as a possible strategy to reduce neurotransmitter release at hyperactive synapses in the BG in parkinsonian state, notably the corticostriatal and striatopallidal projections (Figs. 3.1b and 3.2).

At the level of GPe, GPi, and SNr, mGlu4 and mGlu7 receptors are abundantly expressed in striatal GABAergic terminals (Kinoshita et al. 1998; Kosinski et al. 1999; Bradley et al. 1999a, b; Corti et al. 2002; Bogenpohl et al. 2013), which is consistent with the strong inhibitory effects of mGlu4 receptor agonists or allosteric potentiators on GABAergic transmission at striatopallidal synapses (Valenti et al. 2003; Beurrier et al. 2009; Gubellini et al. 2014) (Fig. 3.3b). In addition, significant presynaptic mGlu4 receptor expression has also been reported in the GP and SNr on putative glutamatergic terminals, arising most likely from the STN, of both rodents and nonhuman primates (Bradley et al. 1999a, b; Bogenpohl et al. 2013) (Fig. 3.3b). Thus, group III mGlu receptors are located to subserve a powerful control over GABAergic and glutamatergic transmission in the GP and SNr.

In contrast to other BG nuclei, the levels of group III mGlu receptor mRNA and protein expression in the STN are low (Kosinski et al. 1999; Bradley et al. 1999a, b;

Corti et al. 2002). However, the activation of these receptors causes a presynaptic depression of excitatory transmission in the rat STN (Awad-Granko and Conn 2001), suggesting the presence of functional group III mGlu receptor subtypes on glutamatergic afferents to the STN (Fig. 3.3c).

In the SNc, slice recording studies have shown that mGlu4 receptor (and possibly mGlu8) activation reduces excitatory transmission (Katayama et al. 2003; Valenti et al. 2005). It has been suggested that this downregulation of glutamatergic transmission may contribute to the neuroprotective effects of mGlu4 receptor agonists against toxin-induced degeneration of midbrain DAergic neurons in models of PD (Battaglia et al. 2006; Betts et al. 2012) (see also Masilamoni and Smith, Chap. 6, this book).

3.4 Group I mGlu Receptors and Parkinson's Disease

Early studies showed that mGlu5 receptor antagonism could ameliorate motor dysfunction in animal models of PD (Conn et al. 2005; Phillips et al. 2006; Ossowska et al. 2007). Negative allosteric modulators (NAMs) that exhibit noncompetitive inhibition of mGlu5 receptors, such as MPEP (2-methyl-6-(phenylethynyl)-pyridine) and MTEP (3-((2-Methyl-1,3-thiazol-4-yl)ethynyl)pyridine hydrochloride), were effective in relieving motor symptoms and LID in a variety of rodent and nonhuman primate models of PD (Breysse et al. 2002, 2003; Coccurello et al. 2004; Ambrosi et al. 2010; Hovelso et al. 2012; Gasparini et al. 2013; Morin et al. 2013; Amalric 2015). Anatomical and behavioral studies further showed that mGlu5 receptor blockade normalizes the hyperactive STN and the cortical input onto the "indirect" pathway (Oueslati et al. 2005; Phillips et al. 2006). Antagonism of mGlu5 receptor is also a potential new target in the management of LID (Gasparini et al. 2013; Rascol et al. 2014). Increased postsynaptic mGlu5 receptor density and specific striatal binding with selective mGlu5 receptor ligands were observed in MPTP-lesioned macaques with LID (Samadi et al. 2008b; Rascol et al. 2014) and in postmortem brains of parkinsonian patients with LID (Ouattara et al. 2011). Further evidence for antidyskinetic efficacy of mGlu5 receptor antagonists in LID come from preclinical studies in rodent (6-OHDA lesion) and monkey (MPTP lesion) PD models (Mela et al. 2007; Levandis et al. 2008; Maranis et al. 2012; Gasparini et al. 2013; Rascol et al. 2014). The mGlu5 receptor NAM mavoglurant (AFQ056) was recently tested in PD patients in clinical phase II and III studies (Berg et al. 2011): this drug, as well as another mGlu5 receptor NAM, dipraglurant (ADX-48621), has shown potential antidyskinetic action in PD patients, without reducing the efficacy of antiparkinsonian therapy (Stocchi et al. 2013; Rascol et al. 2014). Adverse effects including dizziness and hallucinations were reported, however, probably involving other structures outside the BG, such as the lateral habenula, the amygdala, the prefrontal cortex and the hippocampus. Inhibition of glutamate activity by acting on mGlu receptors located in these areas may thus prove to be a novel target to reduce LID (Bastide et al. 2014; Engeln et al. 2014).

Another major breakthrough of targeting group I mGlu receptors for PD came from studies showing that their blockade, in particular of mGlu5 (by MPEP or other selective antagonists), could provide neuroprotection in rodent and monkey models of PD (Battaglia et al. 2004; Vernon et al. 2005, 2007a; Aguirre et al. 2005; Armentero et al. 2006; Ambrosi et al. 2010; Chen et al. 2011; Masilamoni et al. 2011; Hsieh et al. 2012; Fuzzati-Armentero et al. 2015).

3.5 Group II mGlu Receptors and Parkinson's Disease

Selective agonists of mGlu2 and mGlu3 receptors potently decrease excitatory transmission at corticostriatal synapses via a presynaptic mechanism (Picconi et al. 2002). The modulation of group II mGlu receptors did not however provide significant benefit in animal models of PD. In particular, agonists of mGlu2 and mGlu3 receptors did not show good evidence of an antiparkinsonian action in rodent models of PD (Rylander et al. 2009) and even worsened motor symptoms (Murray et al. 2002). Group II mGlu receptors also do not play a significant role in LID, and the expression of mGlu2 and mGlu3 receptors in the BG does not change with L-DOPA treatment in parkinsonian primates (Samadi et al. 2008a). The recent development of subtype 2 or 3 selective agonists may help to resolve this issue.

3.6 Group III mGlu Receptors and Parkinson's Disease

Among group III (mGlu4, mGlu7, and mGlu8) mGlu receptors, the targeting of mGlu4 has been the most characterized in PD models using orthosteric agonists or positive allosteric modulators (PAMs). Each one presents its own advantages and shortcomings, in particular orthosteric agonist being less selective and less able to pass the blood-brain barrier compared to PAMs but more effective in activating the receptor. Whether mGlu4 receptor PAMs are preferable or not to orthosteric agonists for potential PD treatments is still a matter of debate (Flor and Acher 2012; Iderberg et al. 2015).

The first drug targeting mGlu4 receptor showing an antiparkinsonian effect was the PAM N-phenyl-7-(hydroxylimino)cyclopropa[b]-chromen-1a-carboxamide (PHCCC), which produced a marked reversal of reserpine-induced akinesia in rats when injected intracerebroventricularly (Marino et al. 2003). More recent, potent, and selective mGlu4 receptor PAMs, such as VU0155041 (Niswender et al. 2008), ML128 (Hopkins et al. 2010), VU0400195 and ML182 (Jones et al. 2011), N-phenyl-7-(hydroxylimino)cyclopropa[b]-chromen-1a-carboxamide (PHCCC) (Broadstock et al. 2012), and VU0364770 (Jones et al. 2012), have been shown to decrease haloperidol-induced catalepsy and/or reserpine-induced akinesia in rats when injected in the brain or administered systemically (orally or by i.p. injection). When tested on more relevant PD models, systemic administration of VU0364770 decreased forelimb asymmetry in rats with unilateral nigral 6-OHDA lesion and reduced premature responses in rats with bilateral striatal 6-OHDA lesions (Jones et al. 2012) [surprisingly, a subsequent study VU0364770 failed to reproduce such effects on forelimb asymmetry, but it could decrease L-DOPA-induced rotations and improve rotarod score (Iderberg et al. 2015)]. Other two mGlu4 PAMs, namely ADX88178 (Le Poul et al. 2012) and Lu AF21934 (Bennouar et al. 2013), could reverse haloperidol-induced catalepsy in rats but had no impact on forelimb akinesia resulting from unilateral 6-OHDA lesion. An interesting feature of VU0364770, ADX88178, and Lu AF21934, however, is that they could significantly reduce forelimb asymmetry in rats with unilateral striatal 6-OHDA lesion when coadministered with low doses of L-DOPA that per se were ineffective (Jones et al. 2012; Le Poul et al. 2012; Bennouar et al. 2013). Such synergic action also allowed to slightly improve LID (Iderberg et al. 2015) and, more interestingly, significantly reduce its occurrence (Bennouar et al. 2013). Overall, these studies suggest that a combined treatment with low doses of L-DOPA and an mGlu4 receptor PAM could be a valuable therapeutic strategy for alleviating PD motor symptoms and minimizing the occurrence of LID. A possible explanation for the little or no effect of these PAMs alone against akinesia in animal PD models could be found in their lack of efficacy at striatopallidal GABAergic synapses (Gubellini et al. 2014) and on the activity of GP neurons (Bogenpohl et al. 2013), which thus would limit their action to the striatum.

Concerning group III orthosteric agonists, it was reported that the administration of (1S,3R,4S)-1-aminocyclopentane-1,3,4-tricarboxylic acid (ACPT-I), O-phospho-L-serine (L-SOP), L-2-amino-4-phosphono-butanoate (L-AP4), or L-AP4 analogs causes complex effects in rats treated with haloperidol or lesioned with 6-OHDA, notably a reduction of motor deficits when injected in the GP and a lesser or opposite effect when injected in the SNr (Lopez et al. 2007, 2008; Sibille et al. 2007; Konieczny et al. 2007; Sun et al. 2013), the latter outcome being also observed with the selective mGlu8 agonist (S)-3,4-dicarboxyphenylglycine (DCPG) (Lopez et al. 2007). Together, these data suggest that the activation of group III mGlu receptors—notably mGlu4—provides benefits in parkinsonian rats by modulating GABAergic neurotransmission in the GP, which mGlu4 PAMs seem unable to achieve (Bogenpohl et al. 2013; Gubellini et al. 2014). Interestingly, chronic systemic injection of ACPT-I in rats with partial bilateral nigrostriatal lesions could alleviate the akinetic deficit evidenced in a reaction time task, and this was associated with the reversal of subthalamonigral overactivity (Lopez et al. 2012). Concerning subtype-selective ligands, the selective mGlu4 agonist LSP1-3081 (or L-AP4) could inhibit both glutamatergic and GABAergic synaptic transmission in the striatum and reduce forelimb asymmetry in rats with unilateral 6-OHDA lesion when injected in the striatum (Cuomo et al. 2009), confirming that this structure of the BG is a main target of drugs acting on group III mGlu receptors for PD treatment. Another mGlu4 agonist, LSP1-2111, could depress striatopallidal synaptic transmission, improve the performance of rats with striatal 6-OHDA lesion in the reaction time test when injected in the GP, as well as reduce haloperidol-induced catalepsy when i.p. injected (Beurrier et al. 2009). Interestingly, systemic administration of LSP1-2111 also produced a reduction of LID in 6-OHDA-lesioned mice (Lopez et al. 2011), similarly to the PAM Lu AF21934 (Bennouar et al. 2013).

Concerning mGlu7 receptor, promising results were provided by the PAM AMN082 (oral or intrastriatal administration), which could reverse haloperidol-induced catalepsy in naïve rats, reduce apomorphine-induced rotations in unilateral 6-OHDA-lesioned rats, and improve reaction time task in rats with bilateral striatal 6-OHDA lesion (Greco et al. 2010). Systemic (i.p.), intrastriatal, or intranigral administration of AMN082 also decreased haloperidol-induced catalepsy (Konieczny and Lenda 2013) and reserpine-induced akinesia (Broadstock et al. 2012). However, no data with this drug are available on primate models of PD.

The mGlu8 receptor agonist DCPG (intracerebroventricular administration in rats) could reduce motor deficits caused by prolonged (18–20 h) but not acute (2 h) haloperidol and reserpine treatments, and improve forelimb use asymmetry caused by unilateral 6-OHDA lesion (Johnson et al. 2013). However, a previous study showed no alleviation of prolonged reserpine-induced akinesia (Broadstock et al. 2012).

Group III agonists such as L-AP4 and L-SOP have also been shown to confer some degree of neuroprotection in PD models. For example, reduction of TH+ cell loss in the SNc and/or striatal DA denervation was reported in 6-OHDA-lesioned rodents administered with these agonists (Vernon et al. 2005, 2007b; Austin et al. 2010; Betts et al. 2012), and decrease of rotenone toxicity on midbrain TH+ neurons in culture was obtained by L-AP4 application (Jiang et al. 2006).

3.7 Concluding Remarks

The three groups of mGlu receptors are widely expressed in the BG nuclei where they act as key regulators of BG function in both normal and diseased conditions, by modulating neuronal excitability as well as synaptic transmission and plasticity. Since the late 1990s, data from various laboratories have provided strong evidence that the targeting of specific mGlu receptors by means of selective drugs could be a possible strategy in PD treatment for restoring synaptic function and neuronal activity in the BG, thus improving motor symptoms. In this context, the spectacular development of subtype-selective mGlu receptor agonists, antagonists and allosteric modulators has provided scientists with a wide range of neuropharmacological tools to test this hypothesis.

In light of the vast anatomical and functional studies of mGlu receptors gathered throughout the BG circuitry, converging evidence shows that antagonism of group I mGlu receptors (in particular mGlu5), as well as stimulation of group III mGlu receptors (in particular mGlu4), can alleviate motor symptoms, reduce LID, and provide neuroprotection in some animal models of PD. Clinical trials confirmed in part such findings, showing that mGlu5 receptor NAMs can improve motor symptoms and LID in parkinsonian patients.

References

Aguirre JA, Kehr J, Yoshitake T, Liu FL, Rivera A, Fernandez-Espinola S, Andbjer B, Leo G, Medhurst AD, Agnati LF, Fuxe K (2005) Protection but maintained dysfunction of nigral dopaminergic nerve cell bodies and striatal dopaminergic terminals in MPTP-lesioned mice after acute treatment with the mGluR5 antagonist MPEP. Brain Res 1033:216–220

Ahlskog JE, Muenter MD (2001) Frequency of levodopa-related dyskinesias and motor fluctuations as estimated from the cumulative literature. Mov Disord 16:448–458

Ambrosi G, Armentero MT, Levandis G, Bramanti P, Nappi G, Blandini F (2010) Effects of early and delayed treatment with an mGluR5 antagonist on motor impairment, nigrostriatal damage and neuroinflammation in a rodent model of Parkinson's disease. Brain Res Bull 82:29–38

Armentero MT, Fancellu R, Nappi G, Bramanti P, Blandini F (2006) Prolonged blockade of NMDA or mGluR5 glutamate receptors reduces nigrostriatal degeneration while inducing selective metabolic changes in the basal ganglia circuitry in a rodent model of Parkinson's disease. Neurobiol Dis 22:1–9

Amalric M. Targeting metabotropic glutamate receptors (mGluRs) in Parkinson's disease. Curr Opin Pharmacol. 2015;20:29–34.

Austin PJ, Betts MJ, Broadstock M, O'Neill MJ, Mitchell SN, Duty S (2010) Symptomatic and neuroprotective effects following activation of nigral group III metabotropic glutamate receptors in rodent models of Parkinson's disease. Br J Pharmacol 160:1741–1753

Awad H, Hubert GW, Smith Y, Levey AI, Conn PJ (2000) Activation of metabotropic glutamate receptor 5 has direct excitatory effects and potentiates NMDA receptor currents in neurons of the subthalamic nucleus. J Neurosci 20:7871–7879

Awad-Granko H, Conn PJ (2001) Activation of groups I or III metabotropic glutamate receptors inhibits excitatory transmission in the rat subthalamic nucleus. Neuropharmacology 41:32–41

Bagetta V, Ghiglieri V, Sgobio C, Calabresi P, Picconi B (2010) Synaptic dysfunction in Parkinson's disease. Biochem Soc Trans 38:493–497

Bamford NS, Robinson S, Palmiter RD, Joyce JA, Moore C, Meshul CK (2004) Dopamine modulates release from corticostriatal terminals. J Neurosci 24:9541–9552

Bastide MF, Dovero S, Charron G, Porras G, Gross CE, Fernagut PO, Bezard E (2014) Immediate-early gene expression in structures outside the basal ganglia is associated to l-DOPA-induced dyskinesia. Neurobiol Dis 62:179–192

Bastide MF et al (2015) Pathophysiology of L-dopa-induced motor and non-motor complications in Parkinson's disease. Prog Neurobiol 132:96–168

Battaglia G, Bruno V, Pisani A, Centonze D, Catania MV, Calabresi P, Nicoletti F (2001) Selective blockade of type-1 metabotropic glutamate receptors induces neuroprotection by enhancing gabaergic transmission. Mol Cell Neurosci 17:1071–1083

Battaglia G, Busceti CL, Molinaro G, Biagioni F, Storto M, Fornai F, Nicoletti F, Bruno V (2004) Endogenous activation of mGlu5 metabotropic glutamate receptors contributes to the development of nigro-striatal damage induced by 1-methyl-4-phenyl-1,2,3,6-tetrahydropyridine in mice. J Neurosci 24:828–835

Battaglia G, Busceti CL, Molinaro G, Biagioni F, Traficante A, Nicoletti F, Bruno V (2006) Pharmacological activation of mGlu4 metabotropic glutamate receptors reduces nigrostriatal degeneration in mice treated with 1-methyl-4-phenyl-1,2,3,6-tetrahydropyridine. J Neurosci 26:7222–7229

Baude A, Nusser Z, Roberts JD, Mulvihill E, McIlhinney RA, Somogyi P (1993) The metabotropic glutamate receptor (mGluR1 alpha) is concentrated at perisynaptic membrane of neuronal subpopulations as detected by immunogold reaction. Neuron 11:771–787

Benabid AL, Koudsie A, Benazzouz A, Piallat B, Krack P, Limousin-Dowsey P, Lebas JF, Pollak P (2001) Deep brain stimulation for Parkinson's disease. Adv Neurol 86:405–412

Benabid AL, Chabardes S, Mitrofanis J, Pollak P (2009) Deep brain stimulation of the subthalamic nucleus for the treatment of Parkinson's disease. Lancet Neurol 8:67–81

Bennouar KE, Uberti MA, Melon C, Bacolod MD, Jimenez HN, Cajina M, Kerkerian-Le GL, Doller D, Gubellini P (2013) Synergy between L-DOPA and a novel positive allosteric modulator of metabotropic glutamate receptor 4: implications for Parkinson's disease treatment and dyskinesia. Neuropharmacology 66:158–169

Berg D, Godau J, Trenkwalder C, Eggert K, Csoti I, Storch A, Huber H, Morelli-Canelo M, Stamelou M, Ries V, Wolz M, Schneider C, Di PT, Gasparini F, Hariry S, Vandemeulebroecke M, Bi-Saab W, Cooke K, Johns D, Gomez-Mancilla B (2011) AFQ056 treatment of levodopa-induced dyskinesias: results of 2 randomized controlled trials. Mov Disord 26:1243–1250

Betts MJ, O'Neill MJ, Duty S (2012) Allosteric modulation of the group III mGlu(4) receptor provides functional neuroprotection in the 6-hydroxydopamine rat model of Parkinson's disease. Br J Pharmacol 166:2317–2330

Beurrier C, Lopez S, Revy D, Selvam C, Goudet C, Lherondel M, Gubellini P, Kerkerian-Legoff L, Acher F, Pin JP, Amalric M (2009) Electrophysiological and behavioral evidence that modulation of metabotropic glutamate receptor 4 with a new agonist reverses experimental parkinsonism. FASEB J 23:3619–3628

Blandini F, Nappi G, Tassorelli C, Martignoni E (2000) Functional changes of the basal ganglia circuitry in Parkinson's disease. Prog Neurobiol 62:63–88

Bogenpohl J, Galvan A, Hu X, Wichmann T, Smith Y (2013) Metabotropic glutamate receptor 4 in the basal ganglia of parkinsonian monkeys: ultrastructural localization and electrophysiological effects of activation in the striatopallidal complex. Neuropharmacology 66:242–252

Bonsi P, Cuomo D, Picconi B, Sciamanna G, Tscherter A, Tolu M, Bernardi G, Calabresi P, Pisani A (2007) Striatal metabotropic glutamate receptors as a target for pharmacotherapy in Parkinson's disease. Amino Acids 32:189–195

Bradley SR, Standaert DG, Levey AI, Conn PJ (1999a) Distribution of group III mGluRs in rat basal ganglia with subtype-specific antibodies. Ann N Y Acad Sci 868:531–534

Bradley SR, Standaert DG, Rhodes KJ, Rees HD, Testa CM, Levey AI, Conn PJ (1999b) Immunohistochemical localization of subtype 4a metabotropic glutamate receptors in the rat and mouse basal ganglia. J Comp Neurol 407:33–46

Bradley SR, Marino MJ, Wittmann M, Rouse ST, Awad H, Levey AI, Conn PJ (2000) Activation of group II metabotropic glutamate receptors inhibits synaptic excitation of the substantia Nigra pars reticulata. J Neurosci 20:3085–3094

Breysse N, Baunez C, Spooren W, Gasparini F, Amalric M. Chronic but not acute treatment with a metabotropic glutamate 5 receptor antagonist reverses the akinetic deficits in a rat model of parkinsonism. J Neurosci. 2002;22:5669–5678.

Breysse N, Amalric M, Salin P. Metabotropic glutamate 5 receptor blockade alleviates akinesia by normalizing activity of selective basal-ganglia structures in parkinsonian rats. J Neurosci. 2003;23:8302–8309.

Brichta L, Greengard P, Flajolet M (2013) Advances in the pharmacological treatment of Parkinson's disease: targeting neurotransmitter systems. Trends Neurosci 36:543–554

Broadstock M, Austin PJ, Betts MJ, Duty S (2012) Antiparkinsonian potential of targeting group III metabotropic glutamate receptor subtypes in the rodent substantia nigra pars reticulata. Br J Pharmacol 165:1034–1045

Buttery PC, Barker RA (2014) Treating Parkinson's disease in the 21st century: can stem cell transplantation compete? J Comp Neurol 522:2802–2816

Calabresi P, Picconi B, Tozzi A, Ghiglieri V, Di FM (2014) Direct and indirect pathways of basal ganglia: a critical reappraisal. Nat Neurosci 17:1022–1030

Chase TN (1998) Levodopa therapy: consequences of the nonphysiologic replacement of dopamine. Neurology 50:S17–S25

Chase TN, Oh JD (2000) Striatal dopamine- and glutamate-mediated dysregulation in experimental parkinsonism. Trends Neurosci 23:S86–S91

Chase TN, Bibbiani F, Oh JD (2003) Striatal glutamatergic mechanisms and extrapyramidal movement disorders. Neurotox Res 5:139–146

Chen L, Liu J, Ali U, Gui ZH, Wang Y, Wang T, Hou C, Fan LL (2011) Blockade of mGluR5 reverses abnormal firing of subthalamic nucleus neurons in 6-hydroxydopamine partially lesioned rats. Chin J Phys 54:303–309

Chuhma N, Choi WY, Mingote S, Rayport S (2009) Dopamine neuron glutamate cotransmission: frequency-dependent modulation in the mesoventromedial projection. Neuroscience 164:1068–1083

Coccurello R, Breysse N, Amalric M. Simultaneous blockade of adenosine A2A and metabotropic glutamate mGlu5 receptors increase their efficacy in reversing Parkinsonian deficits in rats. Neuropsychopharmacology. 2004;29:1451–1461.

Conn PJ, Pin JP (1997) Pharmacology and functions of metabotropic glutamate receptors. Annu Rev Pharmacol Toxicol 37:205–237

Conn PJ, Battaglia G, Marino MJ, Nicoletti F (2005) Metabotropic glutamate receptors in the basal ganglia motor circuit. Nat Rev Neurosci 6:787–798

Corti C, Aldegheri L, Somogyi P, Ferraguti F (2002) Distribution and synaptic localisation of the metabotropic glutamate receptor 4 (mGluR4) in the rodent CNS. Neuroscience 110:403–420

Cuomo D, Martella G, Barabino E, Platania P, Vita D, Madeo G, Selvam C, Goudet C, Oueslati N, Pin JP, Acher F, Pisani A, Beurrier C, Melon C, Kerkerian-Le GL, Gubellini P (2009) Metabotropic glutamate receptor subtype 4 selectively modulates both glutamate and GABA transmission in the striatum: implications for Parkinson's disease treatment. J Neurochem 109:1096–1105

DeLong MR, Wichmann T (2007) Circuits and circuit disorders of the basal ganglia. Arch Neurol 64:20–24

Engeln M, De DP, Li Q, Bezard E (2014) Fernagut PO. Widespread Monoaminergic Dysregulation of Both Motor and Non-Motor Circuits in Parkinsonism and Dyskinesia, Cereb Cortex

Fasano A, Daniele A, Albanese A (2012) Treatment of motor and non-motor features of Parkinson's disease with deep brain stimulation. Lancet Neurol 11:429–442

Flor PJ, Acher FC (2012) Orthosteric versus allosteric GPCR activation: the great challenge of group-III mGluRs. Biochem Pharmacol 84:414–424

Fuzzati-Armentero MT, Cerri S, Levandis G, Ambrosi G, Montepeloso E, Antoninetti G, Blandini F, Baqi Y, Muller CE, Volpini R, Costa G, Simola N, Pinna A (2015) Dual target strategy: combining distinct non-dopaminergic treatments reduces neuronal cell loss and synergistically modulates l-DOPA-induced rotational behavior in a rodent model of Parkinson's disease. J Neurochem 134:740–747

Galvan A, Kuwajima M, Smith Y (2006) Glutamate and GABA receptors and transporters in the basal ganglia: what does their subsynaptic localization reveal about their function? Neuroscience 143:351–375

Gasparini F, Di PT, Gomez-Mancilla B (2013) Metabotropic glutamate receptors for Parkinson's disease therapy. Parkinsons Dis 2013:196028

Gonzales KK, Smith Y (2015) Cholinergic interneurons in the dorsal and ventral striatum: anatomical and functional considerations in normal and diseased conditions. Ann N Y Acad Sci 1349:1–45

Gonzales KK, Pare JF, Wichmann T, Smith Y (2013) GABAergic inputs from direct and indirect striatal projection neurons onto cholinergic interneurons in the primate putamen. J Comp Neurol 521:2502–2522

Greco B, Lopez S, van der PH FPJ, Amalric M (2010) Metabotropic glutamate 7 receptor subtype modulates motor symptoms in rodent models of Parkinson's disease. J Pharmacol Exp Ther 332:1064–1071

Guatteo E, Mercuri NB, Bernardi G, Knopfel T (1999) Group I metabotropic glutamate receptors mediate an inward current in rat substantia nigra dopamine neurons that is independent from calcium mobilization. J Neurophysiol 82:1974–1981

Gubellini P, Saulle E, Centonze D, Bonsi P, Pisani A, Bernardi G, Conquet F, Calabresi P (2001) Selective involvement of mGlu1 receptors in corticostriatal LTD. Neuropharmacology 40:839–846

Gubellini P, Saulle E, Centonze D, Costa C, Tropepi D, Bernardi G, Conquet F, Calabresi P (2003) Corticostriatal LTP requires combined mGluR1 and mGluR5 activation. Neuropharmacology 44:8–16

Gubellini P, Pisani A, Centonze D, Bernardi G, Calabresi P (2004) Metabotropic glutamate receptors and striatal synaptic plasticity: implications for neurological diseases. Prog Neurobiol 74:271–300

Gubellini P, Salin P, Kerkerian-Le GL, Baunez C (2009) Deep brain stimulation in neurological diseases and experimental models: from molecule to complex behavior. Prog Neurobiol 89:79–123

Gubellini P, Melon C, Dale E, Doller D, Kerkerian-Le GL (2014) Distinct effects of mGlu4 receptor positive allosteric modulators at corticostriatal vs. striatopallidal synapses may differentially contribute to their antiparkinsonian action. Neuropharmacology 85C:166–177

Hanson JE, Smith Y (1999) Group I metabotropic glutamate receptors at GABAergic synapses in monkeys. J Neurosci 19:6488–6496

Hnasko TS, Chuhma N, Zhang H, Goh GY, Sulzer D, Palmiter RD, Rayport S, Edwards RH (2010) Vesicular glutamate transport promotes dopamine storage and glutamate corelease in vivo. Neuron 65:643–656

Hopkins CR, Niswender CM, Lewis LM, Weaver CD, Lindsley CW. Discovery of a potent, selective and in vivo active mGluR4 positive allosteric modulator. 2010.

Hovelso N, Sotty F, Montezinho LP, Pinheiro PS, Herrik KF, Mork A (2012) Therapeutic potential of metabotropic glutamate receptor modulators. Curr Neuropharmacol 10:12–48

Hsieh MH, Ho SC, Yeh KY, Pawlak CR, Chang HM, Ho YJ, Lai TJ, Wu FY (2012) Blockade of metabotropic glutamate receptors inhibits cognition and neurodegeneration in an MPTP-induced Parkinson's disease rat model. Pharmacol Biochem Behav 102:64–71

Hubert GW, Smith Y (2004) Age-related changes in the expression of axonal and glial group I metabotropic glutamate receptor in the rat substantia nigra pars reticulata. J Comp Neurol 475:95–106

Hubert GW, Paquet M, Smith Y (2001) Differential subcellular localization of mGluR1a and mGluR5 in the rat and monkey Substantia nigra. J Neurosci 21:1838–1847

Iderberg H, Maslava N, Thompson AD, Bubser M, Niswender CM, Hopkins CR, Lindsley CW, Conn PJ, Jones CK, Cenci MA (2015) Pharmacological stimulation of metabotropic glutamate receptor type 4 in a rat model of Parkinson's disease and l-DOPA-induced dyskinesia: Comparison between a positive allosteric modulator and an orthosteric agonist. Neuropharmacology 95:121–129

Iskhakova L, Smith Y (2015) MGluR4-containing corticostriatal terminals: Synaptic interactions with direct and indirect pathway neurons in mice. Brain Struct Funct 221(9):4589–4599

Jankovic J (2008) Parkinson's disease: clinical features and diagnosis. J Neurol Neurosurg Psychiatry 79:368–376

Jiang Q, Yan Z, Feng J (2006) Activation of group III metabotropic glutamate receptors attenuates rotenone toxicity on dopaminergic neurons through a microtubule-dependent mechanism. J Neurosci 26:4318–4328

Johnson KA, Niswender CM, Conn PJ, Xiang Z (2011) Activation of group II metabotropic glutamate receptors induces long-term depression of excitatory synaptic transmission in the substantia nigra pars reticulata. Neurosci Lett 504:102–106

Johnson KA, Jones CK, Tantawy MN, Bubser M, Marvanova M, Ansari MS, Baldwin RM, Conn PJ, Niswender CM (2013) The metabotropic glutamate receptor 8 agonist (S)-3,4-DCPG reverses motor deficits in prolonged but not acute models of Parkinson's disease. Neuropharmacology 66:187–195

Jones CK, Engers DW, Thompson AD, Field JR, Blobaum AL, Lindsley SR, Zhou Y, Gogliotti RD, Jadhav S, Zamorano R, Bogenpohl J, Smith Y, Morrison R, Daniels JS, Weaver CD, Conn PJ, Lindsley CW, Niswender CM, Hopkins CR (2011) Discovery, Synthesis, and Structure-Activity Relationship Development of a Series of N-4-(2,5-Dioxopyrrolidin-1-yl)phenylpicolinamides (VU0400195, ML182): Characterization of a Novel Positive Allosteric Modulator of the Metabotropic Glutamate Receptor 4 (mGlu(4)) with Oral Efficacy in an Antiparkinsonian Animal Model. J Med Chem 54:7639–7647

Jones CK, Bubser M, Thompson AD, Dickerson JW, Turle-Lorenzo N, Amalric M, Blobaum AL, Bridges TM, Morrison RD, Jadhav S, Engers DW, Italiano K, Bode J, Daniels JS, Lindsley CW, Hopkins CR, Conn PJ, Niswender CM (2012) The metabotropic glutamate receptor 4-positive allosteric modulator VU0364770 produces efficacy alone and in combination with L-DOPA or an adenosine 2A antagonist in preclinical rodent models of Parkinson's disease. J Pharmacol Exp Ther 340:404–421

Kahn L, Alonso G, Robbe D, Bockaert J, Manzoni OJ (2001) Group 2 metabotropic glutamate receptors induced long term depression in mouse striatal slices. Neurosci Lett 316:178–182

Kaneda K, Tachibana Y, Imanishi M, Kita H, Shigemoto R, Nambu A, Takada M (2005) Down-regulation of metabotropic glutamate receptor 1alpha in globus pallidus and substantia nigra of parkinsonian monkeys. Eur J Neurosci 22:3241–3254

Kaneda K, Kita T, Kita H (2007) Repetitive activation of glutamatergic inputs evokes a long-lasting excitation in rat globus pallidus neurons in vitro. J Neurophysiol 97:121–133

Karas PJ, Mikell CB, Christian E, Liker MA, Sheth SA (2013) Deep brain stimulation: a mechanistic and clinical update. Neurosurg Focus 35:E1

Katayama J, Akaike N, Nabekura J (2003) Characterization of pre- and post-synaptic metabotropic glutamate receptor-mediated inhibitory responses in substantia nigra dopamine neurons. Neurosci Res 45:101–115

Kinoshita A, Shigemoto R, Ohishi H (1998) van der PH, Mizuno N. Immunohistochemical localization of metabotropic glutamate receptors, mGluR7a and mGluR7b, in the central nervous system of the adult rat and mouse: a light and electron microscopic study. J Comp Neurol 393:332–352

Konieczny J, Lenda T (2013) Contribution of the mGluR7 receptor to antiparkinsonian-like effects in rats: a behavioral study with the selective agonist AMN082. Pharmacol Rep 65:1194–1203

Konieczny J, Wardas J, Kuter K, Pilc A, Ossowska K (2007) The influence of group III metabotropic glutamate receptor stimulation by (1S,3R,4S)-1-aminocyclo-pentane-1,3,4-tricarboxylic acid on the parkinsonian-like akinesia and striatal proenkephalin and prodynorphin mRNA expression in rats. Neuroscience 145:611–620

Kosinski CM, Risso BS, Conn PJ, Levey AI, Landwehrmeyer GB, Penney JB Jr, Young AB, Standaert DG (1999) Localization of metabotropic glutamate receptor 7 mRNA and mGluR7a protein in the rat basal ganglia. J Comp Neurol 415:266–284

Kupferschmidt DA, Lovinger DM (2015) Inhibition of presynaptic calcium transients in cortical inputs to the dorsolateral striatum by metabotropic GABA(B) and mGlu2/3 receptors. J Physiol 593:2295–2310

Kuramoto E, Fujiyama F, Unzai T, Nakamura K, Hioki H, Furuta T, Shigemoto R, Ferraguti F, Kaneko T (2007) Metabotropic glutamate receptor 4-immunopositive terminals of medium-sized spiny neurons selectively form synapses with cholinergic interneurons in the rat neostriatum. J Comp Neurol 500:908–922

Kuwajima M, Hall RA, Aiba A, Smith Y (2004) Subcellular and subsynaptic localization of group I metabotropic glutamate receptors in the monkey subthalamic nucleus. J Comp Neurol 474:589–602

Kuwajima M, Dehoff MH, Furuichi T, Worley PF, Hall RA, Smith Y (2007) Localization and expression of group I metabotropic glutamate receptors in the mouse striatum, globus pallidus, and subthalamic nucleus: regulatory effects of MPTP treatment and constitutive Homer deletion. J Neurosci 27:6249–6260

Le Poul E, Bolea C, Girard F, Poli S, Charvin D, Campo B, Bortoli J, Bessif A, Luo B, Koser AJ, Hodge LM, Smith KM, DiLella AG, Liverton N, Hess F, Browne SE, Reynolds IJ (2012) A potent and selective metabotropic glutamate receptor 4 positive allosteric modulator improves movement in rodent models of Parkinson's disease. J Pharmacol Exp Ther 343:167–177

Levandis G, Bazzini E, Armentero MT, Nappi G, Blandini F (2008) Systemic administration of an mGluR5 antagonist, but not unilateral subthalamic lesion, counteracts l-DOPA-induced dyskinesias in a rodent model of Parkinson's disease. Neurobiol Dis 29:161–168

Levy R, Hazrati LN, Herrero MT, Vila M, Hassani OK, Mouroux M, Ruberg M, Asensi H, Agid Y, Feger J, Obeso JA, Parent A, Hirsch EC (1997) Re-evaluation of the functional anatomy of the basal ganglia in normal and Parkinsonian states. Neuroscience 76:335–343

Li X, Gardner EL, Xi ZX (2008) The metabotropic glutamate receptor 7 (mGluR7) allosteric agonist AMN082 modulates nucleus accumbens GABA and glutamate, but not dopamine, in rats. Neuropharmacology 54:542–551

Lopez S, Turle-Lorenzo N, Acher F, De LE, Mele A, Amalric M (2007) Targeting group III metabotropic glutamate receptors produces complex behavioral effects in rodent models of Parkinson's disease. J Neurosci 27:6701–6711

Lopez S, Turle-Lorenzo N, Johnston TH, Brotchie JM, Schann S, Neuville P, Amalric M (2008) Functional interaction between adenosine A2A and group III metabotropic glutamate receptors to reduce parkinsonian symptoms in rats. Neuropharmacology 55:483–490

Lopez S, Bonito-Oliva A, Pallottino S, Acher F, Fisone G (2011) Activation of metabotropic glutamate 4 receptors decreases L-DOPA-induced dyskinesia in a mouse model of Parkinson's disease. J Parkinsons Dis 1:339–346

Lopez S, Jouve L, Turle-Lorenzo N, Kerkerian-Legoff L, Salin P, Amalric M (2012) Antiparkinsonian action of a selective group III mGlu receptor agonist is associated with reversal of subthalamonigral overactivity. Neurobiol Dis 46:69–77

Lovinger DM (2010) Neurotransmitter roles in synaptic modulation, plasticity and learning in the dorsal striatum. Neuropharmacology 58:951–961

Lovinger DM, McCool BA (1995) Metabotropic glutamate receptor-mediated presynaptic depression at corticostriatal synapses involves mGLuR2 or 3. J Neurophysiol 73:1076–1083

Mansouri M, Kasugai Y, Fukazawa Y, Bertaso F, Raynaud F, Perroy J, Fagni L, Kaufmann WA, Watanabe M, Shigemoto R, Ferraguti F (2015) Distinct subsynaptic localization of type 1 metabotropic glutamate receptors at glutamatergic and GABAergic synapses in the rodent cerebellar cortex. Eur J Neurosci 41:157–167

Maranis S, Stamatis D, Tsironis C, Konitsiotis S (2012) Investigation of the antidyskinetic site of action of metabotropic and ionotropic glutamate receptor antagonists. Intracerebral infusions in 6-hydroxydopamine-lesioned rats with levodopa-induced dyskinesia. Eur J Pharmacol 683:71–77

Marino MJ, Wittmann M, Bradley SR, Hubert GW, Smith Y, Conn PJ (2001) Activation of group I metabotropic glutamate receptors produces a direct excitation and disinhibition of GABAergic projection neurons in the substantia nigra pars reticulata. J Neurosci 21:7001–7012

Marino MJ, Wad-Granko H, Ciombor KJ, Conn PJ (2002) Haloperidol-induced alteration in the physiological actions of group I mGlus in the subthalamic nucleus and the substantia nigra pars reticulata. Neuropharmacology 43:147–159

Marino MJ, Williams DL Jr, O'Brien JA, Valenti O, McDonald TP, Clements MK, Wang R, DiLella AG, Hess JF, Kinney GG, Conn PJ (2003) Allosteric modulation of group III metabotropic glutamate receptor 4: a potential approach to Parkinson's disease treatment. Proc Natl Acad Sci U S A 100:13668–13673

Marti M, Paganini F, Stocchi S, Bianchi C, Beani L, Morari M (2001) Presynaptic group I and II metabotropic glutamate receptors oppositely modulate striatal acetylcholine release. Eur J Neurosci 14:1181–1184

Masilamoni GJ, Bogenpohl JW, Alagille D, Delevich K, Tamagnan G, Votaw JR, Wichmann T, Smith Y. Metabotropic glutamate receptor 5 antagonist protects dopaminergic and noradrenergic neurons from degeneration in MPTP-treated monkeys. Brain. 2011;134:2057–2073.

Mela F, Marti M, Dekundy A, Danysz W, Morari M, Cenci MA (2007) Antagonism of metabotropic glutamate receptor type 5 attenuates l-DOPA-induced dyskinesia and its molecular and neurochemical correlates in a rat model of Parkinson's disease. J Neurochem 101:483–497

Messenger MJ, Dawson LG, Duty S (2002) Changes in metabotropic glutamate receptor 1-8 gene expression in the rodent basal ganglia motor loop following lesion of the nigrostriatal tract. Neuropharmacology 43:261–271

Mitrano DA, Smith Y (2007) Comparative analysis of the subcellular and subsynaptic localization of mGluR1a and mGluR5 metabotropic glutamate receptors in the shell and core of the nucleus accumbens in rat and monkey. J Comp Neurol 500:788–806

Morin N, Morissette M, Gregoire L, Gomez-Mancilla B, Gasparini F, Di PT (2013) Chronic treatment with MPEP, an mGlu5 receptor antagonist, normalizes basal ganglia glutamate neurotransmission in L-DOPA-treated parkinsonian monkeys. Neuropharmacology 73:216–231

Murray TK, Messenger MJ, Ward MA, Woodhouse S, Osborne DJ, Duty S, O'Neill MJ (2002) Evaluation of the mGluR2/3 agonist LY379268 in rodent models of Parkinson's disease. Pharmacol Biochem Behav 73:455–466

Niswender CM, Johnson KA, Weaver CD, Jones CK, Xiang Z, Luo Q, Rodriguez AL, Marlo JE, de Paulis T, Thompson AD, Days EL, Nalywajko T, Austin CA, Williams MB, Ayala JE, Williams R, Lindsley CW, Conn PJ (2008) Discovery, characterization, and antiparkinsonian effect of novel positive allosteric modulators of metabotropic glutamate receptor 4. Mol Pharmacol 74:1345–1358

Nusser Z, Mulvihill E, Streit P, Somogyi P (1994) Subsynaptic segregation of metabotropic and ionotropic glutamate receptors as revealed by immunogold localization. Neuroscience 61:421–427

Obeso JA, Rodriguez-Oroz MC, Rodriguez M, Lanciego JL, Artieda J, Gonzalo N, Olanow CW (2000) Pathophysiology of the basal ganglia in Parkinson's disease. Trends Neurosci 23:S8–19

Obeso JA, Rodriguez-Oroz MC, Benitez-Temino B, Blesa FJ, Guridi J, Marin C, Rodriguez M (2008) Functional organization of the basal ganglia: therapeutic implications for Parkinson's disease. Mov Disord 23(Suppl 3):S548–S559

Olanow CW (2007) The pathogenesis of cell death in Parkinson's disease. Mov Disord 22(Suppl 17):S335–S342

Olanow CW, Stern MB, Sethi K (2009) The scientific and clinical basis for the treatment of Parkinson disease. Neurology 72:S1–136

Ossowska K, Konieczny J, Wardas J, Pietraszek M, Kuter K, Wolfarth S, Pilc A (2007) An influence of ligands of metabotropic glutamate receptor subtypes on parkinsonian-like symptoms and the striatopallidal pathway in rats. Amino Acids 32:179–188

Ouattara B, Gregoire L, Morissette M, Gasparini F, Vranesic I, Bilbe G, Johns DR, Rajput A, Hornykiewicz O, Rajput AH, Gomez-Mancilla B, Di PT (2011) Metabotropic glutamate receptor type 5 in levodopa-induced motor complications. Neurobiol Aging 32:1286–1295

Oueslati A, Breysse N, Amalric M, Kerkerian-Le GL, Salin P (2005) Dysfunction of the cortico-basal ganglia-cortical loop in a rat model of early parkinsonism is reversed by metabotropic glutamate receptor 5 antagonism. Eur J Neurosci 22:2765–2774

Paquet M, Smith Y (2003) Group I metabotropic glutamate receptors in the monkey striatum: subsynaptic association with glutamatergic and dopaminergic afferents. J Neurosci 23:7659–7669

Petralia RS, Wang YX, Niedzielski AS, Wenthold RJ (1996) The metabotropic glutamate receptors, mGluR2 and mGluR3, show unique postsynaptic, presynaptic and glial localizations. Neuroscience 71:949–976

Phillips JM, Lam HA, Ackerson LC, Maidment NT (2006) Blockade of mGluR glutamate receptors in the subthalamic nucleus ameliorates motor asymmetry in an animal model of Parkinson's disease. Eur J Neurosci 23:151–160

Picconi B, Pisani A, Centonze D, Battaglia G, Storto M, Nicoletti F, Bernardi G, Calabresi P (2002) Striatal metabotropic glutamate receptor function following experimental parkinsonism and chronic levodopa treatment. Brain 125:2635–2645

Picconi B, Piccoli G, Calabresi P (2012) Synaptic dysfunction in Parkinson's disease. Adv Exp Med Biol 970:553–572

Pin JP, Acher F (2002) The metabotropic glutamate receptors: structure, activation mechanism and pharmacology. Curr Drug Targets CNS Neurol Disord 1:297–317

Pisani A, Bonsi P, Centonze D, Bernardi G, Calabresi P (2001a) Functional coexpression of excitatory mGluR1 and mGluR5 on striatal cholinergic interneurons. Neuropharmacology 40:460–463

Pisani A, Gubellini P, Bonsi P, Conquet F, Picconi B, Centonze D, Bernardi G, Calabresi P (2001b) Metabotropic glutamate receptor 5 mediates the potentiation of N-methyl-D-aspartate responses in medium spiny striatal neurons. Neuroscience 106:579–587

Pisani A, Bonsi P, Catania MV, Giuffrida R, Morari M, Marti M, Centonze D, Bernardi G, Kingston AE, Calabresi P (2002) Metabotropic glutamate 2 receptors modulate synaptic inputs and calcium signals in striatal cholinergic interneurons. J Neurosci 22:6176–6185

Pisani A, Bonsi P, Centonze D, Gubellini P, Bernardi G, Calabresi P (2003) Targeting striatal cholinergic interneurons in Parkinson's disease: focus on metabotropic glutamate receptors. Neuropharmacology 45:45–56

Pisani A, Bernardi G, Ding J, Surmeier DJ (2007) Re-emergence of striatal cholinergic interneurons in movement disorders. Trends Neurosci 30:545–553

Poisik OV, Mannaioni G, Traynelis S, Smith Y, Conn PJ (2003) Distinct functional roles of the metabotropic glutamate receptors 1 and 5 in the rat globus pallidus. J Neurosci 23:122–130

Poisik O, Raju DV, Verreault M, Rodriguez A, Abeniyi OA, Conn PJ, Smith Y (2005) Metabotropic glutamate receptor 2 modulates excitatory synaptic transmission in the rat globus pallidus. Neuropharmacology 49(Suppl 1):57–69

Poisik OV, Smith Y, Conn PJ (2007) D1- and D2-like dopamine receptors regulate signaling properties of group I metabotropic glutamate receptors in the rat globus pallidus. Eur J Neurosci 26:852–862

Prisco S, Natoli S, Bernardi G, Mercuri NB (2002) Group I metabotropic glutamate receptors activate burst firing in rat midbrain dopaminergic neurons. Neuropharmacology 42:289–296

Rascol O, Fox S, Gasparini F, Kenney C, Di PT, Gomez-Mancilla B (2014) Use of metabotropic glutamate 5-receptor antagonists for treatment of levodopa-induced dyskinesias. Parkinsonism Relat Disord 20:947–956

Rizzone MG, Fasano A, Daniele A, Zibetti M, Merola A, Rizzi L, Piano C, Piccininni C, Romito LM, Lopiano L, Albanese A (2014) Long-term outcome of subthalamic nucleus DBS in Parkinson's disease: from the advanced phase towards the late stage of the disease? Parkinsonism Relat Disord 20:376–381

Rosa M, Giannicola G, Marceglia S, Fumagalli M, Barbieri S, Priori A (2012) Neurophysiology of deep brain stimulation. Int Rev Neurobiol 107:23–55

Rylander D, Recchia A, Mela F, Dekundy A, Danysz W, Cenci MA (2009) Pharmacological modulation of glutamate transmission in a rat model of L-DOPA-induced dyskinesia: effects on motor behavior and striatal nuclear signaling. J Pharmacol Exp Ther 330:227–235

Samadi P, Gregoire L, Morissette M, Calon F, Hadj TA, Belanger N, Dridi M, Bedard PJ, Di PT (2008a) Basal ganglia group II metabotropic glutamate receptors specific binding in non-human primate model of L-Dopa-induced dyskinesias. Neuropharmacology 54:258–268

Samadi P, Gregoire L, Morissette M, Calon F, Hadj TA, Dridi M, Belanger N, Meltzer LT, Bedard PJ, Di PT (2008b) mGluR5 metabotropic glutamate receptors and dyskinesias in MPTP monkeys. Neurobiol Aging 29:1040–1051

Schapira AH, Jenner P (2011) Etiology and pathogenesis of Parkinson's disease. Mov Disord 26:1049–1055

Schapira AH, Olanow CW, Greenamyre JT, Bezard E (2014) Slowing of neurodegeneration in Parkinson's disease and Huntington's disease: future therapeutic perspectives. Lancet 384:545–555

Sergeeva OA, Doreulee N, Chepkova AN, Kazmierczak T, Haas HL (2007) Long-term depression of cortico-striatal synaptic transmission by DHPG depends on endocannabinoid release and nitric oxide synthesis. Eur J Neurosci 26:1889–1894

Shen KZ, Johnson SW (2003) Group II metabotropic glutamate receptor modulation of excitatory transmission in rat subthalamic nucleus. J Physiol 553:489–496

Shen KZ, Johnson SW (2013) Group I mGluRs evoke K-ATP current by intracellular Ca2+ mobilization in rat subthalamus neurons. J Pharmacol Exp Ther 345:139–150
Sibille P, Lopez S, Brabet I, Valenti O, Oueslati N, Gaven F, Goudet C, Bertrand HO, Neyton J, Marino MJ, Amalric M, Pin JP, Acher FC (2007) Synthesis and biological evaluation of 1-amino-2-phosphonomethylcyclopropanecarboxylic acids, new group III metabotropic glutamate receptor agonists. J Med Chem 50:3585–3595
Smith Y, Charara A, Hanson JE, Paquet M, Levey AI (2000) GABA(B) and group I metabotropic glutamate receptors in the striatopallidal complex in primates. J Anat 196(Pt 4):555–576
Smith Y, Charara A, Paquet M, Kieval JZ, Pare JF, Hanson JE, Hubert GW, Kuwajima M, Levey AI (2001) Ionotropic and metabotropic GABA and glutamate receptors in primate basal ganglia. J Chem Neuroanat 22:13–42
Smith Y, Wichmann T, Factor SA, DeLong MR (2012) Parkinson's disease therapeutics: new developments and challenges since the introduction of levodopa. Neuropsychopharmacology 37:213–246
Stocchi F, Rascol O, Destee A, Hattori N, Hauser RA, Lang AE, Poewe W, Stacy M, Tolosa E, Gao H, Nagel J, Merschhemke M, Graf A, Kenney C, Trenkwalder C (2013) AFQ056 in Parkinson patients with levodopa-induced dyskinesia: 13-week, randomized, dose-finding study. Mov Disord 28:1838–1846
Sun XR, Chen L, Chen WF, Yung WH (2012) Electrophysiological and behavioral effects of group I metabotropic glutamate receptors on pallidal neurons in rats. Brain Res 1477:1–9
Sun XR, Chen L, Chen WF, Xue Y, Yung WH (2013) Electrophysiological and behavioral effects of group III metabotropic glutamate receptors on pallidal neurons in normal and parkinsonian rats. Synapse 67:831–838
Sung KW, Choi S, Lovinger DM (2001) Activation of group I mGluRs is necessary for induction of long-term depression at striatal synapses. J Neurophysiol 86:2405–2412
Tallaksen-Greene SJ, Kaatz KW, Romano C, Albin RL (1998) Localization of mGluR1a-like immunoreactivity and mGluR5-like immunoreactivity in identified populations of striatal neurons. Brain Res 780:210–217
Tamaru Y, Nomura S, Mizuno N, Shigemoto R (2001) Distribution of metabotropic glutamate receptor mGluR3 in the mouse CNS: differential location relative to pre- and postsynaptic sites. Neuroscience 106:481–503
Testa CM, Standaert DG, Young AB, Penney JB Jr (1994) Metabotropic glutamate receptor mRNA expression in the basal ganglia of the rat. J Neurosci 14:3005–3018
Testa CM, Standaert DG, Landwehrmeyer GB, Penney JB Jr, Young AB (1995) Differential expression of mGluR5 metabotropic glutamate receptor mRNA by rat striatal neurons. J Comp Neurol 354:241–252
Testa CM, Friberg IK, Weiss SW, Standaert DG (1998) Immunohistochemical localization of metabotropic glutamate receptors mGluR1a and mGluR2/3 in the rat basal ganglia. J Comp Neurol 390:5–19
Trudeau LE, Hnasko TS, Wallen-Mackenzie A, Morales M, Rayport S, Sulzer D (2014) The multilingual nature of dopamine neurons. Prog Brain Res 211:141–164
Valenti O, Conn PJ, Marino MJ (2002) Distinct physiological roles of the Gq-coupled metabotropic glutamate receptors Co-expressed in the same neuronal populations. J Cell Physiol 191:125–137
Valenti O, Marino MJ, Wittmann M, Lis E, DiLella AG, Kinney GG, Conn PJ (2003) Group III metabotropic glutamate receptor-mediated modulation of the striatopallidal synapse. J Neurosci 23:7218–7226
Valenti O, Mannaioni G, Seabrook GR, Conn PJ, Marino MJ (2005) Group III metabotropic glutamate-receptor-mediated modulation of excitatory transmission in rodent substantia nigra pars compacta dopamine neurons. J Pharmacol Exp Ther 313:1296–1304
Vernon AC, Palmer S, Datla KP, Zbarsky V, Croucher MJ, Dexter DT (2005) Neuroprotective effects of metabotropic glutamate receptor ligands in a 6-hydroxydopamine rodent model of Parkinson's disease. Eur J Neurosci 22:1799–1806

Vernon AC, Zbarsky V, Datla KP, Croucher MJ, Dexter DT (2007a) Subtype selective antagonism of substantia nigra pars compacta Group I metabotropic glutamate receptors protects the nigrostriatal system against 6-hydroxydopamine toxicity in vivo. J Neurochem 103:1075–1091

Vernon AC, Zbarsky V, Datla KP, Dexter DT, Croucher MJ (2007b) Selective activation of group III metabotropic glutamate receptors by L-(+)-2-amino-4-phosphonobutryic acid protects the nigrostriatal system against 6-hydroxydopamine toxicity in vivo. J Pharmacol Exp Ther 320:397–409

Wichmann T, Dostrovsky JO (2011) Pathological basal ganglia activity in movement disorders. Neuroscience 198:232–244

Wittmann M, Hubert GW, Smith Y, Conn PJ (2001) Activation of metabotropic glutamate receptor 1 inhibits glutamatergic transmission in the substantia nigra pars reticulata. Neuroscience 105:881–889

Wittmann M, Marino MJ, Conn PJ (2002) Dopamine modulates the function of group II and group III metabotropic glutamate receptors in the substantia nigra pars reticulata. J Pharmacol Exp Ther 302:433–441

Yin HH, Lovinger DM (2006) Frequency-specific and D2 receptor-mediated inhibition of glutamate release by retrograde endocannabinoid signaling. Proc Natl Acad Sci U S A 103:8251–8256

Chapter 4
Metabotropic Glutamate 2 (mGlu2) Receptors and Schizophrenia Treatment

Javier González-Maeso

Abstract Although it is not without controversy, recent preclinical and clinical studies suggest that activation of metabotropic glutamate 2 (mGlu2) receptors by orthosteric mGlu2/3 receptor agonists or allosteric mGlu2 receptor agonists represents a potential new approach to treat schizophrenia and other psychotic disorders. Using rodent models of psychosis, it has been shown recently that expression of the serotonin 5-HT$_{2A}$ receptor is at least in part necessary for the mGlu2-dependent antipsychotic-like behavioral effects of the mGlu2/3 receptor agonist LY379268. Additionally, chronic treatment with atypical antipsychotic drugs epigenetically represses the promoter activity of the *mGlu2* (*Grm2*) gene in mouse and human frontal cortex through a molecular mechanism that requires 5-HT$_{2A}$ receptor-dependent signaling. In the present book chapter, we describe the basic signaling and epigenetic mechanisms whereby serotonin and glutamate system cross talk through these two neurotransmitter receptors and discuss some of the molecular underpinnings of this framework, ranging from pharmacological interaction and biophysical mechanisms to epigenetic pathways.

Keywords G protein-coupled receptor (GPCR) • Metabotropic glutamate receptor (mGlu) • Schizophrenia • Antipsychotic • Epigenetic

4.1 Schizophrenia

Schizophrenia is one of the most incapacitating mental disorders that affects almost 1% of the world's population and is among the most costly of all mental illnesses, with an estimated annual per person direct treatment cost that is approximately twofold higher than the cost of major depression and fourfold higher than an anxiety disorder (Sawa and Snyder 2002; Freedman 2003; Adam 2013). The available

J. González-Maeso, Ph.D. (✉)
Department of Physiology and Biophysics, Virginia Commonwealth University Medical School, Richmond, VA 23298, USA
e-mail: jgmaeso@vcu.edu

R.T. Ngomba et al. (eds.), *mGLU Receptors*, The Receptors 31,
DOI 10.1007/978-3-319-56170-7_4

symptomatic treatment is only partially successful, and, therefore, the development of rational therapeutics based on better understanding of the basic etiological mechanisms and pathogenesis of this psychiatric disorder is imperative. Advances in our knowledge of schizophrenia and its treatment have been limited by a number of factors, including heterogeneity of the schizophrenia phenotype and the lack of clear neuroanatomical and neurodegenerative processes such as those that have provided reference points in the study of Alzheimer's disease, Parkinson's disease, and other neurological disorders. Additionally, the anti-schizophrenia effects of both first-generation or typical antipsychotic drugs, such as chlorpromazine and haloperidol, and second-generation or atypical antipsychotic drugs, such as clozapine, olanzapine, and risperidone, were discovered as secondary effects of drugs originally tested to target a different battery of disorders (Kapur and Remington 2001; Miyamoto et al. 2012; Meltzer 2013). As an example, the first antipsychotic drug chlorpromazine was serendipitously discovered in 1952 as an antihistaminic drug that decreased psychosis in schizophrenia patients. Haloperidol was originally tested as a pain reliever, and clozapine was described in 1958 as a "tricyclic antidepressant with antipsychotic properties." Following these early discoveries, the molecular and neuronal circuit mechanisms of antipsychotic drug action as well as the basic biological causes underlying schizophrenia have been the focus of much attention in molecular psychopharmacology and translational neuroscience fields.

Schizophrenia is among the most heritable common disorders, with first-degree relatives of an affected individual having 8–10 times the risk of the corresponding disorder in the general population which is, as mentioned above, about 1%. Thus, twin and adoption studies consistently demonstrate a genetic influence on schizophrenia risk, confirming work that started in the 1930s (Burmeister et al. 2008). For example, third-degree relatives (e.g., first cousins) share about 12.5% of their genes and show a risk of 2% for developing schizophrenia. Second-degree relatives (e.g., half-siblings) share about 25% of their genes and show a risk of 6%. Most first-degree relatives (e.g., siblings and dizygotic twins) share about 50% of their genes and show a risk of about 9%. Monozygotic twins share 100% of their genes and show risks near 50% (Cardno and Gottesman 2000; Gottesman and Erlenmeyer-Kimling 2001). This underlines the importance of genetic factors in the development of schizophrenia. At the same time, it also demonstrates that genetic factors are not sufficient for the development of this psychiatric disorder.

Within this context, it has been demonstrated that environmental insults during pregnancy, such as maternal infection and maternal severe stress significantly increase the risk of schizophrenia in the adult offspring. Thus, it was shown in 1919 that infection with the influenza virus during pregnancy was associated with the schizophrenia risk (Menninger 1919). These interesting findings have been validated with influenza virus (Brown et al. 2004) as well as with a variety of infectious agents rubella (Brown et al. 2001), genital and reproductive infections (Babulas et al. 2006), and the protozoan parasite *Toxoplasma* (Brown et al. 2005). Epidemiological studies also point toward a statistically significant link between severe maternal adverse life event during pregnancy and schizophrenia in the newborns. These prenatal environmental insults include war (van Os and Selten 1998;

Malaspina et al. 2008), famine (Susser et al. 2008), and death of a relative (Khashan et al. 2008). Together with genome-wide association studies (GWAS) (see Stefansson et al. 2008, 2009; Walsh et al. 2008; Purcell et al. 2009, 2014; Consortium 2008, 2014), these findings suggest that schizophrenia is a multifactorial psychiatric disorder, caused by interactions of several genes, neurodevelopmental alterations, and environmental events. However, the primary target of currently used antipsychotic medications is restricted to monoaminergic neurotransmitter systems: all antipsychotic drugs bind with high-affinity dopamine, serotonin, and norepinephrine receptors (Miyamoto et al. 2005, 2012; Meltzer 2013; Snyder and Murphy 2008; Lieberman et al. 2008). Interestingly, the glutamatergic system has recently emerged as one of the most prominent targets in schizophrenia treatment. Thus, recent large GWAS show that some of the genetic loci associated with schizophrenia are involved in glutamatergic neurotransmission (Consortium 2014). Similarly, the use of preclinical models led to the hypothesis that activation of metabotropic glutamate 2 (mGlu2) receptor might represent a new pharmacological approach to treat schizophrenia, hypothesis that was translated into clinical work (for review, see Moreno et al. 2009). In the present book chapter, we review recent work focused on the potential role of mGlu2 receptors as antipsychotic targets.

4.2 Metabotropic Glutamate Receptors

Glutamate is the neurotransmitter involved in the majority of fast excitatory synapses in the CNS. Until relatively recently, the functional responses induced by glutamate were assumed to be mediated through the opening of sodium/calcium permeable ligand-gated ion channels that were classified as N-methyl-D-aspartate (NMDA) and non-NMDA quisqualate/AMPA and kainate receptor subtypes (Krystal et al. 2003). It was also accepted that dopamine, acetylcholine, serotonin, and norepinephrine modulated glutamate neurotransmission by activation of G protein-coupled receptors (GPCRs). It was in the mid-1980s, however, when two independent groups showed that glutamate was able to release intracellular calcium via activation of the PLC pathway. In 1991, two groups independently cloned the rat mGlu1 receptor subtype after which additional mGlu receptor subtypes where cloned and classified based on their primary sequence, signaling properties, and pharmacological characteristics. Currently, there are eight known subtypes of mGlu receptors, which have been classified into three groups. Group I mGlu receptors, which include mGlu1 and mGlu5, activate PLC via $G_{q/11}$ protein coupling. Group II mGlu receptors (mGlu2 and mGlu3) and group III (mGlu4, mGlu6, mGlu7 and mGlu8) are coupled to PTX-sensitive $G_{i/o}$ proteins and inhibit cAMP formation. This general information of mGlu receptors has been revised previously elsewhere (Conn and Pin 1997; Schoepp 2001; Pin et al. 2003, 2004).

GPCRs are characterized by a central core domain composed of seven transmembrane alpha-helices, with an extracellular N-terminus and an intracellular carboxyl tail (Rosenbaum et al. 2009; Audet and Bouvier 2012). Two unique properties of family III mGlu receptors include a large N-terminal extracellular domain termed Venus flytrap, where the endogenous neurotransmitter glutamate as well as orthosteric agonists and antagonists bind to, as well as their functional properties as homodimeric structural units (Pin et al. 2003, 2004). In this context, GPCRs were assumed to be expressed at the plasma membrane as monomeric entities. However, it was demonstrated that expression of a cloned cDNA for family III gamma-aminobutyric acid B ($GABA_B$) receptor 1 ($GABA_B$-R1) in living cells led to a protein that did not show functional properties and that was located intracellularly (Kaupmann et al. 1997; Couve et al. 1998), which was opposite to previously described GPCRs. Follow-up work showed that $GABA_B$-R1 reached the plasma membrane protein only when expressed as a receptor heterocomplex with the $GABA_B$-R2 (Jones et al. 1998; Kaupmann et al. 1998; White et al. 1998) and that $GABA_B$-R1 alone was retained at the endoplasmic reticulum via an element that controls endoplasmic reticulum retention signal located within its C-terminal domain. Thus co-expression of $GABA_B$-R1 and $GABA_B$-R2 led to a functional heteromeric $GABA_B$ receptor complex via their coiled-coil domain that reaches the cell surface (Margeta-Mitrovic et al. 2000, 2001). This was the first demonstration that heteromeric formation was necessary for GPCR receptor function.

Based on the combination of time-resolved FRET and SNAP-tag approaches to quantitatively analyze protein-protein interactions at the surface of living cells, it was demonstrated that mGlu receptors are expressed as strict dimers (Maurel et al. 2008), although the structural mechanism whereby mGlu receptors are assembled as homodimeric receptor complexes is not the same as that for the $GABA_B$-R1-$GABA_B$-R2 receptor heterodimer. Thus, mutagenesis studies have shown that homodimeric mGlu receptors are stable receptor complexes in part because a disulfide bridge formed by the ninth cysteine at the cysteine-rich domain in lobe 2 of the Venus flytrap. Recent findings further define biophysical mechanisms whereby inter- and intra-subunit rearrangements during activation of homodimeric mGlu1 receptors. By inserting eYFP and CFP in the second intracellular loop and the C-terminus of the mGlu1 receptor, it was shown that in the presence of orthosteric agonists inter-subunit FRET signal was rapidly increased. However, immediately after this inter-subunit movement, the intra-subunit FRET was found decreased, which suggests conformational changes upon inter-subunit movements within each individual subunit. The use of chimeric constructs that affected G protein coupling also suggested that only a single mGlu1 receptor subunit achieves an active state in an mGlu1 receptor homodimers (Hlavackova et al. 2012; El Moustaine et al. 2012; Xue et al. 2015).

Although it has been demonstrated that the formation of disulfide bonds between cysteine residues in the Venus flytrap domains of mGlu protomers is responsible for the formation of stable and functionally active mGlu homodimers, the transmembrane domains (TMs) located at the homodimeric interface of mGlu receptor homodimers remain a topic of much debate. Thus, upon deletion of the extracellular Venus flytrap domain, determination of crystal structure of the human mGlu1 receptor seven TM domain at a resolution of 2.8 Å suggests a parallel 7TM dimer mediated via TM1-TM1 homodimeric interface (Wu et al. 2014). It is worth mentioning, however, that the Venus flytrap domains, and hence disulfide bonds between the two mGlu1 protomers, were excluded and that oligomeric structures obtained after crystallization protocols might not represent those under physiological conditions as it is demonstrated by the crystallization of rhodopsin revealing an antiparallel dimer with respect to the plane of the membrane. Further work in heterologous expression systems (HEK293 cells) suggests that the main mGlu2 receptor homodimeric interface is formed by TM4 and TM5 (Xue et al. 2015). Nevertheless, the cysteine cross-linking approach used to define the mGlu2-mGlu2 homodimeric interface required mutation of all cysteines into alanines, including those that form disulfide bonds between at the Venus flytrap domains, which might lead to structural changes in the relative location of the mGlu2 protomers within the mutated mGlu2 homodimer (Xue et al. 2015). Additional studies are therefore still pending to fully understand the molecular mechanism by which the two mGlu protomers within an mGlu homodimer cross talk.

Especially interesting are the findings that monomeric mGlu2 receptors, either without the Venus flytrap domain as an isolated TM or in full length, purified and reconstituted into nanodisks activate G proteins in the presence of positive allosteric modulators (El Moustaine et al. 2012). However, only a reconstituted full-length dimeric mGlu2 receptor activates G proteins upon orthosteric agonist binding (El Moustaine et al. 2012). These data further support previous evidence that a single 7TM is sufficient to activate G proteins in nanodisks (Whorton et al. 2007, 2008; Kuszak et al. 2009), whereas a homodimeric mGlu2 structure is necessary for G protein coupling. Additional work will be necessary to validate these structural findings in living cells.

4.3 Preclinical and Clinical Schizophrenia Assays with mGlu2/3 Receptor Agonists

Hallucinogenic drugs, such as lysergic acid diethylamide (LSD) and psilocin (Hoch et al. 1952; Schmid et al. 2015; Gouzoulis-Mayfrank et al. 2005; Kometer et al. 2012, 2013; Young 1974; Vollenweider et al. 1998), and dissociative drugs such as ketamine and phencyclidine (PCP) (Gouzoulis-Mayfrank et al. 2005; Lahti et al. 2001; Umbricht et al. 2000) cause psychotic and cognitive symptoms in healthy volunteers that resemble some of those observed in schizophrenia patients. Based on this, although animal models of schizophrenia and other psychiatric disorders have severe limitations (Hanks and Gonzalez-Maeso 2013), LSD-like and PCP-like drugs have been used in preclinical models to test therapeutic effects of compounds on schizophrenia-related phenotypes of psychosis and cognitive deficits (Hanks and Gonzalez-Maeso 2013; Moreno and Gonzalez-Maeso 2013a). PCP-like drugs induce hyperlocomotor activity in rodents, and this is prevented by administration of atypical antipsychotic drugs such as clozapine and risperidone (Nestler and Hyman 2010). Using this paradigm of antipsychotic-like behavior, it was shown in

1998 that the orthosteric mGlu2/3 receptor agonist LY354740 prevented the locomotion activity and stereotypical behavior induced by PCP in rats (Moghaddam and Adams 1998). It was also demonstrated that LY354740 attenuated the effects of PCP on cortical glutamate efflux without affecting dopamine hyperactivity (Moghaddam and Adams 1998). These findings represented the first preclinical demonstration that non-dopaminergic strategies might serve as a new therapeutic approach for schizophrenia treatment. This was followed up by a complementary behavioral study showing that LY354740 suppressed head-twitch behavior induced by the hallucinogenic 5-HT_{2A} receptor agonist DOI in rats (Gewirtz and Marek 2000), which led to the demonstration of a multitude of cellular, electrophysiological, and behavioral antipsychotic-like behaviors of both orthosteric mGlu2/3 receptor agonists and allosteric mGlu2 receptor agonists (for review, see Moreno et al. 2009). Additionally, the use of mGlu2 and mGlu3 receptor knockout mice provided convincing evidence that the effects of the mGlu2/3 agonists LY314582 (Spooren et al. 2000), LY379268 (Woolley et al. 2008), and LY404039 (Fell et al. 2008) diminishing hyperlocomotor activity induced by PCP-like drugs in mice were mGlu2 receptor dependent and not mGlu3 receptor dependent. Positive allosteric modulations of mGlu receptors represent an attractive pharmacological tool because they allow a fine-tuning of the responses induced by orthosteric agonists (Ellaithy et al. 2015). Thus, they modulate the responses induced by the endogenously expressed neurotransmitter receptors, which implicate that they do not induce physiological responses in cellular types or neuronal circuits without the presence of the endogenous mGlu agonist glutamate. Considering the higher differences in amino acid sequence of the TM domains between mGlu receptor subtypes as compared to that of their Venus flytrap, binding sites of positive allosteric modulators are also more attractive from a targetable perspective, and certainly the number of commercially available positive and negative allosteric modulators that bind selectively to a particular mGlu receptor subtypes has increased substantially, whereas this pharmacological advance did not occur with mGlu orthosteric agonists and antagonists, which up to now rarely bind specifically to a particular mGlu receptor subtype. Using selective positive allosteric modulators of the mGlu2 receptor, such as LY487379 (Galici et al. 2005) and BINA (Benneyworth et al. 2007), it has been demonstrated that activation of mGlu2 function reverses psychosis-like states induced by psychedelic drugs, including LSD-like such as DOB as well as PCP and amphetamine. Consequently numerous research groups showed convincing evidence that either orthosteric activation of mGlu2 receptors or potentiation of mGlu2-dependent function by positive allosteric modulators repress cellular and electrophysiological responses induced by LSD-like and PCP-like drugs (Ellaithy et al. 2015; Moreno and Gonzalez-Maeso 2013b). Among the cellular responses modulated via activation of mGlu2 receptors some of particular interest include expression of immediately early genes, such as *Egr-1* and *Egr-2* (Gonzalez-Maeso et al. 2008; Moreno et al. 2011a, 2013). Considering that these genes are involved in mechanisms related to synaptic plasticity and cognitive processes (Veyrac et al. 2013; O'Donovan et al. 1999; Jones et al. 2001), it is therefore tempting to speculate that the mGlu2 receptor and its signaling effects might affect how the brain processes sensory information and how these processes might be abnormal in human subjects with hallucinatory symptoms, such schizophrenia.

From a clinical perspective, it was exceptionally promising when Eli Lilly published in 2007 that treatment with 40 mg of LY2140023 (prodrug of the mGlu2/3 receptor orthosteric agonist LY404039) twice daily or with 15 mg of olanzapine for 4 weeks resulted in statistically significant improvements in the Positive and Negative Syndrome Scale (PANSS) total score (Patil et al. 2007). Additionally, treatment with LY2140023 was well tolerated and did not affect prolactin elevation, extrapyramidal side effects, or weight gain (Patil et al. 2007). Thus, the results of these studies, which enrolled 193 schizophrenia patients and were conducted at 10 psychiatric hospitals in Russia, pointed toward a translational validation in schizophrenia patients of the previously suggested antipsychotic-like effects of mGlu2/3 receptor agonists in rodent schizophrenia models (Patil et al. 2007). However, additional clinical work with LY2140023 showed either inconclusive results (Kinon et al. 2011) or clinical effects that did not separate from placebo (Adams et al. 2013). Thus, the first follow-up study published in 2011 showed that LY2140023 given twice daily for 4 weeks at four different doses (5, 20, 40, or 80 mg/kg), or placebo, to schizophrenia patients did not separate from placebo on the PANSS total score (number of patients per group = 120–122) (Kinon et al. 2011). Notably, olanzapine (n = 62) did not separate significantly from placebo, and there was also a significant effect (14.6-point improvement on the PANSS total score) in the placebo group (Kinon et al. 2011). The results of this study are considered inconclusive as neither LY2140023 nor olanzapine is statistically separated from placebo (Kinon et al. 2011). A further and larger study (open-label 24 weeks) randomized 262 schizophrenia patients to either LY2140023 or standard of care (SOC: olanzapine, risperidone, or aripiprazole) treatment groups (Adams et al. 2013). Doses of LY2140023 were 20, 40, or 80 mg (twice daily). Doses of olanzapine were 10, 15, or 20 mg (one daily). Doses of risperidone were 2, 4, or 6 mg (once or twice daily). Doses of aripiprazole were 10, 20, or 30 mg (once daily). The incidence of serious adverse events was comparable between groups. Improvement in PANSS total score over the initial 6–8 weeks of treatment was similar between groups, but improvement was significantly greater in the SOC group compared to the LY2140023 group at 24-week endpoint (Adams et al. 2013). These negative findings preceded the press release by Eli Lilly and Company with the decision to stop the clinical trials with LY2140023 for the treatment of schizophrenia. Importantly, more recent preclinical data and findings in postmortem human brain samples of schizophrenic subjects and controls provided the basis for a post hoc analysis that provides new insights into why certain schizophrenic subjects do not respond to treatment with LY2140023.

In 2008, it was demonstrated that chronic treatment with the atypical antipsychotic clozapine induces both downregulation of mGlu2/3 receptor density as defined by radioligand binding saturation curves with [^{3}H]LY341495 in mouse frontal cortex and decreased *mGlu2*, but not *mGlu3*, mRNA expression (Gonzalez-Maeso et al. 2008). These findings provided the basis for additional work that showed an epigenetic explanation of the effect of chronic antipsychotic treatment

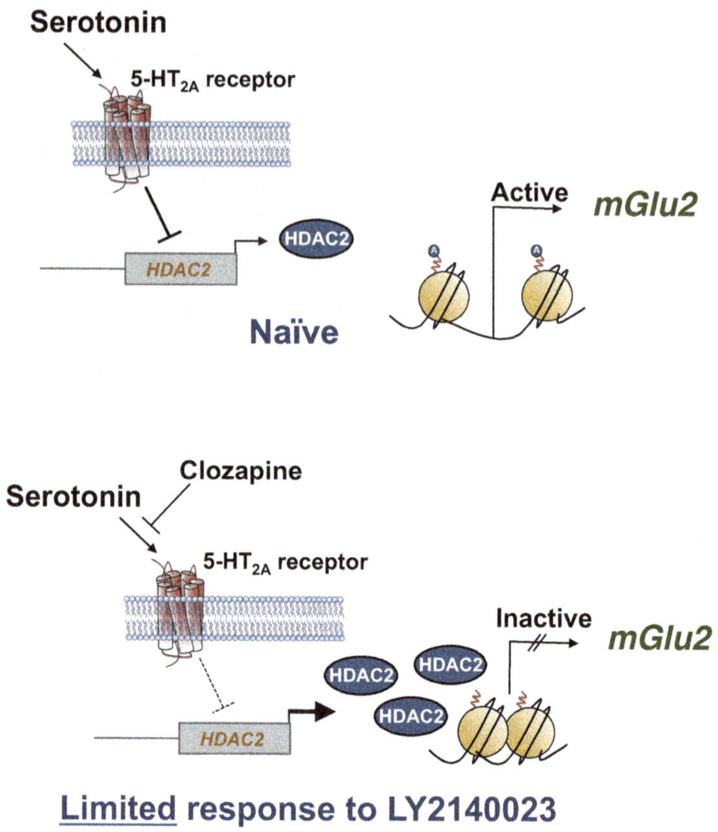

Fig. 4.1 Schematic model of the epigenetic mechanism by which previous long-lasting treatment with atypical antipsychotic drugs limits the mGlu2-dependent therapeutic effects of the mGlu2/3 agonist LY2140023 in schizophrenia patients (see Kurita et al. 2012)

on mGlu2 expression (Kurita et al. 2012, 2013a). Thus, it was demonstrated that chronic treatment (21 days) with atypical antipsychotic drugs, such as olanzapine and risperidone, but not with the typical antipsychotic drug haloperidol, induced histone modifications at the promoter region of the *mGlu2* gene that correlated with transcriptional repression (Fig. 4.1). These included decreased acetylation of histone H3 (H3ac), decreased acetylation of histone H4 (H4ac), and increased tri-methylation of histone H3 at lysine 27 (H3K27me3). Neither histone H3 methylation at lysine 4 (H3K4me1, H3K4me2, and/or H3K4me3), marker of gene activation, nor histone H3 tri-methylation at lysine 9 (H3K9me3), marker of transcriptional repression, was affected after antipsychotic treatment. Additionally, sub-chronic treatment (2 days) with clozapine, olanzapine, or haloperidol did not induce repressive histone modifications at the *mGlu2* promoter. The authors also demonstrated that these epigenetic changes induced at the *mGlu2* promoter in mouse frontal cortex were also observed in postmortem frontal cortex of schizophrenic subjects previously treated with atypical antipsychotic drugs but not in drug-free schizophrenic subjects. These data suggested that decreased acetylation of histone H3 at the mGlu2 promoter and downregulation of mGlu2 expression in frontal cortex is a consequence of atypical antipsychotic drug treatment and not an epigenetic mark of schizophrenia in postmortem tissue samples. Mechanistically, it was also demonstrated that this epigenetic change at the *mGlu2* promoter occurred in association with a serotonin 5-HT_{2A} receptor-dependent upregulation and increased binding of HDAC2 to the mGlu2 promoter (Kurita et al. 2012, 2013a). These observations suggested that previous treatment with atypical antipsychotic drugs, but not with haloperidol, would prevent at least part of the therapeutic effects induced by mGlu2/3 receptor agonists such as LY2140023 (Kurita et al. 2012, 2013a, b).

Importantly, this hypothesis has recently been supported by results for a post hoc analysis showing that a subgroup of schizophrenia patients who were previously exposed to haloperidol showed a therapeutic response to LY2140023 that was comparable to that of risperidone, whereas efficacy of LY2140023 did not differ from placebo in patients previously treated with atypical antipsychotic drugs (Kinon et al. 2015). Together, these findings support the use of preclinical studies in the basic investigation of the molecular and neuronal circuit mechanisms involved in the mechanism of action of antipsychotic drugs and suggest that adjunctive administration of drugs that inhibit HDAC2 may prevent the repressive histone modifications induced at the mGlu2 promoter by chronic atypical antipsychotic drug treatment and hence improve the antipsychotic properties of orthosteric mGlu2/3 agonists and allosteric mGlu2 agonists (Kurita et al. 2013b). This cross talk and genetic and epigenetic levels between the serotonin and glutamatergic systems via 5-HT_{2A} and mGlu2 receptors are further suggested by the repressive histone modifications observed at the mGlu2 promoter in the frontal cortex of 5-HT_{2A} knockout mice (Kurita et al. 2013a) and by single nucleotide polymorphisms (SNPs) located along the 5-HT_{2A} receptor (*Htr2A*) gene that were significantly associated with the therapeutic effects of LY2140023 (Liu et al. 2010). Further work is needed to fully understand the mechanisms and consequences of these functional, genetic, and epigenetic cross talks.

4.4 Environmental Events and mGlu2 Expression

Behavioral assays in mGlu2 and mGlu3 knockout mice suggest that mGlu2 is the principal target responsible for the therapeutic effects induced by orthosteric mGlu2/3 receptor agonists (see above). This is further supported by the behavioral outcomes achieved by allosteric mGlu2 receptor agonists (see above). Thus, although the role of mGlu3 in some of the schizophrenia-related phenotypes observed in rodents cannot be excluded (Walker et al. 2015), these findings suggest that activation of mGlu2 receptor-dependent signaling leads to antipsychotic-like

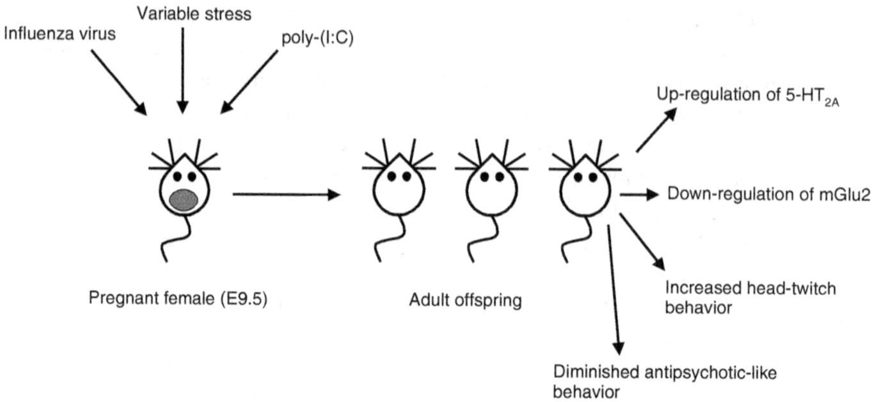

Fig. 4.2 In mouse models, prenatal insults that include maternal influenza virus infection with the mouse-adapted influenza A/WSN/33 (H1N1) virus, maternal variable and unpredictable stress, and activation of the maternal immune system via intraperitoneal injection of poly-(I:C) leads to alterations in expression and behavioral function of mGlu2 and 5-HT$_{2A}$ receptors in frontal cortex in the adult offspring. These behavioral phenotypes include increased head-twitch behavior induced by hallucinogenic 5-HT2A receptor agonists and decreased antipsychotic-like effect of the mGlu2/3 agonist LY379268 on hyperlocomotor responses induced by injection of the PCP-like drug MK801 (see Moreno et al. 2011a, b; Holloway et al. 2013)

effects. Since mGlu2/3 receptor density and mGlu2, but not mGlu3, mRNA expression was downregulated in postmortem human brain of schizophrenic subjects (Gonzalez-Maeso et al. 2008), together, these findings indicate that changes in expression of *mGlu2* might be responsible for psychosis and/or cognitive deficits in schizophrenia patients. From a translational perspective, it would therefore be interesting to generate animal models that recreate this particular biochemical alteration, with the final goal of defining whether downregulation of *mGlu2* expression in frontal cortex is responsible for behavioral changes that model schizophrenia symptoms in mice.

As discussed above, environmental insults during pregnancy, such as maternal viral infection and maternal severe stress, increase the risk of schizophrenia in the adult offspring. Notably, recent findings suggest that maternal infection with the mouse-adapted influenza A/WSN/33 (H1N1) virus causes downregulation of mGlu2/3 binding and decreased *mGlu2*, but not *mGlu3*, mRNA expression in frontal cortex in the adult offspring (Moreno et al. 2011b) (Fig. 4.2). Additionally, the behavioral responses induced by the mGlu2/3 receptor agonist LY379268 were affected by maternal influenza virus infection (Moreno et al. 2011b) (Fig. 4.2). Thus, adult mice born to influenza virus-infected mothers during pregnancy showed decreased effect of LY379268 on the hyperlocomotor activity induced by the noncompetitive NMDA receptor antagonist MK801 (Moreno et al. 2011b) (Fig. 4.2). Additional findings obtained by independent groups testing the effects of several environmental factors, such as variable and unpredictable stress during pregnancy (Holloway et al. 2013; Wang et al. 2015), maternal immune activation during pregnancy with poly (I:C) (Holloway et al. 2013) or lipopolysaccharide (Wischhof et al. 2015), perinatal exposure to bisphenol A (Zhou et al. 2015), prenatal exposure to kynurenic acid (Pershing et al. 2015), comparison of Roman low- and high-avoidance (RLA-I versus RHA-I) rat strains (Klein et al. 2014), and repeated administration of methamphetamine (Chiu et al. 2014), validate that adverse environmental effects induce downregulation of mGlu2 in frontal cortex and dysregulation of mGlu2 receptor-dependent signaling and behavioral function (Table 4.1). This downregulation of mGlu2 has not been reported in knock-in mice with a *tryptophan hydroxylase 2* (*Tph2*) R439H mutation (Jorgensen et al. 2013). However, because either upregulation of 5-HT_{2A} or downregulation of mGlu2 in rodent frontal cortex after prenatal and postnatal environmental indults has been validated in numerous experimental models (Table 4.1), it is tempting to speculate that dysfunction of expression and function of either of these two receptors, or both, play a fundamental role in schizophrenia-related phenotypes. Further investigation will be necessary to understand the molecular and epigenetic mechanisms responsible for the effects of prenatal/perinatal insults on mGlu2 and 5-HT_{2A} receptor densities and their role, if any, on schizophrenia-related structural and behavioral phenotypes.

4.5 Expression of mGlu2 and 5-HT_{2A} as a GPCR Heteromer in Schizophrenia Treatment

An alternative, but not mutually exclusive mechanism, that affects the cross talk between mGlu2 and 5-HT_{2A} is related to their expression in close molecular proximity at the plasma membrane in living cells. GPCRs have been thought to function as monomers—a model of receptor signaling that is further supported by observations based on assays that measured agonist binding and G protein coupling of a single purified monomeric family A GPCR, including β_2-adrenergic receptor, rhodopsin, and μ-opioid receptor (Whorton et al. 2007, 2008; Kuszak et al. 2009). Nevertheless, convincing evidence supports the existence of GPCR homo- and heteromeric complexes that affect pharmacology, trafficking, and receptor function (for review, see Gonzalez-Maeso 2011, 2014). Interestingly, it has been confirmed in heterologous expression systems that at least a fraction of the populations of mGlu2 and 5-HT_{2A} receptors constitute part of the same protein complex as defined by co-immunoprecipitation assays (Moreno et al. 2012; Rives et al. 2009). It has also been demonstrated that these two receptors are expressed in close molecular proximity as defined independently by biophysical assays such as BRET (Gonzalez-Maeso et al. 2008), FRET (Moreno et al. 2012), antibody-based time-resolved (TR)-FRET (Rives et al. 2009), and combination of TR-FRET and the SNAP-tag approach (Delille et al. 2012). In mouse frontal cortex, it was demonstrated that the $G_{i/o}$-coupled mGlu2 and the $G_{q/11}$-coupled 5-HT_{2A} receptors form part of the same

Table 4.1 Summary of published data showing dysregulation of mGlu2 and/or 5-HT$_{2A}$ receptor density and their behavioral function in rodent models of prenatal and postnatal environmental insults

Authors	Paradigm	Rodent	Brain region	Approach	mGlu2 expression	5-HT$_{2A}$ expression	References
Moreno et al. (2011)	Maternal influenza virus	Mouse	FC	Binding	Downregulation	Upregulation	Moreno et al. (2011b)
Jorgensen et al. (2013)	Tph2-R439H mice	Mouse	FC	Binding	Unchanged	Upregulation	Jorgensen et al. (2013)
Holloway et al. (2013)	Maternal stress	Mouse	FC	Binding	Downregulation	Upregulation	Holloway et al. (2013)
Klein et al. (2014)	RLA-I versus RHA-I	Rat	FC	Binding	Downregulation	Upregulation	Klein et al. (2014)
Malkova et al. (2014)	Maternal injection with poly-(I:C)	Mouse	FC	qRT-PCR	Not tested	Upregulation	Malkova et al. (2014)
Chiu et al. (2014)	Repeated methamphetamine	Mouse	FC	WB	Downregulation	Upregulation	Chiu et al. (2014)
Wang et al. (2015)	Mild stress	Rat	FC	WB	Downregulation	Not tested	Wang et al. (2015)
Akatsu et al. (2015)	Maternal stress	Mouse	FC	qRT-PCR	Not tested	Upregulation	Akatsu et al. (2015)
Wischhof et al. (2015)	Maternal injection with LPS	Rat	FC	WB	Unchanged	Upregulation	Wischhof et al. (2015)
Pershing et al. (2015)	Maternal injection with kynurenic acid	Rat	FC	qRT-PCR	Downregulation	Not tested	Pershing et al. (2015)
Maple et al. (2015)	Sleep deprivation	Mouse	FC	qRT-PCR	Not tested	Upregulation	Maple et al. (2015)

Abbreviations: *Binding* radioligand binding assays in either plasma membrane preparations or tissue sections, *FC* frontal cortex, *LPS* lipopolysaccharide, *poly(I:C)* polyinosinic:polycytidylic acid, *qRT-PCR* quantitative real-time polymerase chain reaction, *RHA-I* Roman high-avoidance rat strain, *RLA-I* Roman low-avoidance rat strain, *WB* Western blot

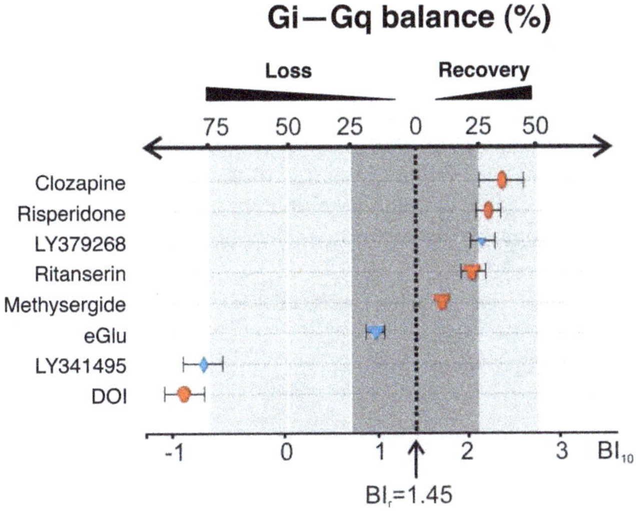

Fig. 4.3 The 5-HT_{2A}-mGlu2 complex-dependent changes in $G_{i/o}$ and $G_{q/11}$ activity predict the psychoactive behavioral effects of a variety of serotonergic (*red*) and glutamatergic (*blue*) compounds, including antipsychotics (clozapine, risperidone, and LY379268), neutral antagonists (ritanserin, methysergide, and eGlu), and propsychotics (LY341495 and DOI). These observations provide a mechanistic insight into antipsychotic drug action (see Fribourg et al. 2011)

protein complex (Gonzalez-Maeso et al. 2008; Fribourg et al. 2011) and are located in close physical proximity at cortical synaptic junctions at the electron microscopy level (Moreno et al. 2012). Acting through the 5-HT_{2A}-mGlu2 receptor heterocomplex, it was reported that both serotonergic and glutamatergic ligands balance $G_{q/11}$- and $G_{i/o}$-protein-dependent signaling (Fribourg et al. 2011) and that this pattern of G protein coupling induced by drugs bound to one receptor of the heteromer in the presence of the other receptor's endogenous ligand (i.e., serotonin or glutamate) predicts their propsychotic-like and antipsychotic-like behavioral effects (Fig. 4.3). Additionally, disruption of heteromeric expression of mGlu2 and 5-HT_{2A} receptors in mouse frontal cortex attenuates the antipsychotic-like behaviors induced in mouse by clozapine and LY379269 (Fribourg et al. 2011). Together, these structural and behavioral findings suggest that the 5-HT_{2A}-mGlu2 heteromer is at least in part necessary for the antipsychotic effects induced by atypical antipsychotic drugs and mGlu2/3 receptor agonists.

4.6 Conclusions

Undoubtedly, the study of the basic molecular and cellular mechanism by which modulation of mGlu2-dependent function affects schizophrenia-related phenotypes bears great promise to enhance our knowledge of schizophrenia and other psychiatric disorders. Recent technological advances, combining viral-mediated over-expression protocols and dendritic spine analysis, now make it possible to study regulation of synaptic remodeling in a highly quantitative fashion. Additionally, bivalent ligands that affect specifically the function of a given GPCR heteromer within a region of the CNS are arising as a promising tool in molecular psychiatry research. Complete understanding of the role of mGlu2 receptor in signaling, plasticity, and behavior might lead to the development of better therapeutic strategies to treat psychotic and cognitive symptoms in schizophrenia patients.

Acknowledgments We thank the National Institutes of Health for grant support of our studies of GPCR structure and function.

References

Adam D (2013) On the spectrum. Nature 496:416–418

Adams DH, Kinon BJ, Baygani S, Millen BA, Velona I, Kollack-Walker S, Walling DP (2013) A long-term, phase 2, multicenter, randomized, open-label, comparative safety study of pomaglumetad methionil (LY2140023 monohydrate) versus atypical antipsychotic standard of care in patients with schizophrenia. BMC Psychiatry 13:143

Akatsu S, Ishikawa C, Takemura K, Ohtani A, Shiga T (2015) Effects of prenatal stress and neonatal handling on anxiety, spatial learning and serotonergic system of male offspring mice. Neurosci Res 101:15–23

Audet M, Bouvier M (2012) Restructuring G-protein- coupled receptor activation. Cell 151:14–23

Babulas V, Factor-Litvak P, Goetz R, Schaefer CA, Brown AS (2006) Prenatal exposure to maternal genital and reproductive infections and adult schizophrenia. Am J Psychiatry 163:927–929

Benneyworth MA, Xiang Z, Smith RL, Garcia EE, Conn PJ, Sanders-Bush E (2007) A selective positive allosteric modulator of metabotropic glutamate receptor subtype 2 blocks a hallucinogenic drug model of psychosis. Mol Pharmacol 72:477–484

Brown AS, Cohen P, Harkavy-Friedman J, Babulas V, Malaspina D, Gorman JM, Susser ES (2001) A.E. Bennett research award. Prenatal rubella, premorbid abnormalities, and adult schizophrenia. Biol Psychiatry 49:473–486

Brown AS, Begg MD, Gravenstein S, Schaefer CA, Wyatt RJ, Bresnahan M, Babulas VP, Susser ES (2004) Serologic evidence of prenatal influenza in the etiology of schizophrenia. Arch Gen Psychiatry 61:774–780

Brown AS, Schaefer CA, Quesenberry CP Jr, Liu L, Babulas VP, Susser ES (2005) Maternal exposure to toxoplasmosis and risk of schizophrenia in adult offspring. Am J Psychiatry 162:767–773

Burmeister M, McInnis MG, Zollner S (2008) Psychiatric genetics: progress amid controversy. Nat Rev Genet 9:527–540

Cardno AG, Gottesman II (2000) Twin studies of schizophrenia: from bow-and-arrow concordances to star wars Mx and functional genomics. Am J Med Genet 97:12–17

Chiu HY, Chan MH, Lee MY, Chen ST, Zhan ZY, Chen HH (2014) Long-lasting alterations in 5-HT2A receptor after a binge regimen of methamphetamine in mice. Int J Neuropsychopharmacol:1–12

Conn PJ, Pin JP (1997) Pharmacology and functions of metabotropic glutamate receptors. Annu Rev Pharmacol Toxicol 37:205–237

Consortium TIS (2008) Rare chromosomal deletions and duplications increase risk of schizophrenia. Nature 455:237–241

Consortium SWG (2014) o. t. P. G. Biological insights from 108 schizophrenia-associated genetic loci. Nature 511:421–427

Couve A, Filippov AK, Connolly CN, Bettler B, Brown DA, Moss SJ (1998) Intracellular retention of recombinant GABAB receptors. J Biol Chem 273:26361–26367

Delille HK, Becker JM, Burkhardt S, Bleher B, Terstappen GC, Schmidt M, Meyer AH, Unger L, Marek GJ, Mezler M (2012) Heterocomplex formation of 5-HT(2A)-mGlu(2) and its relevance for cellular signaling cascades. Neuropharmacology:1–8

El Moustaine D, Granier S, Doumazane E, Scholler P, Rahmeh R, Bron P, Mouillac B, Baneres JL, Rondard P, Pin JP (2012) Distinct roles of metabotropic glutamate receptor dimerization in agonist activation and G-protein coupling. Proc Natl Acad Sci U S A 109:16342–16347

Ellaithy A, Younkin J, Gonzalez-Maeso J, Logothetis DE (2015) Positive allosteric modulators of metabotropic glutamate 2 receptors in schizophrenia treatment. Trends Neurosci 38:506–516

Fell MJ, Svensson KA, Johnson BG, Schoepp DD (2008) Evidence for the role of metabotropic glutamate (mGlu)2 not mGlu3 receptors in the preclinical antipsychotic pharmacology of the mGlu2/3 receptor agonist (−)-(1R,4S,5S,6S)-4-amino-2-sulfonylbicyclo[3.1.0]hexane-4,6-dicarboxylic acid (LY404039). J Pharmacol Exp Ther 326:209–217

Freedman R (2003) Schizophrenia. N Engl J Med 349:1738–1749

Fribourg M, Moreno JL, Holloway T, Provasi D, Baki L, Mahajan R, Park G, Adney SK, Hatcher C, Eltit JM, Ruta JD, Albizu L, Li Z, Umali A, Shim J, Fabiato A, Mackerell AD Jr, Brezina V, Sealfon SC, Filizola M, Gonzalez-Maeso J, Logothetis DE (2011) Decoding the signaling of a GPCR heteromeric complex reveals a unifying mechanism of action of antipsychotic drugs. Cell 147:1011–1023

Galici R, Echemendia NG, Rodriguez AL, Conn PJ (2005) A selective allosteric potentiator of metabotropic glutamate (mGlu) 2 receptors has effects similar to an orthosteric mGlu2/3 receptor agonist in mouse models predictive of antipsychotic activity. J Pharmacol Exp Ther 315:1181–1187

Gewirtz JC, Marek GJ (2000) Behavioral evidence for interactions between a hallucinogenic drug and group II metabotropic glutamate receptors. Neuropsychopharmacology 23:569–576

Gonzalez-Maeso J (2011) GPCR oligomers in pharmacology and signaling. Mol Brain 4:20

Gonzalez-Maeso J (2014) Family a GPCR heteromers in animal models. Front Pharmacol 5:226

Gonzalez-Maeso J, Ang RL, Yuen T, Chan P, Weisstaub NV, Lopez-Gimenez JF, Zhou M, Okawa Y, Callado LF, Milligan G, Gingrich JA, Filizola M, Meana JJ, Sealfon SC (2008) Identification of a serotonin/glutamate receptor complex implicated in psychosis. Nature 452:93–97

Gottesman II, Erlenmeyer-Kimling L (2001) Family and twin strategies as a head start in defining prodromes and endophenotypes for hypothetical early-interventions in schizophrenia. Schizophr Res 51:93–102

Gouzoulis-Mayfrank E, Heekeren K, Neukirch A, Stoll M, Stock C, Obradovic M, Kovar KA (2005) Psychological effects of (S)-ketamine and N,N-dimethyltryptamine (DMT): a double-blind, cross-over study in healthy volunteers. Pharmacopsychiatry 38:301–311

Hanks JB, Gonzalez-Maeso J (2013) Animal models of serotonergic psychedelics. ACS Chem Neurosci 4:33–42

Hlavackova V, Zabel U, Frankova D, Batz J, Hoffmann C, Prezeau L, Pin JP, Blahos J, Lohse MJ (2012) Sequential inter- and intrasubunit rearrangements during activation of dimeric metabotropic glutamate receptor 1. Sci Signal 5:ra59

Hoch PH, Cattell JP, Pennes HH (1952) Effects of mescaline and lysergic acid (d-LSD-25). Am J Psychiatry 108:579–584

Holloway T, Moreno JL, Umali A, Rayannavar V, Hodes GE, Russo SJ, Gonzalez-Maeso J (2013) Prenatal stress induces schizophrenia-like alterations of serotonin 2A and metabotropic glutamate 2 receptors in the adult offspring: role of maternal immune system. J Neurosci 33:1088–1098

Jones KA, Borowsky B, Tamm JA, Craig DA, Durkin MM, Dai M, Yao WJ, Johnson M, Gunwaldsen C, Huang LY, Tang C, Shen Q, Salon JA, Morse K, Laz T, Smith KE, Nagarathnam D, Noble

SA, Branchek TA, Gerald C (1998) GABA(B) receptors function as a heteromeric assembly of the subunits GABA(B)R1 and GABA(B)R2. Nature 396:674–679

Jones MW, Errington ML, French PJ, Fine A, Bliss TV, Garel S, Charnay P, Bozon B, Laroche S, Davis S (2001) A requirement for the immediate early gene Zif268 in the expression of late LTP and long-term memories. Nat Neurosci 4:289–296

Jorgensen CV, Jacobsen JP, Caron MG, Klein AB, Knudsen GM, Mikkelsen JD (2013) Cerebral 5-HT2A receptor binding, but not mGluR2, is increased in tryptophan hydroxylase 2 decrease-of-function mice. Neurosci Lett 555:118–122

Kapur S, Remington G (2001) Atypical antipsychotics: new directions and new challenges in the treatment of schizophrenia. Annu Rev Med 52:503–517

Kaupmann K, Huggel K, Heid J, Flor PJ, Bischoff S, Mickel SJ, McMaster G, Angst C, Bittiger H, Froestl W, Bettler B (1997) Expression cloning of GABA(B) receptors uncovers similarity to metabotropic glutamate receptors. Nature 386:239–246

Kaupmann K, Malitschek B, Schuler V, Heid J, Froestl W, Beck P, Mosbacher J, Bischoff S, Kulik A, Shigemoto R, Karschin A, Bettler B (1998) GABA(B)-receptor subtypes assemble into functional heteromeric complexes. Nature 396:683–687

Khashan AS, Abel KM, McNamee R, Pedersen MG, Webb RT, Baker PN, Kenny LC, Mortensen PB (2008) Higher risk of offspring schizophrenia following antenatal maternal exposure to severe adverse life events. Arch Gen Psychiatry 65:146–152

Kinon BJ, Zhang L, Millen BA, Osuntokun OO, Williams JE, Kollack-Walker S, Jackson K, Kryzhanovskaya L, Jarkova N (2011) A multicenter, inpatient, phase 2, double-blind, placebo-controlled dose-ranging study of LY2140023 monohydrate in patients with DSM-IV schizophrenia. J Clin Psychopharmacol 31:349–355

Kinon BJ, Millen BA, Zhang L, McKinzie DL (2015) Exploratory analysis for a targeted patient population responsive to the metabotropic glutamate 2/3 receptor agonist pomaglumetad methionil in schizophrenia. Biol Psychiatry

Klein AB, Ultved L, Adamsen D, Santini MA, Tobena A, Fernandez-Teruel A, Flores P, Moreno M, Cardona D, Knudsen GM, Aznar S, Mikkelsen JD (2014) 5-HT(2A) and mGlu2 receptor binding levels are related to differences in impulsive behavior in the Roman low- (RLA) and high- (RHA) avoidance rat strains. Neuroscience 263:36–45

Kometer M, Schmidt A, Bachmann R, Studerus E, Seifritz E, Vollenweider FX (2012) Psilocybin biases facial recognition, goal-directed behavior, and mood state toward positive relative to negative emotions through different serotonergic subreceptors. Biol Psychiatry

Kometer M, Schmidt A, Jancke L, Vollenweider FX (2013) Activation of serotonin 2A receptors underlies the psilocybin-induced effects on alpha oscillations, N170 visual-evoked potentials, and visual hallucinations. J Neurosci 33:10544–10551

Krystal JH, D'Souza DC, Mathalon D, Perry E, Belger A, Hoffman R (2003) NMDA receptor antagonist effects, cortical glutamatergic function, and schizophrenia: toward a paradigm shift in medication development. Psychopharmacology 169:215–233

Kurita M, Holloway T, Garcia-Bea A, Kozlenkov A, Friedman AK, Moreno JL, Heshmati M, Golden SA, Kennedy PJ, Takahashi N, Dietz DM, Mocci G, Gabilondo AM, Hanks J, Umali A, Callado LF, Gallitano AL, Neve RL, Shen L, Buxbaum JD, Han MH, Nestler EJ, Meana JJ, Russo SJ, Gonzalez-Maeso J (2012) HDAC2 regulates atypical antipsychotic responses through the modulation of mGlu2 promoter activity. Nat Neurosci 15:1245–1254

Kurita M, Moreno JL, Holloway T, Kozlenkov A, Mocci G, Garcia-Bea A, Hanks JB, Neve R, Nestler EJ, Russo SJ, Gonzalez-Maeso J (2013a) Repressive epigenetic changes at the mGlu2 promoter in frontal cortex of 5-HT2A knockout mice. Mol Pharmacol 83:1166–1175

Kurita M, Holloway T, Gonzalez-Maeso J (2013b) HDAC2 as a new target to improve schizophrenia treatment. Expert Rev Neurother 13:1–3

Kuszak AJ, Pitchiaya S, Anand JP, Mosberg HI, Walter NG, Sunahara RK (2009) Purification and functional reconstitution of monomeric mu-opioid receptors: allosteric modulation of agonist binding by Gi2. J Biol Chem 284:26732–26741

Lahti AC, Weiler MA, Tamara Michaelidis BA, Parwani A, Tamminga CA (2001) Effects of ketamine in normal and schizophrenic volunteers. Neuropsychopharmacology 25:455–467

Lieberman JA, Bymaster FP, Meltzer HY, Deutch AY, Duncan GE, Marx CE, Aprille JR, Dwyer DS, Li XM, Mahadik SP, Duman RS, Porter JH, Modica-Napolitano JS, Newton SS, Csernansky JG (2008) Antipsychotic drugs: comparison in animal models of efficacy, neurotransmitter regulation, and neuroprotection. Pharmacol Rev 60:358–403

Liu W, Downing AC, Munsie LM, Chen P, Reed MR, Ruble CL, Landschulz KT, Kinon BJ, Nisenbaum LK (2010) Pharmacogenetic analysis of the mGlu2/3 agonist LY2140023 monohydrate in the treatment of schizophrenia. Pharmacogenomics J

Malaspina D, Corcoran C, Kleinhaus KR, Perrin MC, Fennig S, Nahon D, Friedlander Y, Harlap S (2008) Acute maternal stress in pregnancy and schizophrenia in offspring: a cohort prospective study. BMC Psychiatry 8:71

Malkova NV, Gallagher JJ, Yu CZ, Jacobs RE, Patterson PH (2014) Manganese-enhanced magnetic resonance imaging reveals increased DOI-induced brain activity in a mouse model of schizophrenia. Proc Natl Acad Sci U S A 111:E2492–E2500

Maple AM, Zhao X, Elizalde DI, McBride AK, Gallitano AL (2015) Htr2a expression responds rapidly to environmental stimuli in an Egr3-dependent manner. ACS Chem Neurosci

Margeta-Mitrovic M, Jan YN, Jan LY (2000) A trafficking checkpoint controls GABA(B) receptor heterodimerization. Neuron 27:97–106

Margeta-Mitrovic M, Jan YN, Jan LY (2001) Function of GB1 and GB2 subunits in G protein coupling of GABA(B) receptors. Proc Natl Acad Sci U S A 98:14649–14654

Maurel D, Comps-Agrar L, Brock C, Rives ML, Bourrier E, Ayoub MA, Bazin H, Tinel N, Durroux T, Prezeau L, Trinquet E, Pin JP (2008) Cell-surface protein-protein interaction analysis with time-resolved FRET and snap-tag technologies: application to GPCR oligomerization. Nat Methods 5:561–567

Meltzer HY (2013) Update on typical and atypical antipsychotic drugs. Annu Rev Med 64:393–406

Menninger KA (1919) Psychoses associated with influenza. J Am Med Assoc 72:235–241

Miyamoto S, Duncan GE, Marx CE, Lieberman JA (2005) Treatments for schizophrenia: a critical review of pharmacology and mechanisms of action of antipsychotic drugs. Mol Psychiatry 10:79–104

Miyamoto S, Miyake N, Jarskog LF, Fleischhacker WW, Lieberman JA (2012) Pharmacological treatment of schizophrenia: a critical review of the pharmacology and clinical effects of current and future therapeutic agents. Mol Psychiatry 17:1206–1227

Moghaddam B, Adams BW (1998) Reversal of phencyclidine effects by a group II metabotropic glutamate receptor agonist in rats. Science 281:1349–1352

Moreno JL, Gonzalez-Maeso J (2013a) Preclinical models of antipsychotic drug action. Int J Neuropsychopharmacol 16:2131–2144

Moreno JL, Gonzalez-Maeso J (2013b) Preclinical models of antipsychotic drug action. Int J Neuropsychopharmacol:1–14

Moreno JL, Sealfon SC, Gonzalez-Maeso J (2009) Group II metabotropic glutamate receptors and schizophrenia. Cell Mol Life Sci 66:3777–3785

Moreno JL, Holloway T, Albizu L, Sealfon SC, Gonzalez-Maeso J (2011a) Metabotropic glutamate mGlu2 receptor is necessary for the pharmacological and behavioral effects induced by hallucinogenic 5-HT2A receptor agonists. Neurosci Lett 493:76–79

Moreno JL, Kurita M, Holloway T, Lopez J, Cadagan R, Martinez-Sobrido L, Garcia-Sastre A, Gonzalez-Maeso J (2011b) Maternal influenza viral infection causes schizophrenia-like alterations of 5-HT2A and mGlu2 receptors in the adult offspring. J Neurosci 31:1863–1872

Moreno JL, Muguruza C, Umali A, Mortillo S, Holloway T, Pilar-Cuellar F, Mocci G, Seto J, Callado LF, Neve RL, Milligan G, Sealfon SC, Lopez-Gimenez JF, Meana JJ, Benson DL, Gonzalez-Maeso J (2012) Identification of three residues essential for 5-HT2A-mGlu2 receptor heteromerization and its psychoactive behavioral function. J Biol Chem 287:44301–44319

Moreno JL, Holloway T, Umali A, Rayannavar V, Sealfon SC, Gonzalez-Maeso J (2013) Persistent effects of chronic clozapine on the cellular and behavioral responses to LSD in mice. Psychopharmacology 225:217–226

Nestler EJ, Hyman SE (2010) Animal models of neuropsychiatric disorders. Nat Neurosci 13:1161–1169

O'Donovan KJ, Tourtellotte WG, Millbrandt J, Baraban JM (1999) The EGR family of transcription-regulatory factors: progress at the interface of molecular and systems neuroscience. Trends Neurosci 22:167–173

Patil ST, Zhang L, Martenyi F, Lowe SL, Jackson KA, Andreev BV, Avedisova AS, Bardenstein LM, Gurovich IY, Morozova MA, Mosolov SN, Neznanov NG, Reznik AM, Smulevich AB, Tochilov VA, Johnson BG, Monn JA, Schoepp DD (2007) Activation of mGlu2/3 receptors as a new approach to treat schizophrenia: a randomized phase 2 clinical trial. Nat Med 13:1102–1107

Pershing ML, Bortz DM, Pocivavsek A, Fredericks PJ, Jorgensen CV, Vunck SA, Leuner B, Schwarcz R, Bruno JP (2015) Elevated levels of kynurenic acid during gestation produce neurochemical, morphological, and cognitive deficits in adulthood: implications for schizophrenia. Neuropharmacology 90:33–41

Pin JP, Galvez T, Prezeau L (2003) Evolution, structure, and activation mechanism of family 3/C G-protein-coupled receptors. Pharmacol Ther 98:325–354

Pin JP, Kniazeff J, Goudet C, Bessis AS, Liu J, Galvez T, Acher F, Rondard P, Prezeau L (2004) The activation mechanism of class-C G-protein coupled receptors. Biol Cell 96:335–342

Purcell SM, Wray NR, Stone JL, Visscher PM, O'Donovan MC, Sullivan PF, Sklar P (2009) Common polygenic variation contributes to risk of schizophrenia and bipolar disorder. Nature 460:748–752

Purcell SM, Moran JL, Fromer M, Ruderfer D, Solovieff N, Roussos P, O'Dushlaine C, Chambert K, Bergen SE, Kahler A, Duncan L, Stahl E, Genovese G, Fernandez E, Collins MO, Komiyama NH, Choudhary JS, Magnusson PK, Banks E, Shakir K, Garimella K, Fennell T, DePristo M, Grant SG, Haggarty SJ, Gabriel S, Scolnick EM, Lander ES, Hultman CM, Sullivan PF, McCarroll SA, Sklar P (2014) A polygenic burden of rare disruptive mutations in schizophrenia. Nature 506:185–190

Rives ML, Vol C, Fukazawa Y, Tinel N, Trinquet E, Ayoub MA, Shigemoto R, Pin JP, Prezeau L (2009) Crosstalk between GABAB and mGlu1a receptors reveals new insight into GPCR signal integration. EMBO J 28:2195–2208

Rosenbaum DM, Rasmussen SG, Kobilka BK (2009) The structure and function of G-protein-coupled receptors. Nature 459:356–363

Sawa A, Snyder SH (2002) Schizophrenia: diverse approaches to a complex disease. Science 296:692–695

Schmid Y, Enzler F, Gasser P, Grouzmann E, Preller KH, Vollenweider FX, Brenneisen R, Muller F, Borgwardt S, Liechti ME (2015) Acute effects of lysergic acid diethylamide in healthy subjects. Biol Psychiatry

Schoepp DD (2001) Unveiling the functions of presynaptic metabotropic glutamate receptors in the central nervous system. J Pharmacol Exp Ther 299:12–20

Snyder EM, Murphy MR (2008) Schizophrenia therapy: beyond atypical antipsychotics. Nat Rev Drug Discov 7:471–472

Spooren WP, Gasparini F, van der Putten H, Koller M, Nakanishi S, Kuhn R (2000) Lack of effect of LY314582 (a group 2 metabotropic glutamate receptor agonist) on phencyclidine-induced locomotor activity in metabotropic glutamate receptor 2 knockout mice. Eur J Pharmacol 397:R1–R2

Stefansson H, Rujescu D, Cichon S, Pietilainen OP, Ingason A, Steinberg S, Fossdal R, Sigurdsson E, Sigmundsson T, Buizer-Voskamp JE, Hansen T, Jakobsen KD, Muglia P, Francks C, Matthews PM, Gylfason A, Halldorsson BV, Gudbjartsson D, Thorgeirsson TE, Sigurdsson A, Jonasdottir A, Jonasdottir A, Bjornsson A, Mattiasdottir S, Blondal T, Haraldsson M, Magnusdottir BB, Giegling I, Moller HJ, Hartmann A, Shianna KV, Ge D, Need AC, Crombie C, Fraser G, Walker N, Lonnqvist J, Suvisaari J, Tuulio-Henriksson A, Paunio T, Toulopoulou T, Bramon E, Di Forti M, Murray R, Ruggeri M, Vassos E, Tosato S, Walshe M, Li T, Vasilescu C, Muhleisen TW, Wang AG, Ullum H, Djurovic S, Melle I, Olesen J, Kiemeney LA, Franke B, Sabatti C, Freimer NB, Gulcher JR, Thorsteinsdottir U, Kong A, Andreassen OA, Ophoff RA, Georgi A, Rietschel M, Werge T, Petursson H, Goldstein DB, Nothen MM, Peltonen L, Collier

DA, St Clair D, Stefansson K, Kahn RS, Linszen DH, van Os J, Wiersma D, Bruggeman R, Cahn W, de Haan L, Krabbendam L, Myin-Germeys I (2008) Large recurrent microdeletions associated with schizophrenia. Nature 455:232–236

Stefansson H, Ophoff RA, Steinberg S, Andreassen OA, Cichon S, Rujescu D, Werge T, Pietilainen OP, Mors O, Mortensen PB, Sigurdsson E, Gustafsson O, Nyegaard M, Tuulio-Henriksson A, Ingason A, Hansen T, Suvisaari J, Lonnqvist J, Paunio T, Borglum AD, Hartmann A, Fink-Jensen A, Nordentoft M, Hougaard D, Norgaard-Pedersen B, Bottcher Y, Olesen J, Breuer R, Moller HJ, Giegling I, Rasmussen HB, Timm S, Mattheisen M, Bitter I, Rethelyi JM, Magnusdottir BB, Sigmundsson T, Olason P, Masson G, Gulcher JR, Haraldsson M, Fossdal R, Thorgeirsson TE, Thorsteinsdottir U, Ruggeri M, Tosato S, Franke B, Strengman E, Kiemeney LA, Melle I, Djurovic S, Abramova L, Kaleda V, Sanjuan J, de Frutos R, Bramon E, Vassos E, Fraser G, Ettinger U, Picchioni M, Walker N, Toulopoulou T, Need AC, Ge D, Yoon JL, Shianna KV, Freimer NB, Cantor RM, Murray R, Kong A, Golimbet V, Carracedo A, Arango C, Costas J, Jonsson EG, Terenius L, Agartz I, Petursson H, Nothen MM, Rietschel M, Matthews PM, Muglia P, Peltonen L, St Clair D, Goldstein DB, Stefansson K, Collier DA (2009) Common variants conferring risk of schizophrenia. Nature 460:744–747

Susser E, St Clair D, He L (2008) Latent effects of prenatal malnutrition on adult health: the example of schizophrenia. Ann N Y Acad Sci 1136:185–192

Umbricht D, Schmid L, Koller R, Vollenweider FX, Hell D, Javitt DC (2000) Ketamine-induced deficits in auditory and visual context-dependent processing in healthy volunteers: implications for models of cognitive deficits in schizophrenia. Arch Gen Psychiatry 57:1139–1147

van Os J, Selten JP (1998) Prenatal exposure to maternal stress and subsequent schizophrenia. The May 1940 invasion of The Netherlands. Br J Psychiatry 172:324–326

Veyrac A, Gros A, Bruel-Jungerman E, Rochefort C, Kleine Borgmann FB, Jessberger S, Laroche S (2013) Zif268/egr1 gene controls the selection, maturation and functional integration of adult hippocampal newborn neurons by learning. Proc Natl Acad Sci U S A 110:7062–7067

Vollenweider FX, Vollenweider-Scherpenhuyzen MF, Babler A, Vogel H, Hell D (1998) Psilocybin induces schizophrenia-like psychosis in humans via a serotonin-2 agonist action. Neuroreport 9:3897–3902

Walker AG, Wenthur CJ, Xiang Z, Rook JM, Emmitte KA, Niswender CM, Lindsley CW, Conn PJ (2015) Metabotropic glutamate receptor 3 activation is required for long-term depression in medial prefrontal cortex and fear extinction. Proc Natl Acad Sci U S A 112:1196–1201

Walsh T, McClellan JM, McCarthy SE, Addington AM, Pierce SB, Cooper GM, Nord AS, Kusenda M, Malhotra D, Bhandari A, Stray SM, Rippey CF, Roccanova P, Makarov V, Lakshmi B, Findling RL, Sikich L, Stromberg T, Merriman B, Gogtay N, Butler P, Eckstrand K, Noory L, Gochman P, Long R, Chen Z, Davis S, Baker C, Eichler EE, Meltzer PS, Nelson SF, Singleton AB, Lee MK, Rapoport JL, King MC, Sebat J (2008) Rare structural variants disrupt multiple genes in neurodevelopmental pathways in schizophrenia. Science 320:539–543

Wang Y, Ma Y, Hu J, Cheng W, Jiang H, Zhang X, Li M, Ren J, Li X (2015) Prenatal chronic mild stress induces depression-like behavior and sex-specific changes in regional glutamate receptor expression patterns in adult rats. Neuroscience 301:363–374

White JH, Wise A, Main MJ, Green A, Fraser NJ, Disney GH, Barnes AA, Emson P, Foord SM, Marshall FH (1998) Heterodimerization is required for the formation of a functional GABA(B) receptor. Nature 396:679–682

Whorton MR, Bokoch MP, Rasmussen SG, Huang B, Zare RN, Kobilka B, Sunahara RK (2007) A monomeric G protein-coupled receptor isolated in a high-density lipoprotein particle efficiently activates its G protein. Proc Natl Acad Sci U S A 104:7682–7687

Whorton MR, Jastrzebska B, Park PS, Fotiadis D, Engel A, Palczewski K, Sunahara RK (2008) Efficient coupling of transducin to monomeric rhodopsin in a phospholipid bilayer. J Biol Chem 283:4387–4394

Wischhof L, Irrsack E, Dietz F, Koch M (2015) Maternal lipopolysaccharide treatment differentially affects 5-HT and mGlu2/3 receptor function in the adult male and female rat offspring. Neuropharmacology 97:275–288

Woolley ML, Pemberton DJ, Bate S, Corti C, Jones DN (2008) The mGlu2 but not the mGlu3 receptor mediates the actions of the mGluR2/3 agonist, LY379268, in mouse models predictive of antipsychotic activity. Psychopharmacology 196:431–440

Wu H, Wang C, Gregory KJ, Han GW, Cho HP, Xia Y, Niswender CM, Katritch V, Meiler J, Cherezov V, Conn PJ, Stevens RC (2014) Structure of a class C GPCR metabotropic glutamate receptor 1 bound to an allosteric modulator. Science 344:58–64

Xue L, Rovira X, Scholler P, Zhao H, Liu J, Pin JP, Rondard P (2015) Major ligand-induced rearrangement of the heptahelical domain interface in a GPCR dimer. Nat Chem Biol 11:134–140

Young BG (1974) A phenomenological comparison of LSD and schizophrenic states. Br J Psychiatry 124:64–74

Zhou R, Chen F, Feng X, Zhou L, Li Y, Chen L (2015) Perinatal exposure to low-dose of bisphenol a causes anxiety-like alteration in adrenal axis regulation and behaviors of rat offspring: a potential role for metabotropic glutamate 2/3 receptors. J Psychiatr Res 64:121–129

Chapter 5
mGlu5: A Metabotropic Glutamate Receptor at the Hub of Hippocampal Information Processing, Persistent Synaptic Plasticity, and Long-Term Memory

Hardy Hagena and Denise Manahan-Vaughan

Abstract In the hippocampus, the metabotropic glutamate (mGlu) receptor, mGlu5, plays a very prominent role in synaptic information storage and memory. This receptor enables persistent (>24 h) forms of synaptic plasticity, in the form of long-term potentiation (LTP) and long-term depression (LTD), and is also required for plasticity forms that are directly modulated by spatial learning. mGlu5 supports hippocampal neuronal oscillations that occur during synaptic plasticity events, supports the stabilization of place fields, and regulates the direction of change in synaptic weights in specific synaptic subcompartments of the hippocampus. Furthermore, dysfunctions in this receptor are associated with potent disturbances of hippocampus-dependent cognition. We propose that the mGlu5 receptor lies at the hub of hippocampal information processing and is pivotal to the accurate, long-term, and reliable acquisition and encoding of new spatial experiences and cognitive representations.

Keywords mGlu5 • Synaptic plasticity • Hippocampus • In vivo • Rodent • Learning • Memory • Metabotropic glutamate receptor

5.1 Introduction

One of the most abundant and important neurotransmitters in the mammalian central nervous system (CNS) is glutamate. Glutamate binding to its ionotropic receptors results in the opening of ion channels, whereas glutamate's action on metabotropic receptors triggers a chemical cascade that can activate many downstream targets. Metabotropic glutamate (mGlu) receptors are distributed throughout the brain and show a prominent distribution in the hippocampus (Ferraguti and Shigemoto 2006), an area that is critically involved in distinct forms of memory

H. Hagena, Ph.D. • D. Manahan-Vaughan, Ph.D. (✉)
Medical Faculty, Department of Neurophysiology, Ruhr University Bochum, Universitaetsstrasse 150, 44780 Bochum, Germany
e-mail: hardy.hagena@rub.de; denise.manahan-vaughan@rub.de; dmv-igsn@rub.de

R.T. Ngomba et al. (eds.), *mGLU Receptors*, The Receptors 31,
DOI 10.1007/978-3-319-56170-7_5

and learning. For neuroscientists, these receptors have become increasingly of interest following reports that they are involved in numerous brain diseases, such as schizophrenia and epilepsy (Ure et al. 2006; Conn et al. 2009; Niswender and Conn 2010), as well as in addiction (Bäckström et al. 2004; Cowen et al. 2005; Kenny et al. 2005; McMillen et al. 2005; Schroeder et al. 2005; Bird and Lawrence 2009) and pathological pain (Urban et al. 2003; Zhou et al. 2013). However, mGlu receptors also play an essential role in normal brain function and are crucially involved in fine-tuning of persistent synaptic plasticity, a cellular mechanism that underlies memory formation, and spatial learning (Balschun and Wetzel 1998; Jia et al. 1998; Balschun et al. 1999; Balschun and Wetzel 2002; Naie and Manahan-Vaughan 2004; Manahan-Vaughan and Braunewell 2005; Naie and Manahan-Vaughan 2005; Altinbilek and Manahan-Vaughan 2007; Bikbaev et al. 2008; Altinbilek and Manahan-Vaughan 2009; Popkirov and Manahan-Vaughan 2011; Goh and Manahan-Vaughan 2012a; Zhang and Manahan-Vaughan 2014).

mGlu receptors are classified into three groups based on their G protein coupling and associated biochemical cascades (Conn and Pin 1997). The mGlu5 and the mGlu1 receptor subtypes are group I mGlu receptors that couple positively to inositol phosphate (Abe et al. 1992) and occur in different isoforms (Minakami et al. 1993; Joly et al. 1995). They contribute to bidirectional regulation of synaptic strength, particularly with regard to forms that last for very long periods of time (Mukherjee and Manahan-Vaughan 2012). This is in marked contrast to group II and group III mGlu receptors; these are negatively coupled to adenylyl cyclase (Conn and Pin 1997), largely serve as autoreceptors for glutamate (Harris and Cotman 1986; Kamiya et al. 1996; Macek et al. 1996; Min et al. 1998; Kwon and Castillo 2008), and mostly contribute to synaptic plasticity by changing excitability thresholds or driving changes in synaptic strength toward LTD (Manahan-Vaughan and Reymann 1995a; Manahan-Vaughan 2000; Klausnitzer et al. 2004; Naie et al. 2006).

The mGlu5 receptor (mGlu5) is of particular interest for synaptic plasticity and memory researchers alike, given its very prominent role in plasticity and memory processes (Mukherjee and Manahan-Vaughan 2012) as well as brain diseases that specifically relate to dysfunctions of the receptor, such as Fragile X (Bassell and Warren 2008; Waung and Huber 2009; Dölen et al. 2010; Krueger and Bear 2011). MGlu5 is ubiquitously expressed in the brain (Romano et al. 1995) and is primarily located in extrasynaptic areas of the postsynaptic membrane (Ferraguti and Shigemoto 2006). In terms of the hippocampus, the CA1 region demonstrates the highest expression, followed by the dentate gyrus and CA3 regions (Romano et al. 1995; Shigemoto et al. 1997).

5.2 Role of mGlu5 in Persistent Forms of Hippocampal Synaptic Plasticity and Memory

Robust hippocampus-dependent memory is likely to be enabled by persistent forms of synaptic plasticity such as LTP and LTD (Kemp and Manahan-Vaughan 2007). Persistent LTP and LTD last for days and weeks in freely behaving rodents (Manahan-Vaughan and Braunewell 1999; Abraham et al. 2002; Abraham 2003), and many forms are tightly associated with novel learning and the creation of associative memories (Manahan-Vaughan and Braunewell 1999; Straube et al. 2003; Kemp and Manahan-Vaughan 2004, 2005, 2012; Etkin et al. 2006; Whitlock et al. 2006).

Numerous studies over the last decade have demonstrated that mGlu5 activation is essential for the long-term persistency of both LTP and LTD in rodents, whereby persistent (>24 h) synaptic plasticity in both rats and mice critically depends on mGlu5 (Mukherjee and Manahan-Vaughan 2012; Goh and Manahan-Vaughan 2012a) (Fig. 5.1). Traditionally, persistent LTP and LTD are induced by patterned stimulation of hippocampal afferents, although direct pharmacological activation of mGlu5 has been shown to induce hippocampal LTD (Palmer et al. 1997; Fitzjohn et al. 1999; Huber et al. 2001; Naie et al. 2007), whereas agonists of group I mGlu receptors can lead to slow-onset synaptic potentiation in vitro (Bashir et al. 1993; Chinestra et al. 1994; Bortolotto et al. 1995) or in vivo (Manahan-Vaughan and Reymann 1995b).

5.2.1 Role of mGlu5 in Persistent LTP and LTP-Related Memory

Many years ago, the phases of LTP were subdivided into early LTP (or short-term potentiation, STP) that lasts less than 3 h and late (L)-LTP that lasts beyond 3–6 h (Racine et al. 1983; Krug et al. 1984; Otani et al. 1989; Matthies et al. 1990; Frey

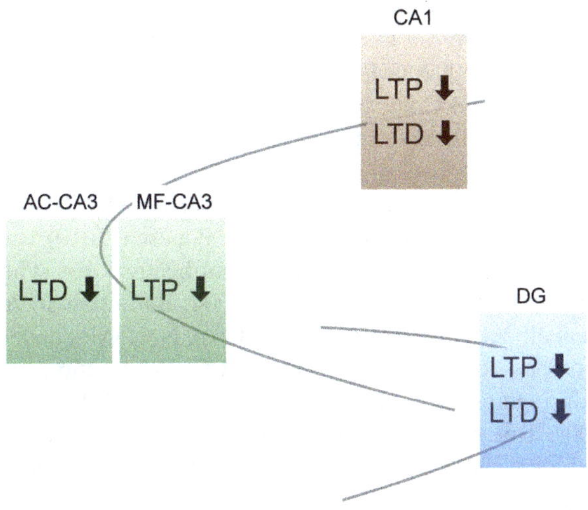

Fig. 5.1 Summary of the effects of mGlu5 antagonism on persistent synaptic plasticity in the rodent hippocampus In freely behaving rats, antagonism of mGlu5 receptors in the dentate gyrus (DG, *blue rectangle*) results in inhibition of persistent (>24 h) LTP and LTD in vivo. LTP is impaired at mossy fiber-CA3 (MF-CA3) synapses, whereas LTD is impaired at commissural/associational-CA3 (AC-CA3) synapses (*green rectangle*). (By contrast, MF-CA3 LTD and AC-CA3 LTP are *not* prevented by antagonism of mGlu5). At Schaffer collateral-CA1 synapses, mGlu5 receptor antagonism inhibits LTP and LTD in both rats and mice (*brown rectangle*)

et al. 1993, 1996; Nguyen et al. 1994; Frey and Morris 1997). Whereas hippocampal E-LTP, at least in the CA1 region, critically depends on NMDAR activation (Harris et al. 1984; Coan and Collingridge 1987; Errington et al. 1987; Bashir et al. 1991; Dudek and Bear 1992; Bear and Malenka 1994; Hrabetova and Sacktor 1997; Hrabetova et al. 2000), LTP depends on protein kinase (Frey et al. 1993) and immediate early gene activation (Abraham et al. 1991; Fazeli et al. 1993; Dragunow 1996). L-LTP requires protein translation (Frey et al. 1988; Otani and Abraham 1989; Bliss and Collingridge 1993) and protein transcription (Nguyen et al. 1994; Frey et al. 1996).

Activation of the N-methyl-D-aspartate receptor (NMDAR) is crucial for long-term changes in synaptic plasticity and leads to Ca^{2+}-influx into the postsynaptic compartment. This initiates the activation of downstream targets, such as Ca^{2+}/calmodulin-dependent protein kinase II (CaMKII) (Malenka and Bear 2004; Kerchner and Nicoll 2008; Lisman et al. 2012; Coultrap et al. 2014), phosphatidylinositol 3-kinase (PI3K) (Perkinton et al. 2002; Zhu et al. 2002), extracellular signal-regulated kinase (ERK) (Kim et al. 2005; Ivanov et al. 2006), as well as protein kinase A (PKA) (Roberson and Sweatt 1996; Banko et al. 2004; Giordano et al. 2005). mGlu5 receptors actively contribute to these processes. In vivo, or in vitro, treatment of the hippocampus with specific antagonists of mGlu5 *prior* to high-frequency stimulation (HFS) to induce LTP, results in an impairment of both the induction and maintenance phases of LTP, whereas treatment post-HFS affects only the late phases (Naie and Manahan-Vaughan 2004; Neyman and Manahan-Vaughan 2008). mGlu5 receptors are positively coupled to phospholipase C via G_q (Tanaka et al. 2000), and glutamate binding results in phosphatidylinositol hydrolysis, Ca^{2+}-release from intracellular stores, and to a potentiation of NMDAR currents (Conn and Pin 1997; Jia et al. 1998; Attucci et al. 2001; Mannaioni et al. 2001; Marino and Conn 2002; O'Brien et al. 2004). Receptor activation also stimulates ERK1/2 (Peavy and Conn 1998) and regulates NMDAR subunit composition. NMDARs are heterotetramers comprising two GluN1 and two GluN2 subunits (Moriyoshi et al. 1991). GluN2A- and GluN2B-containing NMDAR exhibit very different opening thresholds and durations (Chen et al. 1999; Banke and Traynelis 2003; Erreger et al. 2005; Amico-Ruvio and Popescu 2010); thus, a directed change in subunit composition has immense implications for synaptic plasticity and the expression of LTD or LTP (Liu et al. 2004; Bartlett et al. 2007; Morishita et al. 2007; Xu et al. 2009; Ballesteros et al. 2016). NMDARs undergo a switch in their subunit composition from a high GluN2B/GluN2A ratio to a low GluN2B/GluN2A ratio during normal development (Monyer et al. 1994; Sheng et al. 1994; Sans et al. 2000). This switch is mediated by activation of NMDAR and mGlu5 receptors (Matta et al. 2011). MGlu5 knockouts, for example, do not express the developmental and experience-dependent switch from GluN2B to GluN2A-containing receptors (Matta et al. 2011). The switch from GluN2B to GluN2A is an important factor in regulating the direction of LTP and LTD (Liu et al. 2004; Morishita et al. 2007). Thus, an alteration in the subunit composition and/or relative expression can be expected to change the outcome of persistent plasticity (Ballesteros et al. 2016; Jansen et al. 2017).

Interestingly, although LTP does not depend on NMDAR activation in all hippocampal subfields (Harris and Cotman 1986; Wang et al. 1996; Nicoll and Schmitz 2005), LTP at the perforant path-dentate gyrus (pp-DG) synapse (Naie and Manahan-Vaughan 2004), the mossy fiber-CA3 synapse (Hagena and Manahan-Vaughan 2015), and the Schaffer collateral-CA1 synapse (Popkirov and Manahan-Vaughan 2011) all critically require mGlu5 activation in order for LTP to be sustained for periods of over 24 h. Long-term changes of synaptic plasticity are based on the synthesis and translation of proteins (Frey et al. 1988; Manahan-Vaughan et al. 2000) and subsequent activation of immediate early genes (IEGs) such as c-fos or Arc. All of these processes are influenced by mGlu5 activation. An involvement of mGlu5 was first shown in striatal cells, where binding of glutamate to mGlu5 receptors mediates downstream targets such as JNK, Ca^{2+}/calmodulin-dependent protein kinase (CaMK), and phosphorylation of ERK1/2 and Elk-1 (Jong et al. 2009), which is involved in the activation of the IEG c-fos (Cavigelli et al. 1995). Intracellular mGlu5 receptor activation stimulates activity-regulated cytoskeletal-associated protein, Arc/Arg3.1 (Kumar et al. 2012), another IEG that is upregulated in response to synaptic activity (Lyford et al. 1995; Guzowski et al. 2005). Furthermore, group I mGlu receptors contain an intracellular C-terminal tail that interacts with the IEG Homer 1a, which is also a marker for enhanced synaptic activity (Brakeman et al. 1997; Tu et al. 1998). Homer proteins may regulate the function of mGlu receptors upon synaptic activity (Tu et al. 1998). Different family members of Homer have been described that couple to mGlu5 but exercise different functions. Homer 1b and 1c are constitutively expressed, whereas Homer 1a is upregulated in response to synaptic activity but impairs the coupling of the receptor (Xiao et al. 1998; Yang et al. 2004; Duncan et al. 2005; Mao et al. 2005).

Persistent forms of synaptic plasticity that last for longer than 24 h all require activation of mGlu5 receptors (Table 5.1). The stabilization of synaptic plasticity over long periods of time is likely to be supported by synaptic remodeling: proteins of the postsynaptic density (PSD), such as Homer and Shank interact to increase dendritic spine morphology (Sala et al. 2001) and to connect receptors with their signaling cascades (Scannevin and Huganir 2000; Sheng and Pak 2000; Sheng and Sala 2001; Tomita et al. 2001). These processes in turn engage with mGlu5: Homer proteins enable proximal interactions between mGlu5 receptors with the inositol trisphosphate (IP_3) receptor (Tu et al. 1998; Xiao et al. 2000) that in turn mediates Ca^{2+} release from intracellular Ca^{2+} stores. Furthermore, Homer interacts with another protein termed Shank that associates with the NMDAR/PSD-95 complex (Naisbitt et al. 1999) and mediates the clustering of group I mGlu receptors (Tu et al. 1999). Through this clustering, mGlu receptors are linked with the NMDAR scaffold, connecting both signaling pathways. Following the induction of LTP, mGlu5 receptor expression is upregulated in the dentate gyrus and in the CA1 region and is accompanied by a reduction of mGlu2/3 receptor expression (Manahan-Vaughan et al. 2003). This change in mGlu receptor homeostasis can be expected to positively enhance presynaptic release of, and postsynaptic responsiveness to, glutamate. Enhancements of LTP via environmental enrichment also depend on mGlu5 (Buschler and Manahan-Vaughan 2017), suggesting that this receptor is dynamically regulated by learning experiences.

Table 5.1 Regulation of hippocampal synaptic plasticity by mGlu5 receptors

	DG	MF-CA3	AC-CA3	CA1
LTP	LTP blocked[a–d]	LTP blocked[f]	No effect[f]	Enhancement of LTP following prolonged antagonism[e]
	L-LTP blocked[e]	No effect[d]		LTP blocked[c,d,g]
LTD	LTD blocked[h]	No effect[f]	LTD blocked[f]	LTD blocked[g–k]

Summary of the influence of pharmacological antagonism of mGlu5 receptors or knockout of mGlu5 receptors on long-term potentiation (LTP) and long-term depression (LTD) in the hippocampal perforant path-dentate gyrus (DG), mossy fiber-CA3 (MF-CA3), commissural/associational-CA3 (AC-CA3), or Schaffer collateral-CA1 (CA1) synapses

[a]Naie and Manahan-Vaughan (2004)
[b]Balschun and Wetzel (2002)
[c]Manahan-Vaughan and Braunewell (2005)
[d]Lu et al. (1997)
[e]Bikbaev et al. (2008)
[f]Hagena and Manahan-Vaughan (2015)
[g]Manahan-Vaughan (1997)
[h]Ayala et al. (2009)
[i]Goh and Manahan-Vaughan (2012a)
[j]Popkirov and Manahan-Vaughan (2011)
[k]Dietz and Manahan-Vaughan (2017)

LTP is associated with specific aspects of spatial and associative memory including learning about novel or global changes in space (Kemp and Manahan-Vaughan 2004, 2007), some aspects of spatial reference memory (Morris et al. 1986), and aversive associative memories such as conditioned fear (Rogan et al. 1997; Whitlock et al. 2006; Zhou et al. 2009). Studies using rodents that lack mGlu5 or have received antagonist treatment to block mGlu5 corroborate, at least indirectly, that LTP may enable these forms of memory (Table 5.2). Thus, antagonist treatment that prevents persistent LTP also prevents spatial reference memory (Balschun and Wetzel 2002; Naie and Manahan-Vaughan 2004). Conversely, positive allosteric modulation of mGlu5 promotes spatial memory (Balschun et al. 2006; Ayala et al. 2009; Xu et al. 2013) and persistent forms of LTP (Kroker et al. 2011; Bikbaev and Manahan-Vaughan 2017). Furthermore, mGlu5 is required for the consolidation of conditioned fear (Lu et al. 1997; Handford et al. 2014) and for spatial reference memory in the water maze (Lu et al. 1997; Steckler et al. 2005), and fear conditioning enhances mGlu5 expression in the hippocampus (Riedel et al, 2000). Overall, mGlu5 serves as a prominent common denominator shared by hippocampal LTP and specific forms of spatial memory.

5.2.2 *Role of mGlu5 in Persistent LTD and LTD-Related Memory*

Both LTP and LTD comprise cellular mechanisms through which spatial information is stored by the hippocampus (Manahan-Vaughan and Braunewell 1999; Braunewell and Manahan-Vaughan 2001; Kemp and Manahan-Vaughan 2004;

Table 5.2 Regulation of hippocampus-dependent learning and memory by mGlu5 receptors

	Knockouts	Agonist/PAMs	Antagonist/NAMs
Spatial learning	Deficits in memory performance in the MWM[a]	Enhancement of novel object recognition with lower concentration of CDPPB (10 mg/kg)[b]	Impairment of retention in inhibitory avoidance task[g]
		Improvement in spatial alternation retention in Y-maze[c]	Impaired retention of spatial alternation task; no effect if applied immediately after training[h]
		Enhanced memory abilities in T-maze[d]	
		PAMs improve memory performance in the MWM[c]	
		Enhanced performance in reversal learning and search strategy in Barnes maze[d]	
		Enhanced reversal training in MWM[e]	
		MPEP in high concentrations impaired spatial learning in the MWM[f]	
Working memory and reference memory			Impairment of working and reference memory tasks[i–l]
			Impairment of instrumental learning[l]
			Impairment of working memory performance in radial maze[j]
Extinction learning		No effect on single-session and multi-session fear extinction[e]	Extinction of consolidated context is impaired in a spatial T-maze task[n]
		Improved fear extinction learning after single retrieval trial[e]	
		Extinction of cocaine-associated contextual memory is enhanced[m]	
Fear conditioning	Impairment of processing contextual information in fear-conditioning test[a]		Impairment of fear conditioning[o]
			Attenuation of cue-elicited freezing during fear conditioning[p]

(continued)

Table 5.2 (continued)

Pharmacological antagonism or agonism of mGlu5 potently modulates various forms of hippocampus-dependent learning

CDPPB 3-cyano-N-(1,3-diphenyl-1H-pyrazol-5-yl)benzamide

MPEP 2-methyl-6-(phenylethynyl)pyridine

MWM Morris water maze

PAM positive allosteric modulator

[a] Lu et al. (1997)

[b] Uslaner et al. (2009)

[c] Balschun et al. (2006)

[d] Fowler et al. (2013)

[e] Xu et al. (2013)

[f] Steckler et al. (2005)

[g] Simonyi et al. (2007)

[h] Balschun and Wetzel (2002)

[i] Bikbaev et al. (2008)

[j] Naie and Manahan-Vaughan (2004)

[k] Manahan-Vaughan and Braunewell (2005)

[l] Homayoun et al. (2004)

[m] Gass and Olive (2009)

[n] André et al. (2014)

[o] Gravius et al. (2006)

[p] Handford et al. (2014)

[q] Ayala et al. (2009)

Hagena and Manahan-Vaughan 2011). Whereas LTP is associated with the acquisition of information about global changes in the spatial environment (Kemp and Manahan-Vaughan 2004; Uzakov et al. 2005; Hagena and Manahan-Vaughan 2011) or aversive associative experiences (Rogan et al. 1997; Whitlock et al. 2006; Zhou et al. 2009), LTD is associated with the acquisition of details about the spatial content of an experience (Kemp and Manahan-Vaughan 2004, 2008, 2012; Hagena and Manahan-Vaughan 2011; Goh and Manahan-Vaughan 2012b).

LTD, in contrast to LTP, is induced by slow and prolonged influx of postsynaptic Ca^{2+} (Wickens and Abraham 1991; Artola and Singer 1993). This activates protein phosphatases (Mulkey et al. 1993; Mauna et al. 2011) leading to a dephosphorylation (Bear and Malenka 1994; Roche et al. 1994) and subsequent internalization of AMPA receptors (Lüscher et al. 2000; Lüscher and Malenka 2012). Long-term stabilization of hippocampal LTD is enabled by IEG activation (Kemp et al. 2013) and protein synthesis (Manahan-Vaughan et al. 2000). Although for a time it was assumed that LTD comprised the mechanistic mirror image of LTP (Bear and Malenka 1994; Stevens and Wang 1994), several studies demonstrated that different molecular processes are likely to distinguish these forms of synaptic plasticity (Lee et al. 2000; Braunewell and Manahan-Vaughan 2001; Shonesy et al. 2014).

Intriguingly, pharmacological antagonism of mGlu5 receptors prevents both LTP and LTD (Table 5.1). One mechanism whereby this occurs is likely to comprise suppression of mGlu5 receptor-mediated NMDAR currents (Harney et al. 2006), as well as changes in intracellular Ca^{2+} levels that normally occur following mGlu5 activation (Harney et al. 2006; Naie et al. 2007). Antagonism of mGlu5 receptors may also prevent the activation of downstream proteins necessary for the maintenance of persistent depression (Tu et al. 1999; Kumar et al. 2012) or prevent dendritic protein synthesis that is associated with LTD (Huber et al. 2001).

Strikingly, antagonism of mGlu5 not only prevents LTD that is triggered by spatial learning, it also prevents the memory of the novel spatial experience that would normally facilitate LTD (Goh and Manahan-Vaughan 2012a). Furthermore, mGlu5 receptors are required for LTD that is facilitated by visuospatial (Popkirov and Manahan-Vaughan 2011) and audiospatial learning (Dietz and Manahan-Vaughan 2017), suggesting that its requirement extends to learning through different sensory modalities. These observations reinforce the likelihood that mGlu5 is pivotal to normal information encoding and cognition within the hippocampus.

5.3 Role of mGlu5 in Hippocampal Information Processing

5.3.1 Neuronal Oscillations

Within the CNS and hippocampus, information is transferred across neural networks by means of neuronal oscillations (Harris et al. 2003; Buzsáki and Draguhn 2004; Itskov et al. 2008; Axmacher et al. 2009). Hippocampal oscillations in the theta (5–10 Hz) and gamma (30–100) frequency ranges reflect tightly coupled patterns of network activity (Huerta and Lisman 1993; Whittington et al. 1995, 1997; Huerta and Lisman 1996; Traub et al. 2004) that occur during behavioral learning (Bragin et al. 1995; Chrobak and Buzsáki 1998; Buzsáki 2002, 2005; Csicsvari et al. 2003). mGlu5 is intrinsically involved in enabling appropriate network activity in the hippocampus. Persistent antagonism of mGlu5 in rodents results in a suppression of theta and gamma oscillations in the hippocampus that are associated with a loss of spatial memory ability and changes in the profile of LTP in both the dentate gyrus and CA1 subfields (Bikbaev et al. 2008). Strikingly, the precise pattern of theta-gamma responses that occur during patterned high-frequency afferent stimulation (HFS) to induce hippocampal LTP accurately predicts the success or failure of HFS in enabling LTP (Bikbaev and Manahan-Vaughan 2007, 2008), and this process depends on activation of mGlu5 (Bikbaev and Manahan-Vaughan 2017).

Gamma oscillations occur in synchrony with theta oscillations and exhibit the highest amplitudes in the hippocampus, compared to other CNS areas (Csicsvari et al. 2003; Bartos et al. 2007). They contribute to different cognitive processes (Singer 1993; Llinás et al. 1998; Fries et al. 2001; Kahana 2006) and may enable temporal coding of information, as well as the acquisition and retrieval of memory traces (Lisman and Idiart 1995; Axmacher et al. 2006). Prolonged treatment of rodents with an mGlu5 antagonist prevents the increases in gamma oscillations that typically occur in association with LTP induction (Bikbaev et al. 2008). This may correspond to a loss of precision of granule cell firing in the dentate gyrus (Whittington et al. 1995; Ylinen et al. 1995) that is essential for the optimal encoding of incoming information (Bragin et al. 1995; Csicsvari et al. 2003). This may also form the basis for mGlu5-related disruptions in sensory information processing, cognition, and neuronal excitability that have been reported in diseases such as psychosis (Lipska and Weinberger 2000; Seong et al. 2002; van den Buuse et al. 2005; Arguello and Gogos 2006; Krivoy et al. 2008), Fragile X syndrome (Bassell and Warren 2008; Waung and Huber 2009; Dölen et al. 2010; Krueger and Bear 2011), and epilepsy (Moldrich et al. 2003).

Frequently, neuronal oscillations are disrupted in brain illnesses that affect hippocampus-dependent learning and synaptic plasticity (Dugladze et al. 2007; Treviño et al. 2007). In line with this, the failure of HFS to induce hippocampal LTP in amyloid-beta treated rats is tightly associated with a failure of HFS to evoke changes in theta-gamma responses that are characteristic of subsequent successful LTP (Kalweit et al. 2015). Taken together, the results of the abovementioned studies suggest that the stability or precise characteristics of neuronal oscillations that emerge during learning may be essential for the successful encoding of synaptic information. Furthermore, the observation that mGlu5 activity is positively correlated to theta and gamma activity in the dentate gyrus suggests that mGlu5 may be involved in eliciting a precise temporal encoding of neuronal signals within the hippocampus (Bikbaev et al. 2008). This possibility is rendered all the more likely by reports that antagonists of mGlu5 prevent long-term spatial reference memory (Manahan-Vaughan and Braunewell 2005) and object-place memory (Goh and Manahan-Vaughan 2012a) and disrupt the stability of hippocampal place fields (Zhang and Manahan-Vaughan 2014).

5.3.2 The Role of mGlu5 in Place Fields

Place cells comprise hippocampal neurons that display firing patterns that correlate with the context-dependent location of an animal in a spatial environment (O'Keefe and Dostrovsky 1971; O'Keefe 1976; Olton et al. 1978). The location of the rodent in space in which the cells respond with an increase in firing activity is called the place field (O'Keefe 1979)). Place cells enable orientation and navigation in space and, together with information provided by grid cells in the entorhinal cortex (Fyhn et al. 2004; Hafting et al. 2005) and head direction cells in structures such as the thalamus and subiculum (Taube et al. 1990a, b; Taube 1995; Zhang 1996), enable the creation of a cognitive map of an environment (O'Keefe 1979).

The initiation of place fields depends on activation of NMDAR (McHugh et al. 1996; Kentros et al. 1998), but strikingly, the long-term stability of place fields is dependent on activation of mGlu5 (Zhang and Manahan-Vaughan 2014). This observation creates an interesting link between the involvement of mGlu5 in neuronal oscillations, persistent synaptic plasticity, and place fields. Although a causative interdependency for these three processes has not yet been demonstrated, the dependency of these processes on mGlu5 suggests that it is only a matter of time before this interconnectivity is demonstrated.

5.3.3 The Special Case of the CA3 Region

In the hippocampal CA1 region, persistent LTP and LTD both depend on NMDAR activation (Morris et al. 1986; Davis et al. 1992; Manahan-Vaughan 1997), whereas in the dentate gyrus, not all forms of LTP require NMDAR (Manahan-Vaughan et al. 1998). Persistent forms of dentate gyrus LTD are also NMDAR-independent (Pöschel and Manahan-Vaughan 2007). In the CA3 region, this differentiation is even more distinct: LTP and LTD at mossy fiber (MF)-CA3 afferents, which last for at least 24 h in freely behaving rodents (Hagena and Manahan-Vaughan 2011, 2013), do not depend on NMDAR, whereas these forms of plasticity are NMDAR dependent at commissural associational (AC)-CA3 synapses (Hagena and Manahan-Vaughan 2013). In vitro studies suggest that presynaptic mechanisms are critical for the induction of MF-CA3 LTP (Schmitz et al. 2000; Contractor et al. 2001; Lauri et al. 2001), whereas in vivo *studies* have revealed that early and late phases of MF-CA3 LTP depend on protein translation, but only the late phase requires transcription (Hagena and Manahan-Vaughan 2013).

The MF-CA3 and AC-CA3 synapses convey very different informational content to the CA3 region. Input from dentate gyrus granule cells arrives via MF-CA3 synapses, whereas input from the entorhinal cortex (associational path) as well as from contralateral CA3 afferents (commissural path) arrives via the AC-CA3 synapses to the CA3 region (Blackstad 1956; Ishizuka et al. 1990). In contrast to the dentate gyrus and CA1 regions, where both LTP and LTD critically depend on mGlu5, MF-CA3 LTP *but not* LTD requires mGlu5 activation, whereas AC-CA3 LTD *but not*

LTP depend on mGlu5 (Hagena and Manahan-Vaughan 2015). The CA3 region is believed to contribute to working memory, pattern completion, and pattern separation (Marr 1971; Nakazawa 2002; Kesner 2007, 2013; Leutgeb and Leutgeb 2007; Neunuebel and Knierim 2014). The fact that mGlu5 activation can serve to promote synaptic potentiation in mossy fiber synapse of the CA3 region while mediating synaptic depression in AC-CA3 synapses (Hagena and Manahan-Vaughan 2015) suggests that it may play a very important role in regulating signal-to-noise ratios and the precise nature of information encoding in the CA3 region: the selective promotion of LTP at MF-CA3 synapses, concurrent with synaptic depression at AC-CA3 synapses can be expected to enable a high resolution, input specific and highly tuned retention of the encoded experience. This in turn would facilitate subsequent information processing related to pattern separation or pattern completion. Thus, the synapse-specific regulation by mGlu5 of LTP and LTD in the CA3 region may comprise a key mechanistic determinant of the unique properties of this subfield in learning and memory.

5.4 mGlu5 at the Hub of Hippocampal Function

mGlu5 receptors are critically involved, not only in inducing and maintaining but also, in shaping synaptic plasticity: they regulate the direction of change in synaptic strength through hippocampal synapses and also serve as a switch that toggles the *specific* direction of change in synaptic strength, depending on the hippocampal synaptic subpopulation involved (Hagena and Manahan-Vaughan 2015). Activation of mGlu5 receptors is required for forms of LTP and LTD that last for very long periods of time in rodents (Table 5.1) and are also required for spatial learning, context-dependent extinction learning, and associative fear memory (Table 5.2). Furthermore, disruptions in mGlu5 receptor expression or function are tightly linked to a plethora of brain diseases and cognitive disorders that impact upon the hippocampus (Urban et al. 2003; Bäckström et al. 2004; Cowen et al. 2005; Kenny et al. 2005; McMillen et al. 2005; Schroeder et al. 2005; Ure et al. 2006; Bassell and Warren 2008; Bird and Lawrence 2009; Conn et al. 2009; Waung and Huber 2009; Dölen et al. 2010; Niswender and Conn 2010; Krueger and Bear 2011). No other neurotransmitter receptor has been demonstrated to be so intrinsically involved in so many diverse aspects of hippocampal physiology and pathology and most particularly in the context of synaptic information encoding, hippocampus-dependent learning, and optimal cognitive function. In sum, the results of the studies reported in this article strongly support the likelihood that the mGlu5 receptor lies at the hub of hippocampal information processing and is pivotal to the accurate, long-term, and reliable acquisition and encoding of new spatial experiences and cognitive representations.

Acknowledgments Recent studies from the authors' lab in the area of mGlu5 function, that are mentioned in the text, were supported by grants to DMV from the German Research Foundation (SFB 874/B1, B3; FOR1581/P2; SFB1280/A4).

References

Abe T, Sugihara H, Nawa H, Shigemoto R, Mizuno N, Nakanishi S (1992) Molecular characterization of a novel metabotropic glutamate receptor mGluR5 coupled to inositol phosphate/Ca2+ signal transduction. J Biol Chem 267:13361–13368

Abraham WC (2003) How long will long-term potentiation last? Philos Trans R Soc Lond Ser B Biol Sci 358:735–744

Abraham WC, Dragunow M, Tate WP (1991) The role of immediate early genes in the stabilization of long-term potentiation. Mol Neurobiol 5:297–314

Abraham WC, Logan B, Greenwood JM, Dragunow M (2002) Induction and experience-dependent consolidation of stable long-term potentiation lasting months in the hippocampus. J Neurosci 22:9626–9634

Altinbilek B, Manahan-Vaughan D (2007) Antagonism of group III metabotropic glutamate receptors results in impairment of LTD but not LTP in the hippocampal CA1 region, and prevents long-term spatial memory. Eur J Neurosci 26:1166–1172

Altinbilek B, Manahan-Vaughan D (2009) A specific role for group II metabotropic glutamate receptors in hippocampal long-term depression and spatial memory. Neuroscience 158:149–158

Amico-Ruvio SA, Popescu GK (2010) Stationary gating of GluN1/GluN2B receptors in intact membrane patches. Biophys J 98:1160–1169

André MAE, Güntürkün O, Manahan-Vaughan D (2014) The metabotropic glutamate receptor, mGlu5, is required for extinction learning that occurs in the absence of a context change. Hippocampus 25:149–158

Arguello PA, Gogos JA (2006) Modeling madness in mice: one piece at a time. Neuron 52:179–196

Artola A, Singer W (1993) Long-term depression of excitatory synaptic transmission and its relationship to long-term potentiation. Trends Neurosci 16:480–487

Attucci S, Carlà V, Mannaioni G, Moroni F (2001) Activation of type 5 metabotropic glutamate receptors enhances NMDA responses in mice cortical wedges. Br J Pharmacol 132:799–806

Axmacher N, Mormann F, Fernández G, Elger CE, Fell J (2006) Memory formation by neuronal synchronization. Brain Res Rev 52:170–182

Axmacher N, Draguhn A, Elger CE, Fell J (2009) Memory processes during sleep: beyond the standard consolidation theory. Cell Mol Life Sci 66:2285–2297

Ayala JE, Chen Y, Banko JL, Sheffler DJ, Williams R, Telk AN, Watson NL, Xiang Z, Zhang Y, Jones PJ, Lindsley CW, Olive MF, Conn PJ (2009) mGluR5 positive allosteric modulators facilitate both hippocampal LTP and LTD and enhance spatial learning. Neuropsychopharmacology 34:2057–2071

Bäckström P, Bachteler D, Koch S, Hyytiä P, Spanagel R (2004) mGluR5 antagonist MPEP reduces ethanol-seeking and relapse behavior. Neuropsychopharmacology 29:921–928

Ballesteros JJ, Buschler A, Köhr G, Manahan-Vaughan D (2016) Afferent input selects NMDA receptor subtype to determine the persistency of hippocampal LTP in freely behaving mice. Front Behav Neurosci 8:33. doi:10.3389/fnsyn.2016.00033

Balschun D, Wetzel W (1998) Inhibition of group I metabotropic glutamate receptors blocks spatial learning in rats. Neurosci Lett 249:41–44

Balschun D, Wetzel W (2002) Inhibition of mGluR5 blocks hippocampal LTP in vivo and spatial learning in rats. Pharmacol Biochem Behav 73:375–380

Balschun D, Manahan-Vaughan D, Wagner T, Behnisch T, Reymann KG, Wetzel W (1999) A specific role for group I mGluRs in hippocampal LTP and hippocampus-dependent spatial learning. Learn Mem 6:138–152

Balschun D, Zuschratter W, Wetzel W (2006) Allosteric enhancement of metabotropic glutamate receptor 5 function promotes spatial memory. Neuroscience 142:691–702

Banke TG, Traynelis SF (2003) Activation of NR1/NR2B NMDA receptors. Nat Neurosci 6:144–152

Banko JL, Hou L, Klann E (2004) NMDA receptor activation results in PKA- and ERK-dependent Mnk1 activation and increased eIF4E phosphorylation in hippocampal area CA1. J Neurochem 91:462–470

Bartlett TE, Bannister NJ, Collett VJ, Dargan SL, Massey PV, Bortolotto ZA, Fitzjohn SM, Bashir ZI, Collingridge GL, Lodge D (2007) Differential roles of NR2A and NR2B-containing NMDA receptors in LTP and LTD in the CA1 region of two-week old rat hippocampus. Neuropharmacology 52:60–70

Bartos M, Vida I, Jonas P (2007) Synaptic mechanisms of synchronized gamma oscillations in inhibitory interneuron networks. Nat Rev Neurosci 8:45–56

Bashir ZI, Alford S, Davies SN, Randall AD, Collingridge GL (1991) Long-term potentiation of NMDA receptor-mediated synaptic transmission in the hippocampus. Nature 349:156–158

Bashir ZI, Bortolotto ZA, Davies CH, Berretta N, Irving AJ, Seal AJ, Henley JM, Jane DE, Watkins JC, Collingridge GL (1993) Induction of LTP in the hippocampus needs synaptic activation of glutamate metabotropic receptors. Nature 363:347–350

Bassell GJ, Warren ST (2008) Fragile X syndrome: loss of local mRNA regulation alters synaptic development and function. Neuron 60:201–214

Bear MF, Malenka RC (1994) Synaptic plasticity: LTP and LTD. Curr Opin Neurobiol 4:389–399

Bikbaev A, Manahan-Vaughan D (2007) Hippocampal network activity is transiently altered by induction of long-term potentiation in the dentate gyrus of freely behaving rats. Front Behav Neurosci 1:7

Bikbaev A, Manahan-Vaughan D (2008) Relationship of hippocampal theta and gamma oscillations to potentiation of synaptic transmission. Front Neurosci 2:56–63

Bikbaev A, Manahan-Vaughan D (2017) Metabotropic glutamate receptor, **mGlu5**, regulates hippocampal synaptic plasticity and is required for tetanisation-triggered changes in theta and gamma oscillations. Neuropharmacology 115:20–29. doi:10.1016/j.neuropharm.2016.06.004

Bikbaev A, Neyman S, Ngomba RT, Conn PJ, Conn J, Nicoletti F, Manahan-Vaughan D (2008) MGluR5 mediates the interaction between late-LTP, network activity, and learning. Tsien JZ. PLoS One 3:e2155

Bird MK, Lawrence AJ (2009) Group I metabotropic glutamate receptors: involvement in drug-seeking and drug-induced plasticity. Curr Mol Pharmacol 2:83–94

Blackstad TW (1956) Commissural connections of the hippocampal region in the rat, with special reference to their mode of termination. J Comp Neurol 105:417–537

Bliss TV, Collingridge GL (1993) A synaptic model of memory: long-term potentiation in the hippocampus. Nature 361:31–39

Bortolotto ZA, Bashir ZI, Davies CH, Taira T, Kaila K, Collingridge GL (1995) Studies on the role of metabotropic glutamate receptors in long-term potentiation: some methodological considerations. J Neurosci Methods 59:19–24

Bragin A, Jandó G, Nádasdy Z, van Landeghem M, Buzsáki G (1995) Dentate EEG spikes and associated interneuronal population bursts in the hippocampal hilar region of the rat. J Neurophysiol 73:1691–1705

Brakeman PR, Lanahan AA, O'Brien R, Roche K, Barnes CA, Huganir RL, Worley PF (1997) Homer: a protein that selectively binds metabotropic glutamate receptors. Nature 386:284–288

Braunewell K-H, Manahan-Vaughan D (2001) Long-term depression: a cellular basis for learning? Rev Neurosci 12:121–140

Buschler A, Manahan-Vaughan D (2017) Metabotropic glutamate receptor, mGlu5, mediates enhancements of hippocampal long-term potentiation after environmental enrichment in young and old mice. Neuropharmacology 115:42–50. doi:10.1016/j.neuropharm.2016.06.003

Buzsáki G (2002) Theta oscillations in the hippocampus. Neuron 33:325–340

Buzsáki G (2005) Theta rhythm of navigation: link between path integration and landmark navigation, episodic and semantic memory. Hippocampus 15:827–840

Buzsáki G, Draguhn A (2004) Neuronal oscillations in cortical networks. Science 304:1926–1929

Cavigelli M, Dolfi F, Claret FX, Karin M (1995) Induction of c-fos expression through JNK-mediated TCF/elk-1 phosphorylation. EMBO J 14:5957–5964

Chen N, Luo T, Raymond LA (1999) Subtype-dependence of NMDA receptor channel open probability. J Neurosci 19:6844–6854

Chinestra P, Diabira D, Urban NN, Barrionuevo G, Ben-Ari Y (1994) Major differences between long-term potentiation and ACPD-induced slow onset potentiation in hippocampus. Neurosci Lett 182:177–180

Chrobak JJ, Buzsáki G (1998) Gamma oscillations in the entorhinal cortex of the freely behaving rat. J Neurosci 18:388–398

Coan EJ, Collingridge GL (1987) Characterization of an N-methyl-d-aspartate receptor component of synaptic transmission in rat hippocampal slices. Neuroscience 22:1–8

Conn PJ, Pin JP (1997) Pharmacology and functions of metabotropic glutamate receptors. Annu Rev Pharmacol Toxicol 37:205–237

Conn PJ, Lindsley CW, Jones CK (2009) Activation of metabotropic glutamate receptors as a novel approach for the treatment of schizophrenia. Trends Pharmacol Sci 30:25–31

Contractor A, Swanson G, Heinemann SF (2001) Kainate receptors are involved in short- and long-term plasticity at mossy fiber synapses in the hippocampus. Neuron 29:209–216

Coultrap SJ, Freund RK, O'Leary H, Sanderson JL, Roche KW, Dell'Acqua ML, Bayer KU (2014) Autonomous CaMKII mediates both LTP and LTD using a mechanism for differential substrate site selection. Cell Rep 6:431–437

Cowen MS, Djouma E, Lawrence AJ (2005) The metabotropic glutamate 5 receptor antagonist 3-[(2-methyl-1,3-thiazol-4-yl)ethynyl]-pyridine reduces ethanol self-administration in multiple strains of alcohol-preferring rats and regulates olfactory glutamatergic systems. J Pharmacol Exp Ther 315:590–600

Csicsvari J, Jamieson B, Wise KD, Buzsáki G (2003) Mechanisms of gamma oscillations in the hippocampus of the behaving rat. Neuron 37:311–322

Davis S, Butcher SP, Morris RG (1992) The NMDA receptor antagonist D-2-amino-5-phosphonopentanoate (D-AP5) impairs spatial learning and LTP in vivo at intracerebral concentrations comparable to those that block LTP in vitro. J Neurosci 12:21–34

Dietz B, Manahan-Vaughan D (2017) Hippocampal long-term depression is facilitated by the acquisition and updating of memory of spatial auditory content and requires **mGlu5** activation. Neuropharmacology 115:30–41. doi:10.1016/j.neuropharm.2016.02.026

Dölen G, Carpenter RL, Ocain TD, Bear MF (2010) Mechanism-based approaches to treating fragile X. Pharmacol Ther 127:78–93

Dragunow M (1996) A role for immediate-early transcription factors in learning and memory. Behav Genet 26:293–299

Dudek SM, Bear MF (1992) Homosynaptic long-term depression in area CA1 of hippocampus and effects of N-methyl-D-aspartate receptor blockade. Proc Natl Acad Sci U S A 89:4363–4367

Dugladze T, Vida I, Tort AB, Gross A, Otahal J, Heinemann U, Kopell NJ, Gloveli T (2007) Impaired hippocampal rhythmogenesis in a mouse model of mesial temporal lobe epilepsy. Proc Natl Acad Sci U S A 104:17530–17535

Duncan RS, Hwang S-Y, Koulen P (2005) Effects of vesl/Homer proteins on intracellular signaling. Exp Biol Med (Maywood) 230:527–535

Erreger K, Dravid SM, Banke TG, Wyllie DJA, Traynelis SF (2005) Subunit-specific gating controls rat NR1/NR2A and NR1/NR2B NMDA channel kinetics and synaptic signalling profiles. J Physiol 563:345–358

Errington ML, Lynch MA, Bliss TV (1987) Long-term potentiation in the dentate gyrus: induction and increased glutamate release are blocked by D(−)aminophosphonovalerate. Neuroscience 20:279–284

Etkin A, Alarcon JM, Weisberg SP, Touzani K, Huang Y-Y, Nordheim A, Kandel ER (2006) A role in learning for SRF: deletion in the adult forebrain disrupts LTD and the formation of an immediate memory of a novel context. Neuron 50:127–143

Fazeli MS, Corbet J, Dunn MJ, Dolphin AC, Bliss TV (1993) Changes in protein synthesis accompanying long-term potentiation in the dentate gyrus in vivo. J Neurosci 13:1346–1353

Ferraguti F, Shigemoto R (2006) Metabotropic glutamate receptors. Cell Tissue Res 326:483–504

Fitzjohn SM, Kingston AE, Lodge D, Collingridge GL (1999) DHPG-induced LTD in area CA1 of juvenile rat hippocampus; characterisation and sensitivity to novel mGlu receptor antagonists. Neuropharmacology 38:1577–1583

Fowler SW, Walker JM, Klakotskaia D, Will MJ, Serfozo P, Simonyi A, Schachtman TR (2013) Effects of a metabotropic glutamate receptor 5 positive allosteric modulator, CDPPB, on spatial learning task performance in rodents. Neurobiol Learn Mem 99:25–31

Frey U, Morris RG (1997) Synaptic tagging and long-term potentiation. Nature 385:533–536

Frey U, Krug M, Reymann KG, Matthies H (1988) Anisomycin, an inhibitor of protein synthesis, blocks late phases of LTP phenomena in the hippocampal CA1 region in vitro. Brain Res 452:57–65

Frey U, Huang YY, Kandel ER (1993) Effects of cAMP simulate a late stage of LTP in hippocampal CA1 neurons. Science 260:1661–1664

Frey U, Frey S, Schollmeier F, Krug M (1996) Influence of actinomycin D, a RNA synthesis inhibitor, on long-term potentiation in rat hippocampal neurons in vivo and in vitro. J Physiol 490(Pt 3):703–711

Fries P, Neuenschwander S, Engel AK, Goebel R, Singer W (2001) Rapid feature selective neuronal synchronization through correlated latency shifting. Nat Neurosci 4:194–200

Fyhn M, Molden S, Witter MP, Moser EI, Moser M-B (2004) Spatial representation in the entorhinal cortex. Science 305:1258–1264

Gass JT, Olive MF (2009) Positive allosteric modulation of mGluR5 receptors facilitates extinction of a cocaine contextual memory. Biol Psychiat 65(8):717–720

Giordano G, Sánchez-Pérez AM, Montoliu C, Berezney R, Malyavantham K, Costa LG, Calvete JJ, Felipo V (2005) Activation of NMDA receptors induces protein kinase A-mediated phosphorylation and degradation of matrin 3. Blocking these effects prevents NMDA-induced neuronal death. J Neurochem 94:808–818

Goh JJ, Manahan-Vaughan D (2012a) Endogenous hippocampal LTD that is enabled by spatial object recognition requires activation of NMDA receptors and the metabotropic glutamate receptor, mGlu5. Hippocampus 23:129–138

Goh JJ, Manahan-Vaughan D (2012b) Spatial object recognition enables endogenous LTD that curtails LTP in the mouse hippocampus. Cereb Cortex 23:1118–1125

Gravius A, Barberi C, Schäfer D, Schmidt WJ, Danysz W (2006) The role of group I metabotropic glutamate receptors in acquisition and expression of contextual and auditory fear conditioning in rats – a comparison. Neuropharmacology *51*(7-8):1146–1155

Guzowski JF, Timlin JA, Roysam B, McNaughton BL, Worley PF, Barnes CA (2005) Mapping behaviorally relevant neural circuits with immediate-early gene expression. Curr Opin Neurobiol 15:599–606

Hafting T, Fyhn M, Molden S, Moser M-B, Moser EI (2005) Microstructure of a spatial map in the entorhinal cortex. Nature 436:801–806

Hagena H, Manahan-Vaughan D (2011) Learning-facilitated synaptic plasticity at CA3 mossy fiber and commissural-associational synapses reveals different roles in information processing. Cereb Cortex 21:2442–2449

Hagena H, Manahan-Vaughan D (2013) Differentiation in the protein synthesis-dependency of persistent synaptic plasticity in mossy fiber and associational/commissural CA3 synapses in vivo. Front Integr Neurosci 7:10

Hagena H, Manahan-Vaughan D (2015) mGlu5 acts as a switch for opposing forms of synaptic plasticity at mossy fiber-CA3 and commissural associational-CA3 synapses. J Neurosci 35:4999–5006

Handford CE, Tan S, Lawrence AJ, Kim JH (2014) The effect of the mGlu5 negative allosteric modulator MTEP and NMDA receptor partial agonist D-cycloserine on pavlovian conditioned fear. Int J Neuropsychopharmacol 17:1521–1532

Harney SC, Rowan M, Anwyl R (2006) Long-term depression of NMDA receptor-mediated synaptic transmission is dependent on activation of metabotropic glutamate receptors and is altered to long-term potentiation by low intracellular calcium buffering. J Neurosci 26:1128–1132

Harris EW, Cotman CW (1986) Long-term potentiation of guinea pig mossy fiber responses is not blocked by N-methyl D-aspartate antagonists. Neurosci Lett 70:132–137

Harris EW, Ganong AH, Cotman CW (1984) Long-term potentiation in the hippocampus involves activation of N-methyl-D-aspartate receptors. Brain Res 323:132–137

Harris KD, Csicsvari J, Hirase H, Dragoi G, Buzsáki G (2003) Organization of cell assemblies in the hippocampus. Nature 424:552–556

Homayoun H, Stefani MR, Adams BW, Tamagan GD, Moghaddam B (2004) Functional interaction between NMDA and mGlu5 receptors: effects on working memory, instrumental learning, motor behaviors, and Dopamine release. Neuropsychopharmacology *29*(7):1259–1269

Hrabetova S, Sacktor TC (1997) Long-term potentiation and long-term depression are induced through pharmacologically distinct NMDA receptors. Neurosci Lett 226:107–110

Hrabetova S, Serrano P, Blace N, Tse HW, Skifter DA, Jane DE, Monaghan DT, Sacktor TC (2000) Distinct NMDA receptor subpopulations contribute to long-term potentiation and long-term depression induction. J Neurosci 20:RC81

Huber KM, Roder JC, Bear MF (2001) Chemical induction of mGluR5- and protein synthesis--dependent long-term depression in hippocampal area CA1. J Neurophysiol 86:321–325

Huerta PT, Lisman JE (1993) Heightened synaptic plasticity of hippocampal CA1 neurons during a cholinergically induced rhythmic state. Nature 364:723–725

Huerta PT, Lisman JE (1996) Synaptic plasticity during the cholinergic theta-frequency oscillation in vitro. Hippocampus 6:58–61

Ishizuka N, Weber J, Amaral DG (1990) Organization of intrahippocampal projections originating from CA3 pyramidal cells in the rat. J Comp Neurol 295:580–623

Itskov V, Pastalkova E, Mizuseki K, Buzsáki G, Buzsaki G, Harris KD (2008) Theta-mediated dynamics of spatial information in hippocampus. J Neurosci 28:5959–5964

Ivanov A, Pellegrino C, Rama S, Dumalska I, Salyha Y, Ben-Ari Y, Medina I (2006) Opposing role of synaptic and extrasynaptic NMDA receptors in regulation of the extracellular signal-regulated kinases (ERK) activity in cultured rat hippocampal neurons. J Physiol 572:789–798

Jansen S, Gottschling, C, Faissner A, Frotscher M, Manahan-Vaughan D (2017) Intrinsic cellular and molecular properties of in vivo hippocampal synaptic plasticity are altered in the absence of key synaptic matrix molecules. Hippocampus. Epub ahead of print. doi:10.1002/hipo.22742

Jia Z, Lu Y, Henderson J, Taverna F, Romano C, Abramow-Newerly W, Wojtowicz JM, Roder J (1998) Selective abolition of the NMDA component of long-term potentiation in mice lacking mGluR5. Learn Mem 5:331–343

Joly C, Gomeza J, Brabet I, Curry K, Bockaert J, Pin JP (1995) Molecular, functional, and pharmacological characterization of the metabotropic glutamate receptor type 5 splice variants: comparison with mGluR1. J Neurosci 15:3970–3981

Jong Y-JI, Kumar V, O'Malley KL (2009) Intracellular metabotropic glutamate receptor 5 (mGluR5) activates signaling cascades distinct from cell surface counterparts. J Biol Chem 284:35827–35838

Kahana MJ (2006) The cognitive correlates of human brain oscillations. J Neurosci 26:1669–1672

Kalweit AN, Yang H, Colitti-Klausnitzer J, Fülöp L, Bozsó Z, Penke B, Manahan-Vaughan D (2015) Acute intracerebral treatment with amyloid-beta (1-42) alters the profile of neuronal oscillations that accompany LTP induction and results in impaired LTP in freely behaving rats. Front Behav Neurosci 9:103

Kamiya H, Shinozaki H, Yamamoto C (1996) Activation of metabotropic glutamate receptor type 2/3 suppresses transmission at rat hippocampal mossy fibre synapses. J Physiol 493(Pt 2):447–455

Kemp A, Manahan-Vaughan D (2004) Hippocampal long-term depression and long-term potentiation encode different aspects of novelty acquisition. Proc Natl Acad Sci U S A 101:8192–8197

Kemp A, Manahan-Vaughan D (2005) The 5-hydroxytryptamine4 receptor exhibits frequency-dependent properties in synaptic plasticity and behavioural metaplasticity in the hippocampal CA1 region in vivo. Cereb Cortex 15:1037–1043

Kemp A, Manahan-Vaughan D (2007) Hippocampal long-term depression: master or minion in declarative memory processes? Trends Neurosci 30:111–118

Kemp A, Manahan-Vaughan D (2008) The hippocampal CA1 region and dentate gyrus differentiate between environmental and spatial feature encoding through long-term depression. Cereb Cortex 18:968–977

Kemp A, Manahan-Vaughan D (2012) Passive spatial perception facilitates the expression of persistent hippocampal long-term depression. Cereb Cortex 22:1614–1621

Kemp A, Tischmeyer W, Manahan-Vaughan D (2013) Learning-facilitated long-term depression requires activation of the immediate early gene, c-fos, and is transcription dependent. Behav Brain Res 254:83–91

Kenny PJ, Boutrel B, Gasparini F, Koob GF, Markou A (2005) Metabotropic glutamate 5 receptor blockade may attenuate cocaine self-administration by decreasing brain reward function in rats. Psychopharmacologia 179:247–254

Kentros C, Hargreaves E, Hawkins RD, Kandel ER, Shapiro M, Muller RV (1998) Abolition of long-term stability of new hippocampal place cell maps by NMDA receptor blockade. Science 280:2121–2126

Kerchner GA, Nicoll RA (2008) Silent synapses and the emergence of a postsynaptic mechanism for LTP. Nat Rev Neurosci 9:813–825

Kesner RP (2007) Behavioral functions of the CA3 subregion of the hippocampus. Learn Mem 14:771–781

Kesner RP (2013) A process analysis of the CA3 subregion of the hippocampus. Front Cell Neurosci 7:78

Kim MJ, Dunah AW, Wang YT, Sheng M (2005) Differential roles of NR2A- and NR2B-containing NMDA receptors in ras-ERK signaling and AMPA receptor trafficking. Neuron 46:745–760

Klausnitzer J, Kulla A, Manahan-Vaughan D (2004) Role of the group III metabotropic glutamate receptor in LTP, depotentiation and LTD in dentate gyrus of freely moving rats. Neuropharmacology 46:160–170

Krivoy A, Fischel T, Weizman A (2008) The possible involvement of metabotropic glutamate receptors in schizophrenia. Eur Neuropsychopharmacol 18:395–405

Kroker KS, Rast G, Rosenbrock H (2011) Differential effect of the mGlu5 receptor positive allosteric modulator ADX-47273 on early and late hippocampal LTP. Neuropharmacology 61:707–714

Krueger DD, Bear MF (2011) Toward fulfilling the promise of molecular medicine in fragile X syndrome. Annu Rev Med 62:411–429

Krug M, Lössner B, Ott T (1984) Anisomycin blocks the late phase of long-term potentiation in the dentate gyrus of freely moving rats. Brain Res Bull 13:39–42

Kumar V, Fahey PG, Jong YJI, Ramanan N, O'Malley KL (2012) Activation of intracellular metabotropic glutamate receptor 5 in striatal neurons leads to up-regulation of genes associated with sustained synaptic transmission including arc/Arg3.1 protein. J Biol Chem 287:5412–5425

Kwon H-B, Castillo PE (2008) Role of glutamate autoreceptors at hippocampal mossy fiber synapses. Neuron 60:1082–1094

Lauri SE, Bortolotto ZA, Bleakman D, Ornstein PL, Lodge D, Isaac JT, Collingridge GL (2001) A critical role of a facilitatory presynaptic kainate receptor in mossy fiber LTP. Neuron 32:697–709

Lee HK, Barbarosie M, Kameyama K, Bear MF, Huganir RL (2000) Regulation of distinct AMPA receptor phosphorylation sites during bidirectional synaptic plasticity. Nature 405:955–959

Leutgeb S, Leutgeb JK (2007) Pattern separation, pattern completion, and new neuronal codes within a continuous CA3 map. Learn Mem 14:745–757

Lipska BK, Weinberger DR (2000) To model a psychiatric disorder in animals: schizophrenia as a reality test. Neuropsychopharmacology 23:223–239

Lisman JE, Idiart MA (1995) Storage of 7 +/− 2 short-term memories in oscillatory subcycles. Science 267:1512–1515

Lisman J, Yasuda R, Raghavachari S (2012) Mechanisms of CaMKII action in long-term potentiation. Nat Rev Neurosci 13:169–182

Liu L, Wong TP, Pozza MF, Lingenhoehl K, Wang Y, Sheng M, Auberson YP, Wang YT (2004) Role of NMDA receptor subtypes in governing the direction of hippocampal synaptic plasticity. Science 304:1021–1024

Llinás R, Ribary U, Contreras D, Pedroarena C (1998) The neuronal basis for consciousness. Philos Trans R Soc Lond Ser B Biol Sci 353:1841–1849

Lu YM, Jia Z, Janus C, Henderson JT, Gerlai R, Wojtowicz JM, Roder JC (1997) Mice lacking metabotropic glutamate receptor 5 show impaired learning and reduced CA1 long-term potentiation (LTP) but normal CA3 LTP. J Neurosci 17:5196–5205

Lüscher C, Malenka RC (2012) NMDA receptor-dependent long-term potentiation and long-term depression (LTP/LTD). Cold Spring Harb Perspect Biol 4:a005710–a005710

Lüscher C, Nicoll RA, Malenka RC, Muller D (2000) Synaptic plasticity and dynamic modulation of the postsynaptic membrane. Nat Neurosci 3:545–550

Lyford GL, Yamagata K, Kaufmann WE, Barnes CA, Sanders LK, Copeland NG, Gilbert DJ, Jenkins NA, Lanahan AA, Worley PF (1995) Arc, a growth factor and activity-regulated gene, encodes a novel cytoskeleton-associated protein that is enriched in neuronal dendrites. Neuron 14:433–445

Macek TA, Winder DG, Gereau RW, Ladd CO, Conn PJ (1996) Differential involvement of group II and group III mGluRs as autoreceptors at lateral and medial perforant path synapses. J Neurophysiol 76:3798–3806

Malenka RC, Bear MF (2004) LTP and LTD: an embarrassment of riches. Neuron 44:5–21

Manahan-Vaughan D (1997) Group 1 and 2 metabotropic glutamate receptors play differential roles in hippocampal long-term depression and long-term potentiation in freely moving rats. J Neurosci 17:3303–3311

Manahan-Vaughan D (2000) Group III metabotropic glutamate receptors modulate long-term depression in the hippocampal CA1 region of two rat strains in vivo. Neuropharmacology 39:1952–1958

Manahan-Vaughan D, Braunewell K-H (1999) Novelty acquisition is associated with induction of hippocampal long-term depression. Proc Natl Acad Sci U S A 96:8739–8744

Manahan-Vaughan D, Braunewell K-H (2005) The metabotropic glutamate receptor, mGluR5, is a key determinant of good and bad spatial learning performance and hippocampal synaptic plasticity. Cereb Cortex 15:1703–1713

Manahan-Vaughan D, Reymann KG (1995a) Regional and developmental profile of modulation of hippocampal synaptic transmission and LTP by AP4-sensitive mGluRs in vivo. Neuropharmacology 34:991–1001

Manahan-Vaughan D, Reymann KG (1995b) 1S,3R-ACPD dose-dependently induces a slow-onset potentiation in the rat hippocampal CA1 region in vivo. Neuropharmacology 34:1103–1105

Manahan-Vaughan D, Braunewell K-H, Reymann KG (1998) Subtype-specific involvement of metabotropic glutamate receptors in two forms of long-term potentiation in the dentate gyrus of freely moving rats. Neuroscience 86:709–721

Manahan-Vaughan D, Kulla A, Frey JU (2000) Requirement of translation but not transcription for the maintenance of long-term depression in the CA1 region of freely moving rats. J Neurosci 20:8572–8576

Manahan-Vaughan D, Ngomba RT, Storto M, Kulla A, Catania MV, Chiechio S, Rampello L, Passarelli F, Capece A, Reymann KG, Nicoletti F (2003) An increased expression of the mGlu5 receptor protein following LTP induction at the perforant path-dentate gyrus synapse in freely moving rats. Neuropharmacology 44:17–25

Mannaioni G, Marino MJ, Valenti O, Traynelis SF, Conn PJ (2001) Metabotropic glutamate receptors 1 and 5 differentially regulate CA1 pyramidal cell function. J Neurosci 21:5925–5934

Mao L, Yang L, Tang Q, Samdani S, Zhang G, Wang JQ (2005) The scaffold protein Homer1b/c links metabotropic glutamate receptor 5 to extracellular signal-regulated protein kinase cascades in neurons. J Neurosci 25:2741–2752

Marino MJ, Conn PJ (2002) Direct and indirect modulation of the N-methyl D-aspartate receptor. Curr Drug Targets CNS Neurol Disord 1:1–16

Marr D (1971) Simple memory: a theory for archicortex. Philos Trans R Soc Lond Ser B Biol Sci 262:23–81

Matta JA, Ashby MC, Sanz-Clemente A, Roche KW, Isaac JTR (2011) mGluR5 and NMDA receptors drive the experience- and activity-dependent NMDA receptor NR2B to NR2A subunit switch. Neuron 70:339–351

Matthies H, Frey U, Reymann K, Krug M, Jork R, Schröeder H (1990) Different mechanisms and multiple stages of LTP. Adv Exp Med Biol 268:359–368

Mauna JC, Miyamae T, Pulli B, Thiels E (2011) Protein phosphatases 1 and 2A are both required for long-term depression and associated dephosphorylation of cAMP response element binding protein in hippocampal area CA1 in vivo. Hippocampus 21:1093–1104

McHugh TJ, Blum KI, Tsien JZ, Tonegawa S, Wilson MA (1996) Impaired hippocampal representation of space in CA1-specific NMDAR1 knockout mice. Cell 87:1339–1349

McMillen BA, Crawford MS, Kulers CM, Williams HL (2005) Effects of a metabotropic, mglu5, glutamate receptor antagonist on ethanol consumption by genetic drinking rats. Alcohol Alcohol 40:494–497

Min MY, Rusakov DA, Kullmann DM (1998) Activation of AMPA, kainate, and metabotropic receptors at hippocampal mossy fiber synapses: role of glutamate diffusion. Neuron 21:561–570

Minakami R, Katsuki F, Sugiyama H (1993) A variant of metabotropic glutamate receptor subtype 5: an evolutionally conserved insertion with no termination codon. Biochem Biophys Res Commun 194:622–627

Moldrich RX, Chapman AG, De Sarro G, Meldrum BS (2003) Glutamate metabotropic receptors as targets for drug therapy in epilepsy. Eur J Pharmacol 476:3–16

Monyer H, Burnashev N, Laurie DJ, Sakmann B, Seeburg PH (1994) Developmental and regional expression in the rat brain and functional properties of four NMDA receptors. Neuron 12:529–540

Morishita W, Lu W, Smith GB, Nicoll RA, Bear MF, Malenka RC (2007) Activation of NR2B-containing NMDA receptors is not required for NMDA receptor-dependent long-term depression. Neuropharmacology 52:71–76

Moriyoshi K, Masu M, Ishii T, Shigemoto R, Mizuno N, Nakanishi S (1991) Molecular cloning and characterization of the rat NMDA receptor. Nature 354:31–37

Morris RG, Anderson E, Lynch GS, Baudry M (1986) Selective impairment of learning and blockade of long-term potentiation by an N-methyl-D-aspartate receptor antagonist, AP5. Nature 319:774–776

Mukherjee S, Manahan-Vaughan D (2012) Role of metabotropic glutamate receptors in persistent forms of hippocampal plasticity and learning. Neuropharmacology 66:65–81

Mulkey RM, Herron CE, Malenka RC (1993) An essential role for protein phosphatases in hippocampal long-term depression. Science 261:1051–1055

Naie K, Manahan-Vaughan D (2004) Regulation by metabotropic glutamate receptor 5 of LTP in the dentate gyrus of freely moving rats: relevance for learning and memory formation. Cereb Cortex 14:189–198

Naie K, Manahan-Vaughan D (2005) Investigations of the protein synthesis dependency of mGluR-induced long-term depression in the dentate gyrus of freely moving rats. Neuropharmacology 49:35–44

Naie K, Gundimi S, Siegmund H, Heinemann U, Manahan-Vaughan D (2006) Group III metabotropic glutamate receptor-mediated, chemically induced long-term depression differentially affects cell viability in the hippocampus. Eur J Pharmacol 535:104–113

Naie K, Tsanov M, Manahan-Vaughan D (2007) Group I metabotropic glutamate receptors enable two distinct forms of long-term depression in the rat dentate gyrus in vivo. Eur J Neurosci 25:3264–3275

Naisbitt S, Kim E, JC T, Xiao B, Sala C, Valtschanoff J, Weinberg RJ, Worley PF, Sheng M (1999) Shank, a novel family of postsynaptic density proteins that binds to the NMDA receptor/PSD-95/GKAP complex and cortactin. Neuron 23:569–582

Nakazawa K (2002) Requirement for hippocampal CA3 NMDA receptors in associative memory recall. Science 297:211–218

Neunuebel JP, Knierim JJ (2014) CA3 retrieves coherent representations from degraded input: direct evidence for CA3 pattern completion and dentate gyrus pattern separation. Neuron 81:416–427

Neyman S, Manahan-Vaughan D (2008) Metabotropic glutamate receptor 1 (mGluR1) and 5 (mGluR5) regulate late phases of LTP and LTD in the hippocampal CA1 region in vitro. Eur J Neurosci 27:1345–1352

Nguyen PV, Abel T, Kandel ER (1994) Requirement of a critical period of transcription for induction of a late phase of LTP. Science 265:1104–1107

Nicoll RA, Schmitz D (2005) Synaptic plasticity at hippocampal mossy fibre synapses. Nat Rev Neurosci 6:863–876

Niswender CM, Conn PJ (2010) Metabotropic glutamate receptors: physiology, pharmacology, and disease. Annu Rev Pharmacol Toxicol 50:295–322

O'Brien JA, Lemaire W, Wittmann M, Jacobson MA, Ha SN, Wisnoski DD, Lindsley CW, Schaffhauser HJ, Rowe B, Sur C, Duggan ME, Pettibone DJ, Conn PJ, Williams DL (2004) A novel selective allosteric modulator potentiates the activity of native metabotropic glutamate receptor subtype 5 in rat forebrain. J Pharmacol Exp Ther 309:568–577

O'Keefe J (1976) Place units in the hippocampus of the freely moving rat. Exp Neurol 51:78–109

O'Keefe J (1979) A review of the hippocampal place cells. Prog Neurobiol 13:419–439

O'Keefe J, Dostrovsky J (1971) The hippocampus as a spatial map. Preliminary evidence from unit activity in the freely-moving rat. Brain Res 34:171–175

Olton DS, Branch M, Best PJ (1978) Spatial correlates of hippocampal unit activity. Exp Neurol 58:387–409

Otani S, Abraham WC (1989) Inhibition of protein synthesis in the dentate gyrus, but not the entorhinal cortex, blocks maintenance of long-term potentiation in rats. Neurosci Lett 106:175–180

Otani S, Marshall CJ, Tate WP, Goddard GV, Abraham WC (1989) Maintenance of long-term potentiation in rat dentate gyrus requires protein synthesis but not messenger RNA synthesis immediately post-tetanization. Neuroscience 28:519–526

Palmer MJ, Irving AJ, Seabrook GR, Jane DE, Collingridge GL (1997) The group I mGlu receptor agonist DHPG induces a novel form of LTD in the CA1 region of the hippocampus. Neuropharmacology 36:1517–1532

Peavy RD, Conn PJ (1998) Phosphorylation of mitogen-activated protein kinase in cultured rat cortical glia by stimulation of metabotropic glutamate receptors. J Neurochem 71:603–612

Perkinton MS, Ip JK, Wood GL, Crossthwaite AJ, Williams RJ (2002) Phosphatidylinositol 3-kinase is a central mediator of NMDA receptor signalling to MAP kinase (Erk1/2), akt/PKB and CREB in striatal neurones. J Neurochem 80:239–254

Popkirov SG, Manahan-Vaughan D (2011) Involvement of the metabotropic glutamate receptor mGluR5 in NMDA receptor-dependent, learning-facilitated long-term depression in CA1 synapses. Cereb Cortex 21:501–509

Pöschel B, Manahan-Vaughan D (2007) Persistent (>24h) long-term depression in the dentate gyrus of freely moving rats is not dependent on activation of NMDA receptors, L-type voltage-gated calcium channels or protein synthesis. Neuropharmacology 52:46–54

Racine RJ, Milgram NW, Hafner S (1983) Long-term potentiation phenomena in the rat limbic forebrain. Brain Res 260:217–231

Riedel G, Casabona G, Platt B, Macphail EM, Nicoletti F (2000) Fear conditioning-induced time- and subregion-specific increase in expression of mGlu5 receptor protein in rat hippocampus. Neuropharmacology 39:1943–1951

Roberson ED, Sweatt JD (1996) Transient activation of cyclic AMP-dependent protein kinase during hippocampal long-term potentiation. J Biol Chem 271:30436–30441

Roche KW, Tingley WG, Huganir RL (1994) Glutamate receptor phosphorylation and synaptic plasticity. Curr Opin Neurobiol 4:383–388

Rogan MT, Stäubli UV, Ledoux JE (1997) Fear conditioning induces associative long-term potentiation in the amygdala. Nature 390:604–607

Romano C, Sesma MA, McDonald CT, O'Malley K, Van den Pol AN, Olney JW (1995) Distribution of metabotropic glutamate receptor mGluR5 immunoreactivity in rat brain. J Comp Neurol 355:455–469

Sala C, Piëch V, Wilson NR, Passafaro M, Liu G, Sheng M (2001) Regulation of dendritic spine morphology and synaptic function by shank and Homer. Neuron 31:115–130

Sans N, Petralia RS, Wang YX, Blahos J, Hell JW, Wenthold RJ (2000) A developmental change in NMDA receptor-associated proteins at hippocampal synapses. J Neurosci 20:1260–1271

Scannevin RH, Huganir RL (2000) Postsynaptic organization and regulation of excitatory synapses. Nat Rev Neurosci 1:133–141

Schmitz D, Frerking M, Nicoll RA (2000) Synaptic activation of presynaptic kainate receptors on hippocampal mossy fiber synapses. Neuron 27:327–338

Schroeder JP, Overstreet DH, Hodge CW (2005) The mGluR5 antagonist MPEP decreases operant ethanol self-administration during maintenance and after repeated alcohol deprivations in alcohol-preferring (P) rats. Psychopharmacologia 179:262–270

Seong E, Seasholtz AF, Burmeister M (2002) Mouse models for psychiatric disorders. Trends Genet 18:643–650

Sheng M, Pak DT (2000) Ligand-gated ion channel interactions with cytoskeletal and signaling proteins. Annu Rev Physiol 62:755–778

Sheng M, Sala C (2001) PDZ domains and the organization of supramolecular complexes. Annu Rev Neurosci 24:1–29

Sheng M, Cummings J, Roldan LA, Jan YN, Jan LY (1994) Changing subunit composition of heteromeric NMDA receptors during development of rat cortex. Nature 368:144–147

Shigemoto R, Kinoshita A, Wada E, Nomura S, Ohishi H, Takada M, Flor PJ, Neki A, Abe T, Nakanishi S, Mizuno N (1997) Differential presynaptic localization of metabotropic glutamate receptor subtypes in the rat hippocampus. J Neurosci 17:7503–7522

Shonesy BC, Jalan-Sakrikar N, Cavener VS, Colbran RJ (2014) CaMKII: a molecular substrate for synaptic plasticity and memory. Prog Mol Biol Transl Sci 122:61–87

Simonyi A, Serfozo P, Shelat PB, Dopheide MM, Coulibaly AP, Schachtman TR (2007) Differential roles of hippocampal metabotropic glutamate receptors 1 and 5 in inhibitory avoidance learning. Neurobiol Learn Mem *88*(3):305–311

Singer W (1993) Synchronization of cortical activity and its putative role in information processing and learning. Annu Rev Physiol 55:349–374

Steckler T, Oliveira AFM, Van Dyck C, Van Craenendonck H, Mateus AMA, Langlois X, Lesage ASJ, Prickaerts J (2005) Metabotropic glutamate receptor 1 blockade impairs acquisition and retention in a spatial water maze task. Behav Brain Res 164:52–60

Stevens CF, Wang Y (1994) Changes in reliability of synaptic function as a mechanism for plasticity. Nature 371:704–707

Straube T, Korz V, Frey JU (2003) Bidirectional modulation of long-term potentiation by novelty-exploration in rat dentate gyrus. Neurosci Lett 344:5–8

Tanaka J, Nakagawa S, Kushiya E, Yamasaki M, Fukaya M, Iwanaga T, Simon MI, Sakimura K, Kano M, Watanabe M (2000) Gq protein alpha subunits galphaq and Galpha11 are localized at postsynaptic extra-junctional membrane of cerebellar purkinje cells and hippocampal pyramidal cells. Eur J Neurosci 12:781–792

Taube JS (1995) Head direction cells recorded in the anterior thalamic nuclei of freely moving rats. J Neurosci 15:70–86

Taube JS, Muller RU, Ranck JB (1990a) Head-direction cells recorded from the postsubiculum in freely moving rats. II. Effects of environmental manipulations. J Neurosci 10:436–447

Taube JS, Muller RU, Ranck JB (1990b) Head-direction cells recorded from the postsubiculum in freely moving rats. I. Description and quantitative analysis. J Neurosci 10:420–435

Tomita S, Nicoll RA, Bredt DS (2001) PDZ protein interactions regulating glutamate receptor function and plasticity. J Cell Biol 153:F19–F24

Traub RD, Bibbig A, LeBeau FEN, Buhl EH, Whittington MA (2004) Cellular mechanisms of neuronal population oscillations in the hippocampus in vitro. Annu Rev Neurosci 27:247–278

Treviño M, Vivar C, Gutiérrez R (2007) Beta/gamma oscillatory activity in the CA3 hippocampal area is depressed by aberrant GABAergic transmission from the dentate gyrus after seizures. J Neurosci 27:251–259

Tu JC, Xiao B, Yuan JP, Lanahan AA, Leoffert K, Li M, Linden DJ, Worley PF (1998) Homer binds a novel proline-rich motif and links group 1 metabotropic glutamate receptors with IP3 receptors. Neuron 21:717–726

Tu JC, Xiao B, Naisbitt S, Yuan JP, Petralia RS, Brakeman P, Doan A, Aakalu VK, Lanahan AA, Sheng M, Worley PF (1999) Coupling of mGluR/Homer and PSD-95 complexes by the shank family of postsynaptic density proteins. Neuron 23:583–592

Urban MO, Hama AT, Bradbury M, Anderson J, Varney MA, Bristow L (2003) Role of metabotropic glutamate receptor subtype 5 (mGluR5) in the maintenance of cold hypersensitivity following a peripheral mononeuropathy in the rat. Neuropharmacology 44:983–993

Ure J, Baudry M, Perassolo M (2006) Metabotropic glutamate receptors and epilepsy. J Neurol Sci 247:1–9
Uslaner JM, Parmentier-Batteur S, Flick RB, Surles NO, Lam JSH, McNaughton CH et al (2009) Dose-dependent effect of CDPPB, the mGluR5 positive allosteric modulator, on recognition memory is associated with GluR1 and CREB phosphorylation in the prefrontal cortex and hippocampus. *Neuropharmacology 57*(5-6):531–538
Uzakov S, Frey JU, Korz V (2005) Reinforcement of rat hippocampal LTP by holeboard training. Learn Mem 12:165–171
van den Buuse M, Garner B, Gogos A, Kusljic S (2005) Importance of animal models in schizophrenia research. Aust N Z J Psychiatry 39:550–557
Wang Y, Wu J, Rowan MJ, Anwyl R (1996) Ryanodine produces a low frequency stimulation-induced NMDA receptor-independent long-term potentiation in the rat dentate gyrus in vitro. J Physiol 495(Pt 3):755–767
Waung MW, Huber KM (2009) Protein translation in synaptic plasticity: mGluR-LTD, fragile X. Curr Opin Neurobiol 19:319–326
Whitlock JR, Heynen AJ, Shuler MG, Bear MF (2006) Learning induces long-term potentiation in the hippocampus. Science 313:1093–1097
Whittington MA, Traub RD, Jefferys JG (1995) Synchronized oscillations in interneuron networks driven by metabotropic glutamate receptor activation. Nature 373:612–615
Whittington MA, Traub RD, Faulkner HJ, Stanford IM, Jefferys JG (1997) Recurrent excitatory postsynaptic potentials induced by synchronized fast cortical oscillations. Proc Natl Acad Sci U S A 94:12198–12203
Wickens JR, Abraham WC (1991) The involvement of L-type calcium channels in heterosynaptic long-term depression in the hippocampus. Neurosci Lett 130:128–132
Xiao B, JC T, Petralia RS, Yuan JP, Doan A, Breder CD, Ruggiero A, Lanahan AA, Wenthold RJ, Worley PF (1998) Homer regulates the association of group 1 metabotropic glutamate receptors with multivalent complexes of homer-related, synaptic proteins. Neuron 21:707–716
Xiao B, JC T, Worley PF (2000) Homer: a link between neural activity and glutamate receptor function. Curr Opin Neurobiol 10:370–374
Xu Z, Chen R-Q, Q-H G, Yan J-Z, Wang S-H, Liu S-Y, Lu W (2009) Metaplastic regulation of long-term potentiation/long-term depression threshold by activity-dependent changes of NR2A/NR2B ratio. J Neurosci 29:8764–8773
Xu J, Zhu Y, Kraniotis S, He Q, Marshall JJ, Nomura T, Stauffer SR, Lindsley CW, Conn PJ, Contractor A (2013) Potentiating mGluR5 function with a positive allosteric modulator enhances adaptive learning. Learn Mem 20:438–445
Yang L, Mao L, Tang Q, Samdani S, Liu Z, Wang JQ (2004) A novel Ca2+−independent signaling pathway to extracellular signal-regulated protein kinase by coactivation of NMDA receptors and metabotropic glutamate receptor 5 in neurons. J Neurosci 24:10846–10857
Ylinen A, Soltész I, Bragin A, Penttonen M, Sík A, Buzsáki G (1995) Intracellular correlates of hippocampal theta rhythm in identified pyramidal cells, granule cells, and basket cells. Hippocampus 5:78–90
Zhang K (1996) Representation of spatial orientation by the intrinsic dynamics of the head-direction cell ensemble: a theory. J Neurosci 16:2112–2126
Zhang S, Manahan-Vaughan D (2014) Place field stability requires the metabotropic glutamate receptor, mGlu5. Hippocampus 24:1330–1340
Zhou M, Conboy L, Sandi C, Joëls M, Krugers HJ (2009) Fear conditioning enhances spontaneous AMPA receptor-mediated synaptic transmission in mouse hippocampal CA1 area. Eur J Neurosci 30:1559–1564
Zhou Q, Wang J, Zhang X, Zeng L, Wang L, Jiang W (2013) Effect of metabotropic glutamate 5 receptor antagonists on morphine efficacy and tolerance in rats with neuropathic pain. Eur J Pharmacol 718:17–23
Zhu D, Lipsky RH, Marini AM (2002) Co-activation of the phosphatidylinositol-3-kinase/akt signaling pathway by N-methyl-D-aspartate and TrkB receptors in cerebellar granule cell neurons. Amino Acids 23:11–17

Chapter 6
Neuroprotective Properties of Glutamate Metabotropic Glutamate Receptors in Parkinson's Disease and Other Brain Disorders

Gunasingh Jeyaraj Masilamoni and Yoland Smith

Abstract Because of their modulatory role, cell-type-specific expression in the CNS, and anti-inflammatory properties, the metabotropic glutamate receptors (mGluRs) have generated significant interest as potential therapeutic targets for various brain disorders. In addition, preclinical studies in animal models of Parkinson's disease have revealed that specific mGluR subtypes mediate significant neuroprotective effects that reduce midbrain dopaminergic neuronal death. Although the underlying mechanisms of these effects remain to be established, there is evidence that intracellular calcium regulation, anti-inflammatory effects, and glutamatergic network regulation contribute to these properties. These protective effects extend beyond midbrain dopaminergic neurons for some mGluRs. In this review, we discuss recent evidence for mGluR-mediated neuroprotection in PD and highlight the challenges to translate these findings into human trials.

Keywords Dopamine • Inflammation • Huntington • Striatum • Substantia nigra • MPTP • Excitotoxicity

6.1 Introduction

Parkinson's disease (PD) is a progressive neurodegenerative disorder clinically characterized by bradykinesia (slowness), rigidity (stiffness), resting tremor, and gait dysfunction with postural instability. The major pathological hallmark of PD is the degeneration of dopaminergic (DA) nigrostriatal neurons in the substantia nigra pars compacta (SNC) and the presence of cytoplasmic α-synuclein-positive inclusions (Lewy bodies) (Fearnley and Lees 1991; Forno 1996; Spillantini et al. 1998; Giasson

G.J. Masilamoni, Ph.D. • Y. Smith, Ph.D. (✉)
Yerkes National Primate Research Center and Department of Neurology, Emory University, 954, Gatewood Rd NE, Atlanta, GA 30329, USA
e-mail: Gjeyara@emory.edu; ysmit01@emory.edu

R.T. Ngomba et al. (eds.), *mGLU Receptors*, The Receptors 31,
DOI 10.1007/978-3-319-56170-7_6

et al. 2000; Dauer and Przedborski 2003). This progressive degeneration of the nigrostriatal system results in complex pathophysiological changes of neuronal activity and neurochemical imbalance in neurotransmitter release throughout the basal ganglia circuitry and their projection targets (Albin et al. 1989; DeLong 1990; Gerfen et al. 1990; Wichmann and DeLong 2006). The onset of parkinsonian motor symptoms appears only when a critical threshold of 50–60% DA neurons loss in SNC and 70–80% degeneration of striatal DA terminals have been reached (Hornykiewicz 1975, 1998). This lag between the development of motor deficits and the protracted extent of the nigrostriatal degenerative process provides an opportunity for neuroprotective intervention that could slow down the degeneration of the dopaminergic system and delay or prevent the development of parkinsonian motor symptoms.

Significant effort has, indeed, been devoted at identifying disease-modifying therapeutics in different animal models of PD. Unfortunately, despite encouraging preclinical data, none of these compounds were found to be neuroprotective when tested in human trials (Siderowf and Stern 2008; Schapira 2009a, b, c; Smith et al. 2012). Various potential explanations have been raised to explain the failure of these trials, including the lack of PD biomarkers and the late start of the neuroprotective therapy (Mandel et al. 2003, Schapira 2004, Siderowf and Stern 2008, Lang 2009, Olanow 2009, Rascol 2009, Schapira 2009a, c). Thus, the need for effective neuroprotective drugs that could alter the progressive degeneration of the dopaminergic nigrostriatal system and the development of PD biomarkers are key interrelated attributes to the successful development of effective neuroprotective therapy in PD.

Although the mechanisms that underlie nigral DA degeneration in PD remain poorly understood, evidence suggests that glutamate excitotoxicity contributes to the cascade of events that lead to DA cell death in the SNc (Fornai et al. 1997; Sonsalla et al. 1998; Greenamyre et al. 1999; Golembiowska et al. 2002; Blandini et al. 2004; Przedborski 2005). Glutamate elicits its action through the activation of ionotropic (iGlu) and metabotropic glutamate receptors (mGluRs). The use of compounds blocking iGlu, particularly of the N-methyl-D-aspartate (NMDA) subtype, has shown significant neuroprotective effects of midbrain dopaminergic neurons in experimental models of parkinsonism (Turski et al. 1991; Zuddas et al. 1991; Storey et al. 1992; Nash et al. 1999; Konitsiotis et al. 2000). However, the use of these compounds failed human trials because of narrow therapeutic windows and marked CNS side effects (Montastruc et al. 1992; Muir 2006). An alternative approach to the development of glutamate-mediated therapies with a good profile of safety and tolerability for brain disorders is the targeting of glutamate receptors that "modulate" rather than "mediate" fast excitatory synaptic transmission. Because the G-protein-coupled mGluRs meet these criteria, they have generated significant interest as new therapeutic targets for brain diseases, including symptomatic and neuroprotective agents in PD (Lovinger et al. 1993, Conn et al. 2005, Bonsi et al. 2007, Ossowska et al. 2007, Gasparini et al. 2008, Niswender and Conn 2010, Nicoletti et al. 2011, Finlay and Duty 2014, Nickols and Conn 2014, Williams and Dexter 2014, Amalric 2015, Nicoletti et al. 2015).

6.2 MGluR Localization in the Basal Ganglia Circuitry

Eight mGluR subtypes (mGluR1–mGluR8) have been cloned and divided into three groups (groups I–III) on the basis of sequence homologies, signal transduction pathways, and ligand-binding patterns (Pin and Duvoisin 1995; Conn and Pin 1997; Nakanishi et al. 1998; Anwyl 1999; Pin and Acher 2002). Members of these three groups of mGluRs are strongly expressed throughout the basal ganglia circuitry, where they regulate neuronal excitability and synaptic transmission via pre- and postsynaptic mechanisms (Calabresi et al. 1992, Lovinger et al. 1993, Lovinger and McCool 1995, Conn and Pin 1997, Pisani et al. 1997, Rouse et al. 2000, Smith et al. 2000, Pisani et al. 2001, Corti et al. 2002, Marino et al. 2002, 2003a, Gubellini et al. 2004, Conn et al. 2005, Johnson et al. 2009, Lovinger 2010). Furthermore, midbrain dopaminergic neurons in the SNc also express specific mGluRs which represent potential therapeutic targets for neuroprotection in PD (see below). Thus, the foundation of preclinical studies that have been performed so far to assess the potential relevance of mGluR-mediated neuroprotection in PD has been based on the assumption that excitotoxic effects upon midbrain dopaminergic neurons can be mediated either directly through overactivation of calcium-dependent postsynaptic glutamate receptors or via overactivity of glutamatergic inputs onto SNc neurons (Battaglia et al. 2002, 2004; Johnson et al. 2009; Masilamoni et al. 2011). In either case, drugs aimed at postsynaptic (i.e., group I) or presynaptic (i.e., groups II and III) mGluRs may have beneficial effects in reducing toxic insults upon midbrain dopaminergic neurons and possibly other vulnerable populations of neurons in PD (Johnson et al. 2009).

Group I mGluRs (mGluRs 1 and 5) are mainly expressed postsynaptically, couple to G_q proteins, and positively modulate neuronal excitability through activation of phospholipase C leading to an increase in intracellular Ca^{2+} and protein kinase C (Kim et al. 1994; Pin and Duvoisin 1995; Conn and Pin 1997; Fiorillo and Williams 1998; Nakanishi et al. 1998; Sala et al. 2005; Zhang et al. 2005), while group II (mGluRs 2 and 3) and group III (mGluRs 4, 6, 7, and 8) mGluRs are mainly localized presynaptically, classically couple to $G_{i/o}$, and negatively modulate neuronal excitability (Conn and Pin 1997). Group I mGluRs are largely expressed postsynaptically in dendrites and spines, where they enhance excitability through various postsynaptic mechanisms that involve regulation of intracellular calcium, functional interactions with NMDA receptors, and modulation of voltage-gated calcium channels (Conn and Pin 1997; Nakanishi et al. 1998; Alagarsamy et al. 1999; Pisani et al. 2001; Sala et al. 2005; Zhang et al. 2005)

On the other hand, group III mGluRs are predominantly located in the presynaptic active zones of glutamatergic and non-glutamatergic synapses, where they act as autoreceptors or heteroreceptors that reduce neurotransmitter release (Cartmell and Schoepp 2000; Ferraguti and Shigemoto 2006). Although mGluRs 4, 7, and 8 are expressed to a variable degree in different brain regions, mGluR6 is restricted to the retina (Vardi et al. 2000).

The two members of the group II mGluR subtypes, mGluR2 and mGluR3, are differentially expressed in neurons and astrocytes, respectively (Mineff and Valtschanoff 1999; Geurts et al. 2003). At the neuronal level, mGluR2 is located presynaptically in glutamatergic and non-glutamatergic terminals, but its pattern of subcellular expression is different from that of group III mGluRs; instead of being aggregated at the active zones close to the transmitter release sites, as is the case from group III mGluRs, mGluR2 is most commonly expressed away from the synapses, often in preterminal axonal segments, of glutamatergic and GABAergic afferents (Schoepp and Conn 1993; Pin and Duvoisin 1995; Conn and Pin 1997; Cartmell and Schoepp 2000; De Blasi et al. 2001; Galvan et al. 2006; Ferraguti et al. 2008; Niswender and Conn 2010). Thus, upon activation, groups II and III mGluRs can reduce glutamatergic signaling and dampen neuronal excitability giving them potential neuroprotective properties (Nicoletti et al. 1996). On the other hand, group I mGluR antagonists display neuroprotective properties, likely through reduced calcium release from intracellular stores and decreased neuroinflammation (Kim et al. 1994; Pin and Duvoisin 1995; Conn and Pin 1997; Nakanishi et al. 1998; Sala et al. 2005; Zhang et al. 2005).

In this review, we will discuss evidence for neuroprotective properties of different subtypes of mGluRs in PD. We will also summarize findings of mGluR-mediated neuroprotection in other brain diseases (Table 6.1). These will be followed by a brief discussion of the subtype-selective mGlu receptor ligands that still hold the promise to become disease-modifying drugs in neurodegenerative disorders.

6.3 Neuroprotective Effects of Group I mGluRs

There is evidence that mice treated with the mGluR5 antagonist 2-methyl-6-(phenylethynyl) pyridine (MPEP) or mice that lack mGluR5 receptors show increased survival of nigrostriatal DA neurons after administration of the dopaminergic neurotoxin 1-methyl-4-phenyl-1,2,3,6-tetrahydropyridine (MPTP) (Battaglia et al. 2004; Aguirre et al. 2005; Armentero et al. 2006; Vernon et al. 2007). mGluR5 receptor antagonists are also protective against nigrostriatal dopamine denervation in the methamphetamine model of parkinsonism (Battaglia et al. 2002). The mGluR5 antagonist, MTEP, was also found to be neuroprotective against the loss of midbrain DA neurons and norepinephrine neurons in the locus coeruleus (LC) in a chronic MPTP-treated monkey model of PD (Masilamoni et al. 2011) (Fig. 6.1). In this study, we showed that neurons in the SNc and LC of monkeys chronically treated with low doses of MPTP over a period of over 20 weeks were significantly spared in animals that received daily administration of MTEP (Masilamoni et al. 2011). Altogether, these rodent and monkey data suggest that the use of mGluR5 receptor antagonists may be a useful strategy to reduce degeneration of catecholaminergic neurons in PD. Taking into consideration evidence that mGluR5 antagonists also have significant anti-dyskinetic effects in rodent and nonhuman primate models of PD (Dekundy et al. 2006; Mela et al. 2007; Gravius et al. 2008; Levandis et al. 2008; Rylander et al. 2009; Johnston et al. 2010; Morin et al. 2010; Rylander et al. 2010; Gregoire et al. 2011) and some antiparkinsonian effects in 6-OHDA-treated rats (Breysse et al. 2002), these findings provide additional support for the potential use of mGluR5 antagonist as PD-relevant therapeutic.

Table 6.1 Studies that showed evidence of mGluR-mediated neuroprotection and anti-inflammatory effects in various brain diseases

Diseases	mGluRs	Species/model	References
Huntington's disease	Group I: mGluR5	Mouse	Popoli et al. (2004), Schiefer et al. (2004), Doria et al. (2013), Ribeiro et al. (2014) and Doria et al. (2015)
		Rat	Souza et al. (2014)
	Group II: GluR2/3	Mouse	Colwell and Levine (1999), Venero et al. (2002), Schiefer et al. (2004) and Reiner et al. (2012)
	Group III	Cell culture	Byrnes et al. (2009b) and Caraci et al. (2011)
Alzheimer's disease	Group I: mGluR5	Mouse	Hamilton et al. (2014)
	Group II: GluR2/3	Rat	Henrich-Noack et al. (2000)
Amyotrophic lateral sclerosis	Group I: mGluR5	Spinal cord motor neurons	D'Antoni et al. (2011)
		Mouse	Rossi et al. (2008)
		Rat	Banasik et al. (2005)
	Group I: mGluR1	Mouse	Milanese et al. (2014)
	Group II: mGluR3	Mouse	Battaglia et al. (2015)
Traumatic brain injury	Group I: mGluR5	Rat primary cortical neuronal culture/rat	Bao et al. (2001) , Movsesyan et al. (2001), Chen et al. (2012a, b) and He et al. (2015)
		Mice	(Loane et al. 2013, 2014)
	Group I: mGluR1	Rat cortical cell culture/ hippocampal slices	Pellegrini-Giampietro et al. (1999) and Kohara et al. (2008)
		Rat	Faden et al. (2001) and Zhou et al. (2009)
	Group I: mGluR1/5	Rat primary cortical neuronal culture/rat	Kingston et al. (1999), Zwienenberg et al. (2001) and Fei et al. (2006)
		Mice	Chiamulera et al. (1992), Pellegrini-Giampietro (2003) and Landucci et al. (2009)
	Group I: mGluR1/5 Group II: mGluR2/3	Rat	Makarewicz et al. (2006), Szydlowska et al. (2007) and Kohara et al. (2008)
	Group II: mGluR2/3	Rat neuronal/glial culture/rat	Pizzi et al. (1996), Faden et al. (1997), Bond et al. (1998, 1999), Henrich-Noack et al. (1998), Lam et al. (1998), Allen et al. (1999), Cai et al. (2002), Stover et al. (2003) and Feng et al. (2012)
		Mice primary neuronal culture/mice	Feng et al. (2012) and Luo et al. (2014)
	Group III: mGluR7	Primary neuronal culture	Sabelhaus et al. (2000) and Domin et al. (2015)

(continued)

Table 6.1 (continued)

Diseases	mGluRs	Species/model	References
Parkinson's disease	Group I: mGluR1/5	Cell culture	Sarnico et al. (2008)
		Mice	Battaglia et al. (2004)
		Rat	Meltzer et al. (1997), Golembiowska et al. (2003), Domenici et al. (2005), Armentero et al. (2006), Chen et al. (2011), Ferrigno et al. (2015) and Fuzzati-Armentero et al. (2015)
		Nonhuman primate	Masilamoni et al. (2011)
	Group I mGluR1/5	Hippocampal cultures	Szydlowska et al. (2007)
		Mice	Battaglia et al. (2002)
		Rat	Vernon et al. (2007, 2008)
	Group I, II and III	Rat	Vernon et al. (2005)
	Group III mGluR4	Human neuroblastoma cell culture	Jantas et al. (2014)
		Cell culture/rat model	Marino et al. (2003b)
		Mice	Maj et al. (2003) and Battaglia et al. (2006)
		Rat	Vernon et al. (2005), Jiang et al. (2006) and Betts et al. (2012)
	Group II: mGluR2/3	Cell culture	Matarredona et al. (2001)
		Mice	Battaglia et al. (2003) and Di Liberto et al. (2010)
		Rat	Murray et al. (2002)
Anti-inflammatory	Group I: mGluR5	Astrocytes/ microglia/neuronal culture	Biber et al. (1999), Byrnes et al. (2009a), Piers et al. (2011) and Xue et al. (2014)
		Rat neuronal cell culture	Degos et al. (2013)
	Group II: mGluR3	Astrocytes	Durand et al. (2013)
	Group II: mGluR2/3	Neuronal and astrocytes mixed culture	Bruno et al. (1997, 1998a)
	Group III	Microglia	Taylor et al. (2002)
	Group III: mGluR7	Cerebellar granule neurons	Lafon-Cazal et al. (1999)

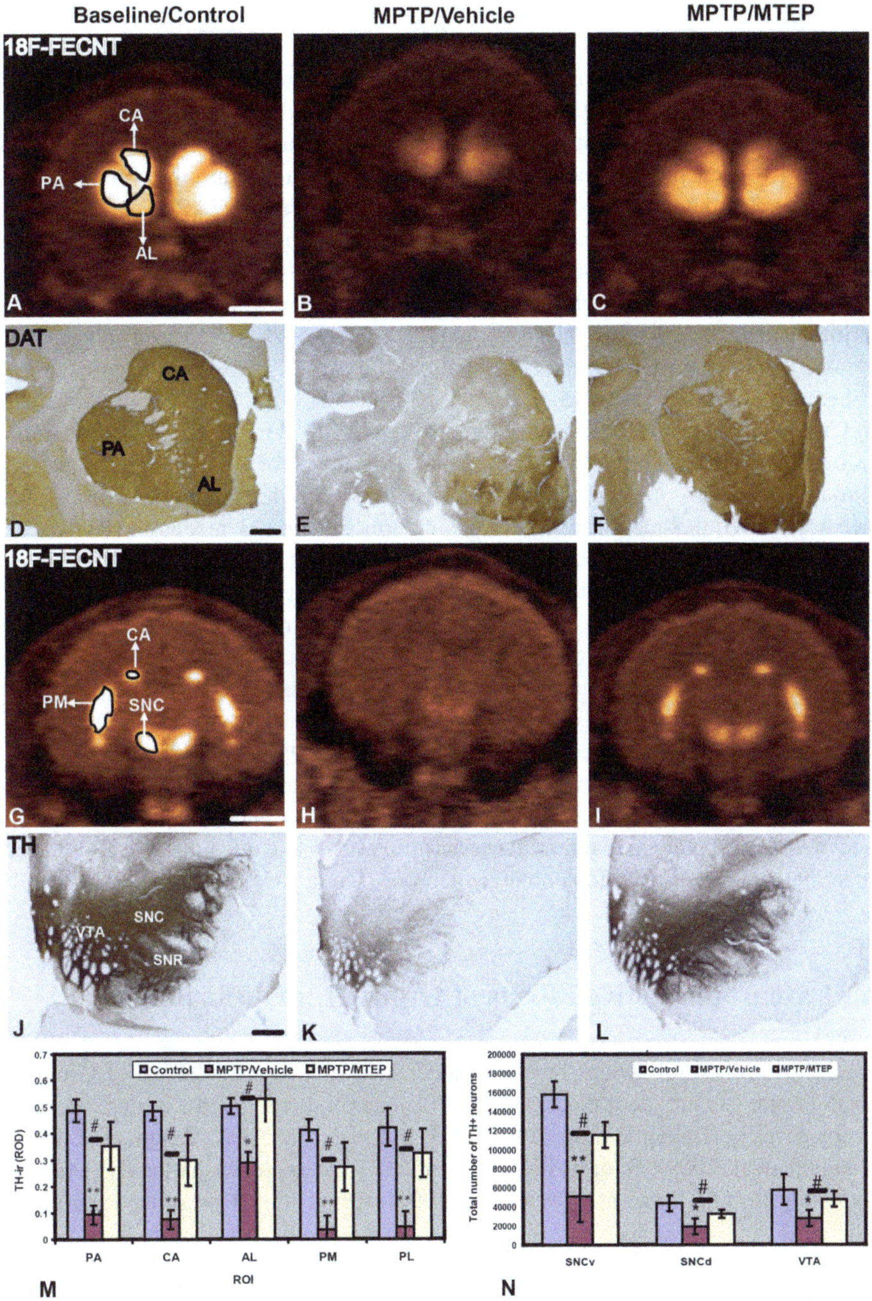

Fig. 6.1 Neuroprotective effects of MTEP towards MPTP-induced neurotoxicity of the midbrain dopaminergic nigrostriatal system in rhesus monkeys. (**a–c**) Coronal sections at the level of the pre-commissural striatum showing binding of the 18F-FECNT dopamine transporter ligand at baseline (**a**) and after chronic treatment with MPTP/vehicle (**b**) or MPTP/MTEP (C). (**d–f**) show corresponding levels of the striatum in the same groups of animals immunostained with a DAT antibody. (**g–l**) Coronal sections showing 18F-FECNT binding (**g–i**) or TH immunostaining (**j–l**) at the level of the

The evidence that various mGluR5 antagonists (AFQ056-mavoglurant; ADX-48621-dipraglurant) currently being tested in human trials as anti-dyskinetic drugs are well tolerated and do not worsen PD motor symptoms (Berg et al. 2011; Stocchi et al. 2013) is encouraging and provides further supports toward the chronic use of mGluR5-related compounds as potential neuroprotective drug in PD.

Although the exact mechanisms by which mGluR5 antagonists mediate their neuroprotective effects upon catecholaminergic neurons in toxin-based models of PD remain to be established, some interesting possibilities have been raised. In light of evidence that calcium dysregulation, mitochondrial respiration impairment, and excitotoxic insults contribute to the degeneration of nigral dopaminergic neurons in PD (Sherer et al. 2002; Maesawa et al. 2004; Wallace et al. 2007; Caudle and Zhang 2009; Chan et al. 2009; Cannon and Greenamyre 2010; Winklhofer and Haass 2010; Van Laar et al. 2011), the blockade of mGluR5 may provide its protective effects through reduction of intracellular calcium levels in dopaminergic neurons via decreased mGluR5-mediated activation of intracellular IP3 receptors (Kim et al. 1994; Pin and Duvoisin 1995; Conn and Pin 1997; Nakanishi et al. 1998; Sala et al. 2005; Zhang et al. 2005) and/or reduction of the potentiating effects of mGluR5 upon NMDA receptor function (Calabresi et al. 1992; Alagarsamy et al. 1999; Tu et al. 1999; Awad et al. 2000). An alternative mechanism could involve the reduction of overactive glutamatergic inputs from the subthalamic nucleus (STN) to SNC neurons in the PD state (Bezard et al. 1998; Rodriguez et al. 1998; Awad et al. 2000; Breysse et al. 2003; Fazal et al. 2003; Shimo and Wichmann 2009; Piallat et al. 2011). Finally, because glial mGluR5 expression is upregulated in some inflammatory processes (Byrnes et al. 2009a; Loane et al. 2009, 2012; Drouin-Ouellet et al. 2011), the blockade of these receptors may reduce these harmful effects and attenuate MPTP-induced toxicity toward midbrain dopaminergic neurons (see further discussion below).

6.4 Neuroprotective Effects of Group II mGluRs in PD

Results from several groups reveal a moderate neuroprotective effect of Group II mGluR agonists in rodent models of PD, the magnitude of which was dependent on lesion severity (Battaglia et al. 2002, 2003, Murray et al. 2002; Vernon et al. 2005; Battaglia et al. 2009). Both mGluR2 and mGluR3 are expressed in the BG circuitry including in STN neurons (Testa et al. 1994). Activation of presynaptic group II

Fig 6.1 (continued) ventral midbrain of the three animal groups used in this study. Abbreviations: *PA* associative putamen, *CA* associative caudate nucleus, *AL* limbic nucleus accumbens, *SNC* substantia nigra, *PM* motor putamen, *CA* associative caudate nucleus. Scale bar: 10 mm (applies to **a–c** and **g–i**) and 5 mm (applies to **d–f** and **j–l**). M: Densitometry analysis N. Stereological estimate of the total number of TH-positive neurons (means ± SD) in SNC-v, SNC-d and VTA regions of control, MPTP/vehicle and MPTP/MTEP treated monkeys. $^{**}p < 0.001$ and $^{*}p < 0.05$ for differences from control and MPTP/vehicle. #, $p < 0.05$ for differences between the vehicle and MTEP-treated animals. There were no significant difference found in the SNCd and VTA between control and MPTP/MTEP treated monkeys (see Masilamoni et al. 2011 for details)

mGluRs at the STN-SNc synapse reduces excitatory postsynaptic current amplitudes in rat SNc neurons (Bradley et al. 2000; Wang et al. 2005). In line with these observations, systemic treatment or intranigral administration of the group II mGluR agonists LY379268 and (2R,4R)-4-aminopyrrolidine-2,4-dicarboxylate (2R,4R-APDC) reduces the extent of 6-OHDA-induced toxicity of dopaminergic neurons in the rat SNc (Murray et al. 2002; Vernon et al. 2005; Chan et al. 2010). In addition, activation of group II mGluRs by dual mGlu2/3 receptor agonist LY379268 or DCG-IV reduces SNc degeneration in mice after intrastriatal MPP+ or systemic MPTP administration, further supporting a role for group II mGluRs as disease-modifying agents in PD (Battaglia et al. 2003). Activation of mGluR2/3 could also contribute to neurorestoration via the production and release of neurotrophic factors from glial cells in vitro (Bruno et al. 1998a; D'Onofrio et al. 2001) and in vivo (Matarredona et al. 2001; Corti et al. 2007; Battaglia et al. 2009). Through the use of the mGluR2/3 receptor agonist LY379268 in mGluR2 and mGluR3 receptor knockout mice, Corti et al. (2007) suggested that the neuroprotective properties of the mGluR2/3 receptor agonist LY379268 in MPTP-treated mice were entirely mediated by activation of the astrocytic mGluR3 receptor and that these effects were amplified in the absence of neuronal mGluR2 receptors, suggesting that mGluR2 activation might be harmful to toxin exposure (Corti et al. 2007). The studies of the neuroprotective properties of the selective mGluR3 receptor agonist, *N*-acetylaspartylglutamate (NAAG) (Orlando et al. 1997; Wroblewska et al. 1997; Bruno et al. 1998a; Bergeron et al. 2005), have been limited by the poor blood-brain barrier permeability for this compound (Westbrook et al. 1986; Sekiguchi et al. 1992).

6.5 Neuroprotective Effects of Group III mGluRs in PD

Several lines of evidence based on cellular and physiological data predict that activation of group III mGluRs in the BG could have neuroprotective effects in PD (Wigmore and Lacey 1998; Valenti et al. 2002, 2003, 2005). These Gi/Go-coupled auto- or heteroreceptors regulate GABAergic and glutamatergic synaptic transmission, most likely through inhibition of voltage-gated calcium entry required for triggering transmitter release (Trombley and Westbrook 1992; Conn and Pin 1997). In regard to neuroprotection in PD, activation of mGluR4 at the striatopallidal synapse reduces the activity of the indirect pathway, which attenuates hyperactivity of the STN, and thereby reduces its excitotoxic effects toward SNc dopaminergic neurons (Valenti et al. 2005). Group III mGluRs activation protects against NMDA-induced toxicity in cultured neurons and in vivo, neuroprotection against excitotoxic insult (Gasparini et al. 1999; Bruno et al. 2000; Flor et al. 2002).

Consistent with the prediction that group III mGluRs activation might have neuroprotective effects in PD, both acute and subchronic intranigral infusion of the group III mGluRs agonist L-AP4 reduce the extent of 6-OHDA toxicity in the rat SNc (Vernon et al. 2005, 2007, 2008; Jiang et al. 2006; Betts et al. 2012). Furthermore, systemic or intrapallidal administration of the mGluR4 allosteric potentiator,

PHCCC (Phenyl-7-(hydroxyimino) cyclopropa [b] chromen-1a-carboxamide), reduces the extent of nigrostriatal MPTP toxicity in wild-type mice, but not in mice lacking mGluR4, further supporting the selective activation of mGluR4 as a neuroprotective approach in PD (Battaglia et al. 2006). The study of mGluR7 and mGluR8 receptor subtypes in neuroprotection awaits the development of specific compounds that selectively regulate these receptors.

6.6 Neuroprotective Effects of mGluR-Related Drugs in Other Brain Diseases

Chronic treatment with the prototypic mGluR5 receptor antagonist, MPEP, attenuates cell death, delays the onset of motor symptoms, and prolongs survival in mutant mice carrying a mutation of SOD1 associated with familiar ALS (Rossi et al. 2008). There is evidence that modulation of mGluR1 and mGluR5 on oligodendrocytes is protective in a rodent model of periventricular leukomalacia during the developmental peak of vulnerability to hypoxia/ischemia, and modulation of these receptors is protective in a rodent model of periventricular leukomalacia (Jantzie et al. 2010). Recent evidence from Huntington's disease (HD) mouse model indicates that the mGluR5 allosteric potentiator, CDPPB, protects striatal neurons against excitotoxic neuronal death, maybe through activation of Akt pathway, without triggering increased intracellular Ca2+ concentration (Ribeiro et al. 2010; Doria et al. 2013, 2015). There is also evidence that the mGluR5 antagonist MPEP might reduce disease progression in HD mouse models (Schiefer et al. 2004; Ribeiro et al. 2011).

6.7 Anti-inflammatory Properties of mGluRs: Implications for Neuroprotection

In the normal brain, microglia, the resident CNS macrophages, are found in a resting state with ramified morphology, but when chronically activated in response to insults, they display an amoeboid morphology (Kreutzberg 1996; Nimmerjahn et al. 2005) and exacerbate neurodegeneration by releasing glutamate and neurotoxic pro-inflammatory cytokines and cytotoxic factors (Barger and Basile 2001; Parker et al. 2002; Cunningham et al. 2005; Barger et al. 2007; McCoy and Tansey 2008; Lee et al. 2009; Tansey and Goldberg 2010). In addition, prolonged activation of microglia prevents them from carrying out their neuro-supportive functions such as the release of key growth factors (Benoit et al. 2008). Microglial toxicity toward midbrain dopaminergic neurons is well documented in a number of cell culture studies and animal models of PD (Gao et al. 2002a, b; Liu and Hong 2003; McCoy and Tansey 2008; Lee et al. 2009; Tansey and Goldberg 2010; Barnum et al. 2014). It has been postulated that both the initiation and progression of PD are triggered by neuroinflammation (Qin et al. 2007a, b; McCoy and Tansey 2008; Lee et al. 2009;

Tansey and Goldberg 2010) and there is a growing body of evidence that in some cases, PD is linked to head trauma, viruses, and infections which subsequently trigger microglial activation (Liu and Hong 2003; Herrera et al. 2005). Inflammation is also heavily implicated in the pathogenesis of other neurodegenerative diseases including AD, ALS, MS, and HD (McGeer and McGeer 2002; Kutzelnigg et al. 2005; Solomon et al. 2006; Gao and Hong 2008; Hickman et al. 2008; Jimenez et al. 2008; Moller 2010).

As discussed above, glia (astrocytes, microglia, and oligodendrocytes) express mGluRs to variable degrees. Although mGluR3 is the most abundant subtype of glial mGluRs, there is evidence that groups I and III mGluRs also display light glial expression. These glial receptors are activated under normal and pathophysiological conditions (Loane et al. 2012). The selective mGluR5 orthosteric agonist, (RS)-2-chloro-5-hydroxyphenylglycine (CHPG), reduces microglial activation and the associated release of pro-inflammatory mediators following stimulation with either lipopolysaccharide (LPS) or interferon-γ (IFNγ) (Byrnes et al. 2009a, b; Loane et al. 2009). CHPG treatment also attenuates NADPH oxidase (NOX2) activity levels and abolishes the neurotoxic potential of activated microglia in microglia/neuron co-culture models. The anti-inflammatory effects of CHPG are abolished in mGluR5 knockout mice or by addition of the mGluR5 antagonist, MTEP, demonstrating selective mGluR5-mediated neuroprotective effects through microglia (Byrnes et al. 2009a; Loane et al. 2009). Recently, mGluR5 positive allosteric modulators were found to significantly attenuate both LPS- and IFNγ-induced nitric oxide (NO) and tumor necrosis factor-α (TNFα) release in cultured microglia with higher potency compared to the orthosteric agonist, CHPG (Xue et al. 2014). These data support the use of mGluR5 PAMs as anti-inflammatory neuroprotective agents. MGluR5 PAMs are currently being investigated as potential treatment for various CNS conditions ranging from schizophrenia and anxiety disorders (Kinney et al. 2003, 2005; Homayoun et al. 2004; Lecourtier et al. 2007; Gravius et al. 2008) to learning and cognition-related problems (Gravius et al. 2008; Ayala et al. 2009; Gass and Olive 2009; Cleva et al. 2010; Reichel et al. 2011). Whether their neuroprotective/anti-inflammatory properties contribute to the potential benefit of these compounds in these disorders remain to be established.

Activation of group III mGluRs also mediates anti-inflammatory effects. For example, LPS-induced microglial activation in vitro is attenuated by the group III mGluR agonists L-AP4 and RS-PPG (Taylor et al. 2003). Activation of glial mGluR4 reduces the production of RANTES, a chemokine involved in neuroinflammation that circulates at higher levels in humans with PD compared with age- and sex-matched controls (Besong et al. 2002; Rentzos et al. 2007). L-AP4 is similarly protective against myelin-induced microglial neurotoxicity in cell culture, a finding attributable to the mGluR-mediated inhibition of soluble toxin production by activated microglia (Pinteaux-Jones et al. 2008). Similarly, oligodendrocyte precursor cells found in lesion sites are particularly vulnerable to cytotoxic and pro-inflammatory factors released by activated microglia, and stimulation of group III mGluRs by L-AP4 prevents microglial-induced inhibition of oligodendrocyte precursor cell proliferation (Taylor et al. 2010). These studies suggest that targeting microglia by group III mGlu receptors has the potential to encourage neuroprotection and regeneration of lost myelin in MS.

Activation of mGluR2/3 stimulates the production and release of neurotrophic factors from glial cells in vitro (Bruno et al. 1998b; D'Onofrio et al. 2001) and in vivo (Matarredona et al. 2001; Corti et al. 2007; Battaglia et al. 2009). Results from several groups reveal a moderate neuroprotective effect of group II mGluR agonists in rodent models of PD, the magnitude of which was dependent on lesion severity (Battaglia et al. 2002, 2003, 2009, Murray et al. 2002; Vernon et al. 2005;).

6.8 mGluR-Mediated Neuroprotection vs Disease Biomarkers

The mGluRs are therapeutic targets of great interest for various brain diseases (Conn et al. 2005; Marino and Conn 2006; Johnson et al. 2009; Niswender and Conn 2010). Their beneficial symptomatic effects are further amplified by their potential neuroprotective properties in PD and other brain diseases. However, caution must be exercised in translating the promising preclinical neuroprotective data discussed in this review to human trials (Olanow 2009; Smith et al. 2012; Schapira et al. 2014). The constant failure of previous neuroprotective human trials in PD patients (Olanow 2009) indicates that the effective assessment of mGluR neuroprotective properties in PD must await the development of biomarkers that will allow to identify at-risk candidates for the disease prior to the appearance of motor symptoms. We hope that the current effort of the Parkinson Progression Marker Initiative (PPMI) multicenter study (Parkinson Progression Marker 2011) will help identify reliable progression biomarkers that could be used toward the future assessment of the disease-modifying properties of mGluRs in PD.

Acknowledgments This research was supported by NIH grant P50NS071669 and a grant from the National Parkinson Disease Foundation.

References

Aguirre JA, Kehr J, Yoshitake T, Liu FL, Rivera A, Fernandez-Espinola S, Andbjer B, Leo G, Medhurst AD, Agnati LF, Fuxe K (2005) Protection but maintained dysfunction of nigral dopaminergic nerve cell bodies and striatal dopaminergic terminals in MPTP-lesioned mice after acute treatment with the mGluR5 antagonist MPEP. Brain Res 1033:216–220

Alagarsamy S, Marino MJ, Rouse ST, Gereau RW, Heinemann SF, Conn PJ (1999) Activation of NMDA receptors reverses desensitization of mGluR5 in native and recombinant systems. Nat Neurosci 2:234–240

Albin RL, Young AB, Penney JB (1989) The functional anatomy of basal ganglia disorders. Trends Neurosci 12:366–375

Allen JW, Ivanova SA, Fan L, Espey MG, Basile AS, Faden AI (1999) Group II metabotropic glutamate receptor activation attenuates traumatic neuronal injury and improves neurological recovery after traumatic brain injury. J Pharmacol Exp Ther 290:112–120

Amalric M (2015) Targeting metabotropic glutamate receptors (mGluRs) in Parkinson's disease. Curr Opin Pharmacol 20:29–34

Anwyl R (1999) Metabotropic glutamate receptors: electrophysiological properties and role in plasticity. Brain Res Brain Res Rev 29:83–120

Armentero MT, Fancellu R, Nappi G, Bramanti P, Blandini F (2006) Prolonged blockade of NMDA or mGluR5 glutamate receptors reduces nigrostriatal degeneration while inducing selective metabolic changes in the basal ganglia circuitry in a rodent model of Parkinson's disease. Neurobiol Dis 22:1–9

Awad H, Hubert GW, Smith Y, Levey AI, Conn PJ (2000) Activation of metabotropic glutamate receptor 5 has direct excitatory effects and potentiates NMDA receptor currents in neurons of the subthalamic nucleus. J Neurosci 20:7871–7879

Ayala JE, Chen Y, Banko JL, Sheffler DJ, Williams R, Telk AN, Watson NL, Xiang Z, Zhang Y, Jones PJ, Lindsley CW, Olive MF, Conn PJ (2009) mGluR5 positive allosteric modulators facilitate both hippocampal LTP and LTD and enhance spatial learning. Neuropsychopharmacology 34:2057–2071

Banasik T, Jasinski A, Pilc A, Majcher K, Brzegowy P (2005) Application of magnetic resonance diffusion anisotropy imaging for the assessment neuroprotecting effects of MPEP, a selective mGluR5 antagonist, on the rat spinal cord injury in vivo. Pharmacol Rep 57:861–866

Bao WL, Williams AJ, Faden AI, Tortella FC (2001) Selective mGluR5 receptor antagonist or agonist provides neuroprotection in a rat model of focal cerebral ischemia. Brain Res 922:173–179

Barger SW, Basile AS (2001) Activation of microglia by secreted amyloid precursor protein evokes release of glutamate by cystine exchange and attenuates synaptic function. J Neurochem 76:846–854

Barger SW, Goodwin ME, Porter MM, Beggs ML (2007) Glutamate release from activated microglia requires the oxidative burst and lipid peroxidation. J Neurochem 101:1205–1213

Barnum CJ, Chen X, Chung J, Chang J, Williams M, Grigoryan N, Tesi RJ, Tansey MG (2014) Peripheral administration of the selective inhibitor of soluble tumor necrosis factor (TNF) XPro(R)1595 attenuates nigral cell loss and glial activation in 6-OHDA hemiparkinsonian rats. J Parkinsons Dis 4:349–360

Battaglia G, Fornai F, Busceti CL, Aloisi G, Cerrito F, De Blasi A, Melchiorri D, Nicoletti F (2002) Selective blockade of mGlu5 metabotropic glutamate receptors is protective against methamphetamine neurotoxicity. J Neurosci 22:2135–2141

Battaglia G, Busceti CL, Pontarelli F, Biagioni F, Fornai F, Paparelli A, Bruno V, Ruggieri S, Nicoletti F (2003) Protective role of group-II metabotropic glutamate receptors against nigrostriatal degeneration induced by 1-methyl-4-phenyl-1,2,3,6-tetrahydropyridine in mice. Neuropharmacology 45:155–166

Battaglia G, Busceti CL, Molinaro G, Biagioni F, Storto M, Fornai F, Nicoletti F, Bruno V (2004) Endogenous activation of mGlu5 metabotropic glutamate receptors contributes to the development of nigro-striatal damage induced by 1-methyl-4-phenyl-1,2,3,6-tetrahydropyridine in mice. J Neurosci 24:828–835

Battaglia G, Busceti CL, Molinaro G, Biagioni F, Traficante A, Nicoletti F, Bruno V (2006) Pharmacological activation of mGlu4 metabotropic glutamate receptors reduces nigrostriatal degeneration in mice treated with 1-methyl-4-phenyl-1,2,3,6-tetrahydropyridine. J Neurosci 26:7222–7229

Battaglia G, Molinaro G, Riozzi B, Storto M, Busceti CL, Spinsanti P, Bucci D, Di Liberto V, Mudo G, Corti C, Corsi M, Nicoletti F, Belluardo N, Bruno V (2009) Activation of mGlu3 receptors stimulates the production of GDNF in striatal neurons. PLoS One 4:e6591

Battaglia G, Riozzi B, Bucci D, Di Menna L, Molinaro G, Pallottino S, Nicoletti F, Bruno V (2015) Activation of mGlu3 metabotropic glutamate receptors enhances GDNF and GLT-1 formation in the spinal cord and rescues motor neurons in the SOD-1 mouse model of amyotrophic lateral sclerosis. Neurobiol Dis 74:126–136

Benoit M, Desnues B, Mege JL (2008) Macrophage polarization in bacterial infections. J Immunol 181:3733–3739

Berg D, Godau J, Trenkwalder C, Eggert K, Csoti I, Storch A, Huber H, Morelli-Canelo M, Stamelou M, Ries V, Wolz M, Schneider C, Di Paolo T, Gasparini F, Hariry S, Vandemeulebroecke M, Abi-Saab W, Cooke K, Johns D, Gomez-Mancilla B (2011) AFQ056 treatment of levodopa-induced dyskinesias: results of 2 randomized controlled trials. Mov Disord 26:1243–1250

Bergeron R, Coyle JT, Tsai G, Greene RW (2005) NAAG reduces NMDA receptor current in CA1 hippocampal pyramidal neurons of acute slices and dissociated neurons. Neuropsychopharmacology 30:7–16

Besong G, Battaglia G, D'Onofrio M, Di Marco R, Ngomba RT, Storto M, Castiglione M, Mangano K, Busceti CL, Nicoletti FR, Bacon K, Tusche M, Valenti O, Conn PJ, Bruno V, Nicoletti F (2002) Activation of group III metabotropic glutamate receptors inhibits the production of RANTES in glial cell cultures. J Neurosci 22:5403–5411

Betts MJ, O'Neill MJ, Duty S (2012) Allosteric modulation of the group III mGlu4 receptor provides functional neuroprotection in the 6-hydroxydopamine rat model of Parkinson's disease. Br J Pharmacol 166:2317–2330

Bezard E, Bioulac B, Gross CE (1998) Glutamatergic compensatory mechanisms in experimental parkinsonism. Prog Neuro-Psychopharmacol Biol Psychiatry 22:609–623

Biber K, Laurie DJ, Berthele A, Sommer B, Tolle TR, Gebicke-Harter PJ, van Calker D, Boddeke HW (1999) Expression and signaling of group I metabotropic glutamate receptors in astrocytes and microglia. J Neurochem 72:1671–1680

Blandini F, Armentero MT, Fancellu R, Blaugrund E, Nappi G (2004) Neuroprotective effect of rasagiline in a rodent model of Parkinson's disease. Exp Neurol 187:455–459

Bond A, O'Neill MJ, Hicks CA, Monn JA, Lodge D (1998) Neuroprotective effects of a systemically active group II metabotropic glutamate receptor agonist LY354740 in a gerbil model of global ischaemia. Neuroreport 9:1191–1193

Bond A, Ragumoorthy N, Monn JA, Hicks CA, Ward MA, Lodge D, O'Neill MJ (1999) LY379268, a potent and selective group II metabotropic glutamate receptor agonist, is neuroprotective in gerbil global, but not focal, cerebral ischaemia. Neurosci Lett 273:191–194

Bonsi P, Cuomo D, Picconi B, Sciamanna G, Tscherter A, Tolu M, Bernardi G, Calabresi P, Pisani A (2007) Striatal metabotropic glutamate receptors as a target for pharmacotherapy in Parkinson's disease. Amino Acids 32:189–195

Bradley SR, Marino MJ, Wittmann M, Rouse ST, Awad H, Levey AI, Conn PJ (2000) Activation of group II metabotropic glutamate receptors inhibits synaptic excitation of the substantia Nigra pars reticulata. J Neurosci 20:3085–3094

Breysse N, Baunez C, Spooren W, Gasparini F, Amalric M (2002) Chronic but not acute treatment with a metabotropic glutamate 5 receptor antagonist reverses the akinetic deficits in a rat model of parkinsonism. J Neurosci 22:5669–5678

Breysse N, Amalric M, Salin P (2003) Metabotropic glutamate 5 receptor blockade alleviates akinesia by normalizing activity of selective basal-ganglia structures in parkinsonian rats. J Neurosci 23:8302–8309

Bruno V, Sureda FX, Storto M, Casabona G, Caruso A, Knopfel T, Kuhn R, Nicoletti F (1997) The neuroprotective activity of group-II metabotropic glutamate receptors requires new protein synthesis and involves a glial-neuronal signaling. J Neurosci 17:1891–1897

Bruno V, Battaglia G, Casabona G, Copani A, Caciagli F, Nicoletti F (1998a) Neuroprotection by glial metabotropic glutamate receptors is mediated by transforming growth factor-beta. J Neurosci 18:9594–9600

Bruno V, Wroblewska B, Wroblewski JT, Fiore L, Nicoletti F (1998b) Neuroprotective activity of N-acetylaspartylglutamate in cultured cortical cells. Neuroscience 85:751–757

Bruno V, Battaglia G, Ksiazek I, van der Putten H, Catania MV, Giuffrida R, Lukic S, Leonhardt T, Inderbitzin W, Gasparini F, Kuhn R, Hampson DR, Nicoletti F, Flor PJ (2000) Selective activation of mGlu4 metabotropic glutamate receptors is protective against excitotoxic neuronal death. J Neurosci 20:6413–6420

Byrnes KR, Stoica B, Loane DJ, Riccio A, Davis MI, Faden AI (2009a) Metabotropic glutamate receptor 5 activation inhibits microglial associated inflammation and neurotoxicity. Glia 57:550–560

Byrnes KR, Stoica B, Riccio A, Pajoohesh-Ganji A, Loane DJ, Faden AI (2009b) Activation of metabotropic glutamate receptor 5 improves recovery after spinal cord injury in rodents. Ann Neurol 66:63–74

Cai Z, Lin S, Rhodes PG (2002) Neuroprotective effects of N-acetylaspartylglutamate in a neonatal rat model of hypoxia-ischemia. Eur J Pharmacol 437:139–145

Calabresi P, Maj R, Pisani A, Mercuri NB, Bernardi G (1992) Long-term synaptic depression in the striatum: physiological and pharmacological characterization. J Neurosci 12:4224–4233

Cannon JR, Greenamyre JT (2010) Neurotoxic in vivo models of Parkinson's disease recent advances. Prog Brain Res 184:17–33

Caraci F, Molinaro G, Battaglia G, Giuffrida ML, Riozzi B, Traficante A, Bruno V, Cannella M, Merlo S, Wang X, Heinz BA, Nisenbaum ES, Britton TC, Drago F, Sortino MA, Copani A, Nicoletti F (2011) Targeting group II metabotropic glutamate (mGlu) receptors for the treatment of psychosis associated with Alzheimer's disease: selective activation of mGlu2 receptors amplifies beta-amyloid toxicity in cultured neurons, whereas dual activation of mGlu2 and mGlu3 receptors is neuroprotective. Mol Pharmacol 79:618-626.

Cartmell J, Schoepp DD (2000) Regulation of neurotransmitter release by metabotropic glutamate receptors. J Neurochem 75:889–907

Caudle WM, Zhang J (2009) Glutamate, excitotoxicity, and programmed cell death in Parkinson disease. Exp Neurol 220:230–233

Chan CS, Gertler TS, Surmeier DJ (2009) Calcium homeostasis, selective vulnerability and Parkinson's disease. Trends Neurosci 32:249–256

Chan CS, Gertler TS, Surmeier DJ (2010) A molecular basis for the increased vulnerability of substantia nigra dopamine neurons in aging and Parkinson's disease. Mov Disord 25(Suppl 1):S63–S70

Chen L, Liu J, Ali U, Gui ZH, Hou C, Fan LL, Wang Y, Wang T (2011) Chronic, systemic treatment with a metabotropic glutamate receptor 5 antagonist produces anxiolytic-like effects and reverses abnormal firing activity of projection neurons in the basolateral nucleus of the amygdala in rats with bilateral 6-OHDA lesions. Brain Res Bull 84:215–223

Chen T, Cao L, Dong W, Luo P, Liu W, Qu Y, Fei Z (2012a) Protective effects of mGluR5 positive modulators against traumatic neuronal injury through PKC-dependent activation of MEK/ERK pathway. Neurochem Res 37:983–990

Chen T, Zhang L, Qu Y, Huo K, Jiang X, Fei Z (2012b) The selective mGluR5 agonist CHPG protects against traumatic brain injury in vitro and in vivo via ERK and Akt pathway. Int J Mol Med 29:630–636

Chiamulera C, Albertini P, Valerio E, Reggiani A (1992) Activation of metabotropic receptors has a neuroprotective effect in a rodent model of focal ischaemia. Eur J Pharmacol 216:335–336

Cleva RM, Gass JT, Widholm JJ, Olive MF (2010) Glutamatergic targets for enhancing extinction learning in drug addiction. Curr Neuropharmacol 8:394–408

Colwell CS, Levine MS (1999) Metabotropic glutamate receptor modulation of excitotoxicity in the neostriatum: role of calcium channels. Brain Res 833:234–241

Conn PJ, Pin JP (1997) Pharmacology and functions of metabotropic glutamate receptors. Annu Rev Pharmacol Toxicol 37:205–237

Conn PJ, Battaglia G, Marino MJ, Nicoletti F (2005) Metabotropic glutamate receptors in the basal ganglia motor circuit. Nat Rev Neurosci 6:787–798

Corti C, Aldegheri L, Somogyi P, Ferraguti F (2002) Distribution and synaptic localisation of the metabotropic glutamate receptor 4 (mGluR4) in the rodent CNS. Neuroscience 110:403–420

Corti C, Battaglia G, Molinaro G, Riozzi B, Pittaluga A, Corsi M, Mugnaini M, Nicoletti F, Bruno V (2007) The use of knock-out mice unravels distinct roles for mGlu2 and mGlu3 metabotropic glutamate receptors in mechanisms of neurodegeneration/neuroprotection. J Neurosci 27:8297–8308

Cunningham C, Wilcockson DC, Campion S, Lunnon K, Perry VH (2005) Central and systemic endotoxin challenges exacerbate the local inflammatory response and increase neuronal death during chronic neurodegeneration. J Neurosci 25:9275–9284

D'Antoni S, Berretta A, Seminara G, Longone P, Giuffrida-Stella AM, Battaglia G, Sortino MA, Nicoletti F, Catania MV (2011) A prolonged pharmacological blockade of type-5 metabotropic glutamate receptors protects cultured spinal cord motor neurons against excitotoxic death. Neurobiol Dis 42:252–264

D'Onofrio M, Cuomo L, Battaglia G, Ngomba RT, Storto M, Kingston AE, Orzi F, De Blasi A, Di Iorio P, Nicoletti F, Bruno V (2001) Neuroprotection mediated by glial group-II metabotropic glutamate receptors requires the activation of the MAP kinase and the phosphatidylinositol-3-kinase pathways. J Neurochem 78:435–445

Dauer W, Przedborski S (2003) Parkinson's disease: mechanisms and models. Neuron 39:889–909
De Blasi A, Conn PJ, Pin J, Nicoletti F (2001) Molecular determinants of metabotropic glutamate receptor signaling. Trends Pharmacol Sci 22:114–120
Degos V, Peineau S, Nijboer C, Kaindl AM, Sigaut S, Favrais G, Plaisant F, Teissier N, Gouadon E, Lombet A, Saliba E, Collingridge GL, Maze M, Nicoletti F, Heijnen C, Mantz J, Kavelaars A, Gressens P (2013) G protein-coupled receptor kinase 2 and group I metabotropic glutamate receptors mediate inflammation-induced sensitization to excitotoxic neurodegeneration. Ann Neurol 73:667–678
Dekundy A, Pietraszek M, Schaefer D, Cenci MA, Danysz W (2006) Effects of group I metabotropic glutamate receptors blockade in experimental models of Parkinson's disease. Brain Res Bull 69:318–326
DeLong MR (1990) Primate models of movement disorders of basal ganglia origin. Trends Neurosci 13:281–285
Di Liberto V, Bonomo A, Frinchi M, Belluardo N, Mudo G (2010) Group II metabotropic glutamate receptor activation by agonist LY379268 treatment increases the expression of brain derived neurotrophic factor in the mouse brain. Neuroscience 165:863–873
Domenici MR, Potenza RL, Martire A, Coccurello R, Pezzola A, Reggio R, Tebano MT, Popoli P (2005) Chronic treatment with the mGlu5R antagonist MPEP reduces the functional effects of the mGlu5R agonist CHPG in the striatum of 6-hydroxydopamine-lesioned rats: possible relevance to the effects of mGlu5R blockade in Parkinson's disease. J Neurosci Res 80:646–654
Domin H, Jantas D, Smialowska M (2015) Neuroprotective effects of the allosteric agonist of metabotropic glutamate receptor 7 AMN082 on oxygen-glucose deprivation- and kainate-induced neuronal cell death. Neurochem Int.
Doria JG, Silva FR, de Souza JM, Vieira LB, Carvalho TG, Reis HJ, Pereira GS, Dobransky T, Ribeiro FM (2013) Metabotropic glutamate receptor 5 positive allosteric modulators are neuroprotective in a mouse model of Huntington's disease. Br J Pharmacol 169:909–921
Doria JG, de Souza JM, Andrade JN, Rodrigues HA, Guimaraes IM, Carvalho TG, Guatimosim C, Dobransky T, Ribeiro FM (2015) The mGluR5 positive allosteric modulator, CDPPB, ameliorates pathology and phenotypic signs of a mouse model of Huntington's disease. Neurobiol Dis 73:163–173
Drouin-Ouellet J, Brownell AL, Saint-Pierre M, Fasano C, Emond V, Trudeau LE, Levesque D, Cicchetti F (2011) Neuroinflammation is associated with changes in glial mGluR5 expression and the development of neonatal excitotoxic lesions. Glia 59:188–199
Durand D, Carniglia L, Caruso C, Lasaga M (2013) mGlu3 receptor and astrocytes: partners in neuroprotection. Neuropharmacology 66:1–11
Faden AI, Ivanova SA, Yakovlev AG, Mukhin AG (1997) Neuroprotective effects of group III mGluR in traumatic neuronal injury. J Neurotrauma 14:885–895
Faden AI, O'Leary DM, Fan L, Bao W, Mullins PG, Movsesyan VA (2001) Selective blockade of the mGluR1 receptor reduces traumatic neuronal injury in vitro and improvesoOutcome after brain trauma. Exp Neurol 167:435–444
Fazal A, Parker F, Palmer AM, Croucher MJ (2003) Characterisation of the actions of group I metabotropic glutamate receptor subtype selective ligands on excitatory amino acid release and sodium-dependent re-uptake in rat cerebrocortical minislices. J Neurochem 86:1346–1358
Fearnley JM, Lees AJ (1991) Ageing and Parkinson's disease: substantia nigra regional selectivity. Brain 114(Pt 5):2283–2301
Fei Z, Zhang X, Bai HM, Jiang XF, Wang XL (2006) Metabotropic glutamate receptor antagonists and agonists: potential neuroprotectors in diffuse brain injury. J Clin Neurosci 13:1023–1027
Feng JF, Zhao X, Gurkoff GG, Van KC, Shahlaie K, Lyeth BG (2012) Post-traumatic hypoxia exacerbates neuronal cell death in the hippocampus. J Neurotrauma 29:1167–1179
Ferraguti F, Shigemoto R (2006) Metabotropic glutamate receptors. Cell Tissue Res 326:483–504
Ferraguti F, Crepaldi L, Nicoletti F (2008) Metabotropic glutamate 1 receptor: current concepts and perspectives. Pharmacol Rev 60:536–581
Ferrigno A, Vairetti M, Ambrosi G, Rizzo V, Richelmi P, Blandini F, Fuzzati-Armentero MT (2015) Selective blockade of mGlu5 metabotropic glutamate receptors is protective against hepatic mitochondrial dysfunction in 6-OHDA lesioned Parkinsonian rats. Clin Exp Pharmacol Physiol 42:695–703

Finlay C, Duty S (2014) Therapeutic potential of targeting glutamate receptors in Parkinson's disease. J Neural Transm 121:861–880

Fiorillo CD, Williams JT (1998) Glutamate mediates an inhibitory postsynaptic potential in dopamine neurons. Nature 394:78–82

Flor PJ, Battaglia G, Nicoletti F, Gasparini F, Bruno V (2002) Neuroprotective activity of metabotropic glutamate receptor ligands. Adv Exp Med Biol 513:197–223

Fornai F, Vaglini F, Maggio R, Bonuccelli U, Corsini GU (1997) Species differences in the role of excitatory amino acids in experimental parkinsonism. Neurosci Biobehav Rev 21:401–415

Forno LS (1996) Neuropathology of Parkinson's disease. J Neuropathol Exp Neurol 55:259–272

Fuzzati-Armentero MT, Cerri S, Levandis G, Ambrosi G, Montepeloso E, Antoninetti G, Blandini F, Baqi Y, Muller CE, Volpini R, Costa G, Simola N, Pinna A (2015) Dual target strategy: combining distinct non-dopaminergic treatments reduces neuronal cell loss and synergistically modulates l-DOPA-induced rotational behavior in a rodent model of Parkinson's disease. J Neurochem 134:740–747

Galvan A, Kuwajima M, Smith Y (2006) Glutamate and GABA receptors and transporters in the basal ganglia: what does their subsynaptic localization reveal about their function? Neuroscience 143:351–375

Gao HM, Hong JS (2008) Why neurodegenerative diseases are progressive: uncontrolled inflammation drives disease progression. Trends Immunol 29:357–365

Gao HM, Hong JS, Zhang W, Liu B (2002a) Distinct role for microglia in rotenone-induced degeneration of dopaminergic neurons. J Neurosci 22:782–790

Gao HM, Jiang J, Wilson B, Zhang W, Hong JS, Liu B (2002b) Microglial activation-mediated delayed and progressive degeneration of rat nigral dopaminergic neurons: relevance to Parkinson's disease. J Neurochem 81:1285–1297

Gasparini F, Bruno V, Battaglia G, Lukic S, Leonhardt T, Inderbitzin W, Laurie D, Sommer B, Varney MA, Hess SD, Johnson EC, Kuhn R, Urwyler S, Sauer D, Portet C, Schmutz M, Nicoletti F, Flor PJ (1999) (R,S)-4-phosphonophenylglycine, a potent and selective group III metabotropic glutamate receptor agonist, is anticonvulsive and neuroprotective in vivo. J Pharmacol Exp Ther 289:1678–1687

Gasparini F, Bilbe G, Gomez-Mancilla B, Spooren W (2008) mGluR5 antagonists: discovery, characterization and drug development. Curr Opin Drug Discov Devel 11:655–665

Gass JT, Olive MF (2009) Positive allosteric modulation of mGluR5 receptors facilitates extinction of a cocaine contextual memory. Biol Psychiatry 65:717–720

Gerfen CR, Engber TM, Mahan LC, Susel Z, Chase TN, Monsma FJ Jr, Sibley DR (1990) D1 and D2 dopamine receptor-regulated gene expression of striatonigral and striatopallidal neurons. Science 250:1429–1432

Geurts JJ, Wolswijk G, Bo L, van der Valk P, Polman CH, Troost D, Aronica E (2003) Altered expression patterns of group I and II metabotropic glutamate receptors in multiple sclerosis. Brain 126:1755–1766

Giasson BI, Duda JE, Murray IV, Chen Q, Souza JM, Hurtig HI, Ischiropoulos H, Trojanowski JQ, Lee VM (2000) Oxidative damage linked to neurodegeneration by selective alpha-synuclein nitration in synucleinopathy lesions. Science 290:985–989

Golembiowska K, Konieczny J, Ossowska K, Wolfarth S (2002) The role of striatal metabotropic glutamate receptors in degeneration of dopamine neurons: review article. Amino Acids 23:199–205

Golembiowska K, Konieczny J, Wolfarth S, Ossowska K (2003) Neuroprotective action of MPEP, a selective mGluR5 antagonist, in methamphetamine-induced dopaminergic neurotoxicity is associated with a decrease in dopamine outflow and inhibition of hyperthermia in rats. Neuropharmacology 45:484–492

Gravius A, Dekundy A, Nagel J, More L, Pietraszek M, Danysz W (2008) Investigation on tolerance development to subchronic blockade of mGluR5 in models of learning, anxiety, and levodopa-induced dyskinesia in rats. J Neural Transm 115:1609–1619

Greenamyre JT, MacKenzie G, Peng TI, Stephans SE (1999) Mitochondrial dysfunction in Parkinson's disease. Biochem Soc Symp 66:85–97

Gregoire L, Morin N, Ouattara B, Gasparini F, Bilbe G, Johns D, Vranesic I, Sahasranaman S, Gomez-Mancilla B, Di Paolo T (2011) The acute antiparkinsonian and antidyskinetic effect

of AFQ056, a novel metabotropic glutamate receptor type 5 antagonist, in L-Dopa-treated parkinsonian monkeys. Parkinsonism Relat Disord 17:270–276
Gubellini P, Pisani A, Centonze D, Bernardi G, Calabresi P (2004) Metabotropic glutamate receptors and striatal synaptic plasticity: implications for neurological diseases. Prog Neurobiol 74:271–300
Hamilton A, Esseltine JL, DeVries RA, Cregan SP, Ferguson SS (2014) Metabotropic glutamate receptor 5 knockout reduces cognitive impairment and pathogenesis in a mouse model of Alzheimer's disease. Mol Brain 7:40
He X, Lakkaraju SK, Hanscom M, Zhao Z, Wu J, Stoica B, MacKerell AD Jr, Faden AI, Xue F (2015) Acyl-2-aminobenzimidazoles: a novel class of neuroprotective agents targeting mGluR5. Bioorg Med Chem 23:2211–2220
Henrich-Noack P, Hatton CD, Reymann KG (1998) The mGlu receptor ligand (S)-4C3HPG protects neurons after global ischaemia in gerbils. Neuroreport 9:985–988
Henrich-Noack P, Flor PJ, Sabelhaus CF, Prass K, Dirnagl U, Gasparini F, Sauter A, Rudin M, Reymann KG (2000) Distinct influence of the group III metabotropic glutamate receptor agonist (R,S)-4-phosphonophenylglycine [(R,S)-PPG] on different forms of neuronal damage. Neuropharmacol 39:911–917
Herrera AJ, Tomas-Camardiel M, Venero JL, Cano J, Machado A (2005) Inflammatory process as a determinant factor for the degeneration of substantia nigra dopaminergic neurons. J Neural Transm 112:111–119
Hickman SE, Allison EK, El Khoury J (2008) Microglial dysfunction and defective beta-amyloid clearance pathways in aging Alzheimer's disease mice. J Neurosci 28:8354–8360
Homayoun H, Stefani MR, Adams BW, Tamagan GD, Moghaddam B (2004) Functional interaction between NMDA and mGlu5 receptors: effects on working memory, instrumental learning, motor behaviors, and dopamine release. Neuropsychopharmacology 29:1259–1269
Hornykiewicz O (1975) Brain monoamines and Parkinsonism. Psychopharmacol Bull 11:34–35
Hornykiewicz O (1998) Biochemical aspects of Parkinson's disease. Neurology 51:S2–S9
Jantas D, Greda A, Golda S, Korostynski M, Grygier B, Roman A, Pilc A, Lason W (2014) Neuroprotective effects of metabotropic glutamate receptor group II and III activators against MPP(+)-induced cell death in human neuroblastoma SH-SY5Y cells: the impact of cell differentiation state. Neuropharmacology 83:36–53
Jantzie LL, Talos DM, Selip DB, An L, Jackson MC, Folkerth RD, Deng W, Jensen FE (2010) Developmental regulation of group I metabotropic glutamate receptors in the premature brain and their protective role in a rodent model of periventricular leukomalacia. Neuron Glia Biol 6:277–288
Jiang Q, Yan Z, Feng J (2006) Activation of group III metabotropic glutamate receptors attenuates rotenone toxicity on dopaminergic neurons through a microtubule-dependent mechanism. J Neurosci 26:4318–4328
Jimenez S, Baglietto-Vargas D, Caballero C, Moreno-Gonzalez I, Torres M, Sanchez-Varo R, Ruano D, Vizuete M, Gutierrez A, Vitorica J (2008) Inflammatory response in the hippocampus of PS1M146L/APP751SL mouse model of Alzheimer's disease: age-dependent switch in the microglial phenotype from alternative to classic. J Neurosci 28:11650–11661
Johnson KA, Conn PJ, Niswender CM (2009) Glutamate receptors as therapeutic targets for Parkinson's disease. CNS Neurol Disord Drug Targets 8:475–491
Johnston TH, Fox SH, McIldowie MJ, Piggott MJ, Brotchie JM (2010) Reduction of L-DOPA-induced dyskinesia by the selective metabotropic glutamate receptor 5 antagonist 3-[(2-methyl-1,3-thiazol-4-yl)ethynyl]pyridine in the 1-methyl-4-phenyl-1,2,3,6-tetrahydropyridine-lesioned macaque model of Parkinson's disease. J Pharmacol Exp Ther 333:865–873
Kim WT, Rioult MG, Cornell-Bell AH (1994) Glutamate-induced calcium signaling in astrocytes. Glia 11:173–184
Kingston AE, O'Neill MJ, Bond A, Bruno V, Battaglia G, Nicoletti F, Harris JR, Clark BP, Monn JA, Lodge D, Schoepp DD (1999) Neuroprotective actions of novel and potent ligands of group I and group II metabotropic glutamate receptors. Ann N Y Acad Sci 890:438–449
Kinney GG, Burno M, Campbell UC, Hernandez LM, Rodriguez D, Bristow LJ, Conn PJ (2003) Metabotropic glutamate subtype 5 receptors modulate locomotor activity and sensorimotor gating in rodents. J Pharmacol Exp Ther 306:116–123

Kinney GG, O'Brien JA, Lemaire W, Burno M, Bickel DJ, Clements MK, Chen TB, Wisnoski DD, Lindsley CW, Tiller PR, Smith S, Jacobson MA, Sur C, Duggan ME, Pettibone DJ, Conn PJ, Williams DL Jr (2005) A novel selective positive allosteric modulator of metabotropic glutamate receptor subtype 5 has in vivo activity and antipsychotic-like effects in rat behavioral models. J Pharmacol Exp Ther 313:199–206

Kohara A, Takahashi M, Yatsugi S, Tamura S, Shitaka Y, Hayashibe S, Kawabata S, Okada M (2008) Neuroprotective effects of the selective type 1 metabotropic glutamate receptor antagonist YM-202074 in rat stroke models. Brain Res 1191:168–179

Konitsiotis S, Blanchet PJ, Verhagen L, Lamers E, Chase TN (2000) AMPA receptor blockade improves levodopa-induced dyskinesia in MPTP monkeys. Neurology 54:1589–1595

Kreutzberg GW (1996) Microglia: a sensor for pathological events in the CNS. Trends Neurosci 19:312–318

Kutzelnigg A, Lucchinetti CF, Stadelmann C, Bruck W, Rauschka H, Bergmann M, Schmidbauer M, Parisi JE, Lassmann H (2005) Cortical demyelination and diffuse white matter injury in multiple sclerosis. Brain 128:2705–2712

Lafon-Cazal M, Fagni L, Guiraud MJ, Mary S, Lerner-Natoli M, Pin JP, Shigemoto R, Bockaert J (1999) mGluR7-like metabotropic glutamate receptors inhibit NMDA-mediated excitotoxicity in cultured mouse cerebellar granule neurons. Eur J Neurosci 11:663–672

Lam AG, Soriano MA, Monn JA, Schoepp DD, Lodge D, McCulloch J (1998) Effects of the selective metabotropic glutamate agonist LY354740 in a rat model of permanent ischaemia. Neurosci Lett 254:121–123

Landucci E, Boscia F, Gerace E, Scartabelli T, Cozzi A, Moroni F, Mannaioni G, Pellegrini-Giampietro DE (2009) Involvement of endocannabinoid signaling in the neuroprotective effects of subtype 1 metabotropic glutamate receptor antagonists in models of cerebral ischemia. Int Rev Neurobiol 85:337–350

Lang AE (2009) When and how should treatment be started in Parkinson disease? Neurology 72:S39–S43

Lecourtier L, Homayoun H, Tamagnan G, Moghaddam B (2007) Positive allosteric modulation of metabotropic glutamate 5 (mGlu5) receptors reverses N-Methyl-D-aspartate antagonist-induced alteration of neuronal firing in prefrontal cortex. Biol Psychiatry 62:739–746

Lee JK, Tran T, Tansey MG (2009) Neuroinflammation in Parkinson's disease. J NeuroImmune Pharmacol 4:419–429

Levandis G, Bazzini E, Armentero MT, Nappi G, Blandini F (2008) Systemic administration of an mGluR5 antagonist, but not unilateral subthalamic lesion, counteracts l-DOPA-induced dyskinesias in a rodent model of Parkinson's disease. Neurobiol Dis 29:161–168

Liu B, Hong JS (2003) Role of microglia in inflammation-mediated neurodegenerative diseases: mechanisms and strategies for therapeutic intervention. J Pharmacol Exp Ther 304:1–7

Loane DJ, Stoica BA, Pajoohesh-Ganji A, Byrnes KR, Faden AI (2009) Activation of metabotropic glutamate receptor 5 modulates microglial reactivity and neurotoxicity by inhibiting NADPH oxidase. J Biol Chem 284:15629–15639

Loane DJ, Stoica BA, Faden AI (2012) Metabotropic glutamate receptor-mediated signaling in neuroglia. Wiley Interdiscip Rev Membr Transp Signal 1:136–150

Loane DJ, Stoica BA, Byrnes KR, Jeong W, Faden AI (2013) Activation of mGluR5 and inhibition of NADPH oxidase improves functional recovery after traumatic brain injury. J Neurotrauma 30:403–412

Loane DJ, Stoica BA, Tchantchou F, Kumar A, Barrett JP, Akintola T, Xue F, Conn PJ, Faden AI (2014) Novel mGluR5 positive allosteric modulator improves functional recovery, attenuates neurodegeneration, and alters microglial polarization after experimental traumatic brain injury. Neurotherapeutics 11:857–869

Lovinger DM (2010) Neurotransmitter roles in synaptic modulation, plasticity and learning in the dorsal striatum. Neuropharmacol 58:951–961

Lovinger DM, McCool BA (1995) Metabotropic glutamate receptor-mediated presynaptic depression at corticostriatal synapses involves mGLuR2 or 3. J Neurophysiol 73:1076–1083

Lovinger DM, Tyler E, Fidler S, Merritt A (1993) Properties of a presynaptic metabotropic glutamate receptor in rat neostriatal slices. J Neurophysiol 69:1236–1244

Luo P, Chen T, Zhao Y, Zhang L, Yang Y, Liu W, Li S, Rao W, Dai S, Yang J, Fei Z (2014) Postsynaptic scaffold protein Homer 1a protects against traumatic brain injury via regulating group I metabotropic glutamate receptors. Cell Death Dis 5:e1174

Maesawa S, Kaneoke Y, Kajita Y, Usui N, Misawa N, Nakayama A, Yoshida J (2004) Long-term stimulation of the subthalamic nucleus in hemiparkinsonian rats: neuroprotection of dopaminergic neurons. J Neurosurg 100:679–687

Maj M, Bruno V, Dragic Z, Yamamoto R, Battaglia G, Inderbitzin W, Stoehr N, Stein T, Gasparini F, Vranesic I, Kuhn R, Nicoletti F, Flor PJ (2003) (-)-PHCCC, a positive allosteric modulator of mGluR4: characterization, mechanism of action, and neuroprotection. Neuropharmacology 45:895–906

Makarewicz D, Duszczyk M, Gadamski R, Danysz W, Lazarewicz JW (2006) Neuroprotective potential of group I metabotropic glutamate receptor antagonists in two ischemic models. Neurochem Int 48:485–490

Mandel S, Weinreb O, Youdim MB (2003) Using cDNA microarray to assess Parkinson's disease models and the effects of neuroprotective drugs. Trends Pharmacol Sci 24:184–191

Marino MJ, Conn PJ (2006) Glutamate-based therapeutic approaches: allosteric modulators of metabotropic glutamate receptors. Curr Opin Pharmacol 6:98–102

Marino MJ, Awad H, Poisik O, Wittmann M, Conn PJ (2002) Localization and physiological roles of metabotropic glutamate receptors in the direct and indirect pathways of the basal ganglia. Amino Acids 23:185–191

Marino MJ, Valenti O, Conn PJ (2003a) Glutamate receptors and Parkinson's disease: opportunities for intervention. Drugs Aging 20:377–397

Marino MJ, Williams DL Jr, O'Brien JA, Valenti O, McDonald TP, Clements MK, Wang R, DiLella AG, Hess JF, Kinney GG, Conn PJ (2003b) Allosteric modulation of group III metabotropic glutamate receptor 4: a potential approach to Parkinson's disease treatment. Proc Natl Acad Sci U S A 100:13668–13673

Masilamoni GJ, Bogenpohl JW, Alagille D, Delevich K, Tamagnan G, Votaw JR, Wichmann T, Smith Y (2011) Metabotropic glutamate receptor 5 antagonist protects dopaminergic and noradrenergic neurons from degeneration in MPTP-treated monkeys. Brain 134:2057–2073

Matarredona ER, Santiago M, Venero JL, Cano J, Machado A (2001) Group II metabotropic glutamate receptor activation protects striatal dopaminergic nerve terminals against MPP+-induced neurotoxicity along with brain-derived neurotrophic factor induction. J Neurochem 76:351–360

McCoy MK, Tansey MG (2008) TNF signaling inhibition in the CNS: implications for normal brain function and neurodegenerative disease. J Neuroinflammation 5:45

McGeer PL, McGeer EG (2002) Inflammatory processes in amyotrophic lateral sclerosis. Muscle Nerve 26:459–470

Mela F, Marti M, Dekundy A, Danysz W, Morari M, Cenci MA (2007) Antagonism of metabotropic glutamate receptor type 5 attenuates l-DOPA-induced dyskinesia and its molecular and neurochemical correlates in a rat model of Parkinson's disease. J Neurochem 101:483–497

Meltzer LT, Serpa KA, Christoffersen CL (1997) Metabotropic glutamate receptor-mediated inhibition and excitation of substantia nigra dopamine neurons. Synapse 26:184–193

Milanese M, Giribaldi F, Melone M, Bonifacino T, Musante I, Carminati E, Rossi PI, Vergani L, Voci A, Conti F, Puliti A, Bonanno G (2014) Knocking down metabotropic glutamate receptor 1 improves survival and disease progression in the SOD1(G93A) mouse model of amyotrophic lateral sclerosis. Neurobiol Dis 64:48–59

Mineff E, Valtschanoff J (1999) Metabotropic glutamate receptors 2 and 3 expressed by astrocytes in rat ventrobasal thalamus. Neurosci Lett 270:95–98

Moller T (2010) Neuroinflammation in Huntington's disease. J Neural Transm 117:1001–1008

Montastruc JL, Rascol O, Senard JM, Rascol A (1992) A pilot study of N-methyl-D-aspartate (NMDA) antagonist in Parkinson's disease. J Neurol Neurosurg Psychiatry 55:630–631

Morin N, Gregoire L, Gomez-Mancilla B, Gasparini F, Di Paolo T (2010) Effect of the metabotropic glutamate receptor type 5 antagonists MPEP and MTEP in parkinsonian monkeys. Neuropharmacol 58:981–986

Movsesyan VA, O'Leary DM, Fan L, Bao W, Mullins PG, Knoblach SM, Faden AI (2001) mGluR5 antagonists 2-methyl-6-(phenylethynyl)-pyridine and (E)-2-methyl-6-(2-phenylethenyl)-pyridine reduce traumatic neuronal injury in vitro and in vivo by antagonizing N-methyl-D-aspartate receptors. J Pharmacol Exp Ther 296:41–47

Muir KW (2006) Glutamate-based therapeutic approaches: clinical trials with NMDA antagonists. Curr Opin Pharmacol 6:53–60

Murray TK, Messenger MJ, Ward MA, Woodhouse S, Osborne DJ, Duty S, O'Neill MJ (2002) Evaluation of the mGluR2/3 agonist LY379268 in rodent models of Parkinson's disease. Pharmacol Biochem Behav 73:455–466

Nakanishi S, Nakajima Y, Masu M, Ueda Y, Nakahara K, Watanabe D, Yamaguchi S, Kawabata S, Okada M (1998) Glutamate receptors: brain function and signal transduction. Brain Res Brain Res Rev 26:230–235

Nash JE, Hill MP, Brotchie JM (1999) Antiparkinsonian actions of blockade of NR2B-containing NMDA receptors in the reserpine-treated rat. Exp Neurol 155:42–48

Nickols HH, Conn PJ (2014) Development of allosteric modulators of GPCRs for treatment of CNS disorders. Neurobiol Dis 61:55–71

Nicoletti F, Bruno V, Copani A, Casabona G, Knopfel T (1996) Metabotropic glutamate receptors: a new target for the therapy of neurodegenerative disorders? Trends Neurosci 19:267–271

Nicoletti F, Bockaert J, Collingridge GL, Conn PJ, Ferraguti F, Schoepp DD, Wroblewski JT, Pin JP (2011) Metabotropic glutamate receptors: from the workbench to the bedside. Neuropharmacol 60:1017–1041

Nicoletti F, Bruno V, Ngomba RT, Gradini R, Battaglia G (2015) Metabotropic glutamate receptors as drug targets: what's new? Curr Opin Pharmacol 20:89–94

Nimmerjahn A, Kirchhoff F, Helmchen F (2005) Resting microglial cells are highly dynamic surveillants of brain parenchyma in vivo. Science 308:1314–1318

Niswender CM, Conn PJ (2010) Metabotropic glutamate receptors: physiology, pharmacology, and disease. Annu Rev Pharmacol Toxicol 50:295–322

Olanow CW (2009) Can we achieve neuroprotection with currently available anti-parkinsonian interventions? Neurology 72:S59–S64

Orlando LR, Luthi-Carter R, Standaert DG, Coyle JT, Penney JB Jr, Young AB (1997) N-acetylaspartylglutamate (NAAG) protects against rat striatal quinolinic acid lesions in vivo. Neurosci Lett 236:91–94

Ossowska K, Konieczny J, Wardas J, Pietraszek M, Kuter K, Wolfarth S, Pilc A (2007) An influence of ligands of metabotropic glutamate receptor subtypes on parkinsonian-like symptoms and the striatopallidal pathway in rats. Amino Acids 32:179–188

Parker MA, Bazan HE, Marcheselli V, Rodriguez de Turco EB, Bazan NG (2002) Platelet-activating factor induces permeability transition and cytochrome c release in isolated brain mitochondria. J Neurosci Res 69:39–50

Parkinson Progression Marker I (2011) The Parkinson Progression Marker Initiative (PPMI). Prog Neurobiol 95:629–635

Pellegrini-Giampietro DE (2003) The distinct role of mGlu1 receptors in post-ischemic neuronal death. Trends Pharmacol Sci 24:461–470

Pellegrini-Giampietro DE, Peruginelli F, Meli E, Cozzi A, Albani-Torregrossa S, Pellicciari R, Moroni F (1999) Protection with metabotropic glutamate 1 receptor antagonists in models of ischemic neuronal death: time-course and mechanisms. Neuropharmacol 38:1607–1619

Piallat B, Polosan M, Fraix V, Goetz L, David O, Fenoy A, Torres N, Quesada JL, Seigneuret E, Pollak P, Krack P, Bougerol T, Benabid AL, Chabardes S (2011) Subthalamic neuronal firing in obsessive-compulsive disorder and Parkinson disease. Ann Neurol 69:793–802

Piers TM, Heales SJ, Pocock JM (2011) Positive allosteric modulation of metabotropic glutamate receptor 5 down-regulates fibrinogen-activated microglia providing neuronal protection. Neurosci Lett 505:140–145

Pin JP, Acher F (2002) The metabotropic glutamate receptors: structure, activation mechanism and pharmacology. Curr Drug Targets CNS Neurol Disord 1:297–317

Pin JP, Duvoisin R (1995) The metabotropic glutamate receptors: structure and functions. Neuropharmacology 34:1–26

Pinteaux-Jones F, Sevastou IG, Fry VA, Heales S, Baker D, Pocock JM (2008) Myelin-induced microglial neurotoxicity can be controlled by microglial metabotropic glutamate receptors. J Neurochem 106:442–454

Pisani A, Calabresi P, Centonze D, Bernardi G (1997) Enhancement of NMDA responses by group I metabotropic glutamate receptor activation in striatal neurones. Br J Pharmacol 120:1007–1014

Pisani A, Gubellini P, Bonsi P, Conquet F, Picconi B, Centonze D, Bernardi G, Calabresi P (2001) Metabotropic glutamate receptor 5 mediates the potentiation of N-methyl-D-aspartate responses in medium spiny striatal neurons. Neuroscience 106:579–587

Pizzi M, Consolandi O, Memo M, Spano PF (1996) Activation of multiple metabotropic glutamate receptor subtypes prevents NMDA-induced excitotoxicity in rat hippocampal slices. Eur J Neurosci 8:1516–1521

Popoli P, Pintor A, Tebano MT, Frank C, Pepponi R, Nazzicone V, Grieco R, Pezzola A, Reggio R, Minghetti L, De Berardinis MA, Martire A, Potenza RL, Domenici MR, Massotti M (2004) Neuroprotective effects of the mGlu5R antagonist MPEP towards quinolinic acid-induced striatal toxicity: involvement of pre- and post-synaptic mechanisms and lack of direct NMDA blocking activity. J Neurochem 89:1479–1489

Przedborski S (2005) Pathogenesis of nigral cell death in Parkinson's disease. Parkinsonism Relat Disord 11(Suppl 1):S3–S7

Qin L, Wu X, Block ML, Liu Y, Breese GR, Hong JS, Knapp DJ, Crews FT (2007a) Systemic LPS causes chronic neuroinflammation and progressive neurodegeneration. Glia 55:453–462

Qin Z, Hu D, Han S, Reaney SH, Di Monte DA, Fink AL (2007b) Effect of 4-hydroxy-2-nonenal modification on alpha-synuclein aggregation. J Biol Chem 282:5862–5870

Rascol O (2009) "Disease-modification" trials in Parkinson disease: target populations, endpoints and study design. Neurology 72:S51–S58

Reichel CM, Schwendt M, McGinty JF, Olive MF, See RE (2011) Loss of object recognition memory produced by extended access to methamphetamine self-administration is reversed by positive allosteric modulation of metabotropic glutamate receptor 5. Neuropsychopharmacol 36:782–792

Reiner A, Wang HB, Del Mar N, Sakata K, Yoo W, Deng YP (2012) BDNF may play a differential role in the protective effect of the mGluR2/3 agonist LY379268 on striatal projection neurons in R6/2 Huntington's disease mice. Brain Res 1473:161–172

Rentzos M, Nikolaou C, Andreadou E, Paraskevas GP, Rombos A, Zoga M, Tsoutsou A, Boufidou F, Kapaki E, Vassilopoulos D (2007) Circulating interleukin-15 and RANTES chemokine in Parkinson's disease. Acta Neurol Scand 116:374–379

Ribeiro FM, Paquet M, Ferreira LT, Cregan T, Swan P, Cregan SP, Ferguson SS (2010) Metabotropic glutamate receptor-mediated cell signaling pathways are altered in a mouse model of Huntington's disease. J Neurosci 30:316–324

Ribeiro FM, Pires RG, Ferguson SS (2011) Huntington's disease and group I metabotropic glutamate receptors. Mol Neurobiol 43:1–11

Ribeiro FM, Hamilton A, Doria JG, Guimaraes IM, Cregan SP, Ferguson SS (2014) Metabotropic glutamate receptor 5 as a potential therapeutic target in Huntington's disease. Expert Opin Ther Targets 18:1293–1304

Rodriguez MC, Obeso JA, Olanow CW (1998) Subthalamic nucleus-mediated excitotoxicity in Parkinson's disease: a target for neuroprotection. Ann Neurol 44:S175–S188

Rossi D, Brambilla L, Valori CF, Roncoroni C, Crugnola A, Yokota T, Bredesen DE, Volterra A (2008) Focal degeneration of astrocytes in amyotrophic lateral sclerosis. Cell Death Differ 15:1691–1700

Rouse ST, Marino MJ, Bradley SR, Awad H, Wittmann M, Conn PJ (2000) Distribution and roles of metabotropic glutamate receptors in the basal ganglia motor circuit: implications for treatment of Parkinson's disease and related disorders. Pharmacol Ther 88:427–435

Rylander D, Recchia A, Mela F, Dekundy A, Danysz W, Cenci MA (2009) Pharmacological modulation of glutamate transmission in a rat model of L-DOPA-induced dyskinesia: effects on motor behavior and striatal nuclear signaling. J Pharmacol Exp Ther 330:227–235

Rylander D, Iderberg H, Li Q, Dekundy A, Zhang J, Li H, Baishen R, Danysz W, Bezard E, Cenci MA (2010) A mGluR5 antagonist under clinical development improves L-DOPA-induced dyskinesia in parkinsonian rats and monkeys. Neurobiol Dis 39:352–361

Sabelhaus CF, Schroder UH, Breder J, Henrich-Noack P, Reymann KG (2000) Neuroprotection against hypoxic/hypoglycaemic injury after the insult by the group III metabotropic glutamate receptor agonist (R, S)-4-phosphonophenylglycine. Br J Pharmacol 131:655–658

Sala C, Roussignol G, Meldolesi J, Fagni L (2005) Key role of the postsynaptic density scaffold proteins Shank and Homer in the functional architecture of Ca2+ homeostasis at dendritic spines in hippocampal neurons. J Neurosci 25:4587–4592

Sarnico I, Boroni F, Benarese M, Sigala S, Lanzillotta A, Battistin L, Spano P, Pizzi M (2008) Activation of NF-kappaB p65/c-Rel dimer is associated with neuroprotection elicited by mGlu5 receptor agonists against MPP(+) toxicity in SK-N-SH cells. J Neural Transm 115:669–676

Schapira AH (2004) Disease modification in Parkinson's disease. Lancet Neurol 3:362–368

Schapira AH (2009a) Molecular and clinical pathways to neuroprotection of dopaminergic drugs in Parkinson disease. Neurology 72:S44–S50

Schapira AH (2009b) Neurobiology and treatment of Parkinson's disease. Trends Pharmacol Sci 30:41–47

Schapira AH (2009c) Neuroprotection in Parkinson's disease. Parkinsonism Relat Disord 15(Suppl 4):S41–S43

Schapira AH, Olanow CW, Greenamyre JT, Bezard E (2014) Slowing of neurodegeneration in Parkinson's disease and Huntington's disease: future therapeutic perspectives. Lancet 384:545–555

Schiefer J, Sprunken A, Puls C, Luesse HG, Milkereit A, Milkereit E, Johann V, Kosinski CM (2004) The metabotropic glutamate receptor 5 antagonist MPEP and the mGluR2 agonist LY379268 modify disease progression in a transgenic mouse model of Huntington's disease. Brain Res 1019:246–254

Schoepp DD, Conn PJ (1993) Metabotropic glutamate receptors in brain function and pathology. Trends Pharmacol Sci 14:13–20

Sekiguchi M, Wada K, Wenthold RJ (1992) N-acetylaspartylglutamate acts as an agonist upon homomeric NMDA receptor (NMDAR1) expressed in Xenopus oocytes. FEBS Lett 311:285–289

Sherer TB, Betarbet R, Stout AK, Lund S, Baptista M, Panov AV, Cookson MR, Greenamyre JT (2002) An in vitro model of Parkinson's disease: linking mitochondrial impairment to altered alpha-synuclein metabolism and oxidative damage. J Neurosci 22:7006–7015

Shimo Y, Wichmann T (2009) Neuronal activity in the subthalamic nucleus modulates the release of dopamine in the monkey striatum. Eur J Neurosci 29:104–113

Siderowf A, Stern MB (2008) Premotor Parkinson's disease: clinical features, detection, and prospects for treatment. Ann Neurol 64(Suppl 2):S139–S147

Smith Y, Charara A, Hanson JE, Paquet M, Levey AI (2000) GABA(B) and group I metabotropic glutamate receptors in the striatopallidal complex in primates. J Anat 196(Pt 4):555–576

Smith Y, Wichmann T, Factor SA, DeLong MR (2012) Parkinson's disease therapeutics: new developments and challenges since the introduction of levodopa. Neuropsychopharmacol 37:213–246

Solomon JN, Lewis CA, Ajami B, Corbel SY, Rossi FM, Krieger C (2006) Origin and distribution of bone marrow-derived cells in the central nervous system in a mouse model of amyotrophic lateral sclerosis. Glia 53:744–753

Sonsalla PK, Albers DS, Zeevalk GD (1998) Role of glutamate in neurodegeneration of dopamine neurons in several animal models of parkinsonism. Amino Acids 14:69–74

Souza LC, Wilhelm EA, Bortolatto CF, Nogueira CW, Boeira SP, Jesse CR (2014) Involvement of mGlu5 receptor in 3-nitropropionic acid-induced oxidative stress in rat striatum. Neurol Res 36:833–840

Spillantini MG, Crowther RA, Jakes R, Cairns NJ, Lantos PL, Goedert M (1998) Filamentous alpha-synuclein inclusions link multiple system atrophy with Parkinson's disease and dementia with Lewy bodies. Neurosci Lett 251:205–208

Stocchi F, Rascol O, Destee A, Hattori N, Hauser RA, Lang AE, Poewe W, Stacy M, Tolosa E, Gao H, Nagel J, Merschhemke M, Graf A, Kenney C, Trenkwalder C (2013) AFQ056 in Parkinson patients with levodopa-induced dyskinesia: 13-week, randomized, dose-finding study. Mov Disord 28:1838–1846

Storey E, Hyman BT, Jenkins B, Brouillet E, Miller JM, Rosen BR, Beal MF (1992) 1-Methyl-4-phenylpyridinium produces excitotoxic lesions in rat striatum as a result of impairment of oxidative metabolism. J Neurochem 58:1975–1978
Stover JF, Sakowitz OW, Beyer TF, Dohse NK, Kroppenstedt SN, Thomale UW, Schaser KD, Unterberg AW (2003) Effects of LY379268, a selective group II metabotropic glutamate receptor agonist on EEG activity, cortical perfusion, tissue damage, and cortical glutamate, glucose, and lactate levels in brain-injured rats. J Neurotrauma 20:315–326
Szydlowska K, Kaminska B, Baude A, Parsons CG, Danysz W (2007) Neuroprotective activity of selective mGlu1 and mGlu5 antagonists in vitro and in vivo. Eur J Pharmacol 554:18–29
Tansey MG, Goldberg MS (2010) Neuroinflammation in Parkinson's disease: its role in neuronal death and implications for therapeutic intervention. Neurobiol Dis 37:510–518
Taylor DL, Diemel LT, Cuzner ML, Pocock JM (2002) Activation of group II metabotropic glutamate receptors underlies microglial reactivity and neurotoxicity following stimulation with chromogranin A, a peptide up-regulated in Alzheimer's disease. J Neurochem 82:1179–1191
Taylor DL, Diemel LT, Pocock JM (2003) Activation of microglial group III metabotropic glutamate receptors protects neurons against microglial neurotoxicity. J Neurosci 23:2150–2160
Taylor DL, Pirianov G, Holland S, McGinnity CJ, Norman AL, Reali C, Diemel LT, Gveric D, Yeung D, Mehmet H (2010) Attenuation of proliferation in oligodendrocyte precursor cells by activated microglia. J Neurosci Res 88:1632–1644
Testa CM, Standaert DG, Young AB, Penney JB Jr (1994) Metabotropic glutamate receptor mRNA expression in the basal ganglia of the rat. J Neurosci 14:3005–3018
Trombley PQ, Westbrook GL (1992) L-AP4 inhibits calcium currents and synaptic transmission via a G-protein-coupled glutamate receptor. J Neurosci 12:2043–2050
Tu JC, Xiao B, Naisbitt S, Yuan JP, Petralia RS, Brakeman P, Doan A, Aakalu VK, Lanahan AA, Sheng M, Worley PF (1999) Coupling of mGluR/Homer and PSD-95 complexes by the Shank family of postsynaptic density proteins. Neuron 23:583–592
Turski L, Bressler K, Rettig KJ, Loschmann PA, Wachtel H (1991) Protection of substantia nigra from MPP+ neurotoxicity by N-methyl-D-aspartate antagonists. Nature 349:414–418
Valenti O, Conn PJ, Marino MJ (2002) Distinct physiological roles of the Gq-coupled metabotropic glutamate receptors Co-expressed in the same neuronal populations. J Cell Physiol 191:125–137
Valenti O, Marino MJ, Conn PJ (2003) Modulation of excitatory transmission onto midbrain dopaminergic neurons of the rat by activation of group III metabotropic glutamate receptors. Ann N Y Acad Sci 1003:479–480
Valenti O, Mannaioni G, Seabrook GR, Conn PJ, Marino MJ (2005) Group III metabotropic glutamate-receptor-mediated modulation of excitatory transmission in rodent substantia nigra pars compacta dopamine neurons. J Pharmacol Exp Ther 313:1296–1304
Van Laar VS, Arnold B, Cassady SJ, Chu CT, Burton EA, Berman SB (2011) Bioenergetics of neurons inhibit the translocation response of Parkin following rapid mitochondrial depolarization. Hum Mol Genet 20:927–940
Vardi N, Duvoisin R, Wu G, Sterling P (2000) Localization of mGluR6 to dendrites of ON bipolar cells in primate retina. J Comp Neurol 423:402–412
Venero JL, Santiago M, Tomas-Camardiel M, Matarredona ER, Cano J, Machado A (2002) DCG-IV but not other group-II metabotropic receptor agonists induces microglial BDNF mRNA expression in the rat striatum Correlation with neuronal injury. Neuroscience 113:857–869
Vernon AC, Palmer S, Datla KP, Zbarsky V, Croucher MJ, Dexter DT (2005) Neuroprotective effects of metabotropic glutamate receptor ligands in a 6-hydroxydopamine rodent model of Parkinson's disease. Eur J Neurosci 22:1799–1806
Vernon AC, Zbarsky V, Datla KP, Croucher MJ, Dexter DT (2007) Subtype selective antagonism of substantia nigra pars compacta group I metabotropic glutamate receptors protects the nigrostriatal system against 6-hydroxydopamine toxicity in vivo. J Neurochem 103:1075–1091
Vernon AC, Croucher MJ, Dexter DT (2008) Additive neuroprotection by metabotropic glutamate receptor subtype-selective ligands in a rat Parkinson's model. Neuroreport 19:475–478
Wallace BA, Ashkan K, Heise CE, Foote KD, Torres N, Mitrofanis J, Benabid AL (2007) Survival of midbrain dopaminergic cells after lesion or deep brain stimulation of the subthalamic nucleus in MPTP-treated monkeys. Brain 130:2129–2145

Wang L, Kitai ST, Xiang Z (2005) Modulation of excitatory synaptic transmission by endogenous glutamate acting on presynaptic group II mGluRs in rat substantia nigra compacta. J Neurosci Res 82:778–787

Westbrook GL, Mayer ML, Namboodiri MA, Neale JH (1986) High concentrations of N-acetylaspartylglutamate (NAAG) selectively activate NMDA receptors on mouse spinal cord neurons in cell culture. J Neurosci 6:3385–3392

Wichmann T, DeLong MR (2006) Basal ganglia discharge abnormalities in Parkinson's disease. J Neural Transm Suppl:21–25

Wigmore MA, Lacey MG (1998) Metabotropic glutamate receptors depress glutamate-mediated synaptic input to rat midbrain dopamine neurones in vitro. Br J Pharmacol 123:667–674

Williams CJ, Dexter DT (2014) Neuroprotective and symptomatic effects of targeting group III mGlu receptors in neurodegenerative disease. J Neurochem 129:4–20

Winklhofer KF, Haass C (2010) Mitochondrial dysfunction in Parkinson's disease. Biochim Biophys Acta 1802:29–44

Wroblewska B, Wroblewski JT, Pshenichkin S, Surin A, Sullivan SE, Neale JH (1997) N-acetylaspartylglutamate selectively activates mGluR3 receptors in transfected cells. J Neurochem 69:174–181

Xue F, Stoica BA, Hanscom M, Kabadi SV, Faden AI (2014) Positive allosteric modulators (PAMs) of metabotropic glutamate receptor 5 (mGluR5) attenuate microglial activation. CNS Neurol Disord Drug Targets 13:558–566

Zhang Y, Rodriguez AL, Conn PJ (2005) Allosteric potentiators of metabotropic glutamate receptor subtype 5 have differential effects on different signaling pathways in cortical astrocytes. J Pharmacol Exp Ther 315:1212–1219

Zhou M, Xu W, Liao G, Bi X, Baudry M (2009) Neuroprotection against neonatal hypoxia/ischemia-induced cerebral cell death by prevention of calpain-mediated mGluR1alpha truncation. Exp Neurol 218:75–82

Zuddas A, Corsini GU, Barker JL, Kopin IJ, Di Porzio U (1991) Specific Reinnervation of Lesioned Mouse Striatum by Grafted Mesencephalic Dopaminergic Neurons. Eur J Neurosci 3:72–85

Zwienenberg M, Gong QZ, Berman RF, Muizelaar JP, Lyeth BG (2001) The effect of groups II and III metabotropic glutamate receptor activation on neuronal injury in a rodent model of traumatic brain injury. Neurosurgery 48:1119–1126. discussion 1126-1117

Chapter 7
Structure, Dynamics, and Modulation of Metabotropic Glutamate Receptors

Philippe Rondard, Xavier Rovira, Cyril Goudet, and Jean-Philippe Pin

Abstract The metabotropic glutamate receptors (mGluRs) are glutamate-activated G-protein-coupled receptors that are expressed throughout the central nervous system. The eight mGluR subtypes serve to modulate transmission at many synapses and are potential therapeutic targets for the treatment of many neurological and psychiatric diseases. In particular, their organization in multiple domains and subunits offers various possibilities for the development of new drugs that modulate mGluR activity. This review will show how recent and original approaches to study the mGluR structure, dynamics, and pharmacology have brought a new view of their mechanism of activation and the possibility of developing innovative ligands, such as light-regulated drugs, to control the physiological activity of mGluRs in animals.

Keywords Allosteric modulator • Partial agonist • Förster resonance energy transfer (FRET) • Lanthanide-based resonance energy transfer (LRET) • Photopharmacology • G-protein-coupled receptor (GPCR)

7.1 Introduction

The eight metabotropic glutamate (mGlu1-8) receptors (mGluRs) are all activated by glutamate, which is the major excitatory neurotransmitter in the central nervous system (Conn and Pin 1997; Niswender and Conn 2010). The mGluRs are thus essential for the regulation of synaptic activity. Accordingly, they are promising drug targets for the treatment of neurological and psychiatric disorders (Gregory et al. 2013; Jacquemont et al. 2011; Patil et al. 2007). Structurally, mGluRs belong to the class C G-protein-coupled receptor (GPCR) family, which also includes the

P. Rondard (✉) • X. Rovira • C. Goudet • J.-P. Pin
CNRS, UMR5203, Institut de Génomique Fonctionnelle,
141 rue de la Cardonille, 34094 Montpellier, France

INSERM U1191, Montpellier, France

Université de Montpellier, Montpellier F-34000, France
e-mail: philippe.rondard@igf.cnrs.fr

R.T. Ngomba et al. (eds.), *mGLU Receptors*, The Receptors 31,
DOI 10.1007/978-3-319-56170-7_7

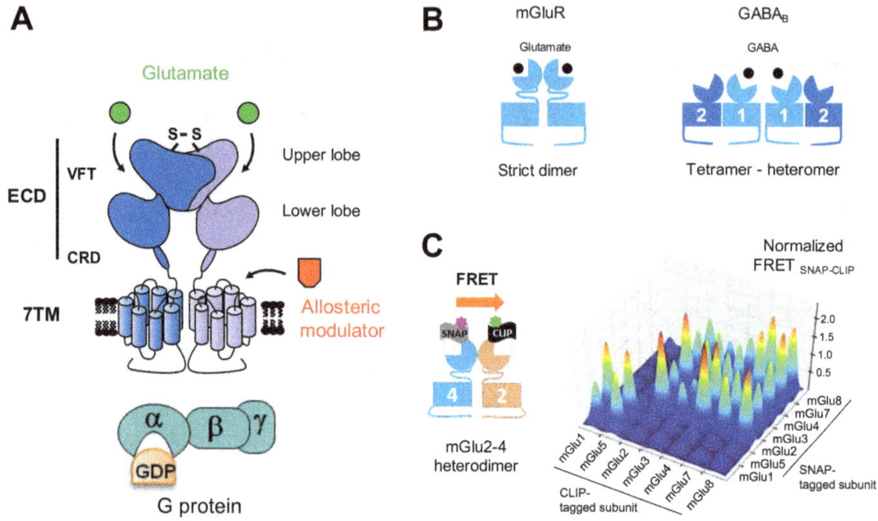

Fig. 7.1 Structural and functional organization of the mGluRs. (**a**) Cartoon illustrating a full-length mGluR dimer stabilized by a disulfide bridge. (**b**) Scheme illustrating the different organization of the mGlu and $GABA_B$ receptors at the surface of living cells. While the mGluR homodimers form strict dimers, the $GABA_B$ receptor, which is composed of $GABA_{B1}$ and $GABA_{B2}$ subunits (named "1" and "2", respectively), forms constitutive and higher-order oligomers (heteromers) through the interaction with the $GABA_{B1}$ subunit, which is responsible for GABA binding (Maurel et al. 2008; Calebiro et al. 2013; Comps-Agrar et al. 2011, 2012; Schwenk et al. 2010). In these heteromers, the $GABA_{B2}$ subunits are responsible for G-protein activation, and they do not interact directly together. (**c**) Specific and functional mGlu heterodimers have been detected at the surface of living cells thanks to the measurement of trFRET signals between the SNAP- and CLIP-tagged mGlu subunits that were co-transfected in the cells (Adapted from Doumazane et al. 2011). The SNAP- and CLIP-tag were covalently labeled with a pair of fluorophores (compatible with trFRET signal detection), as shown for the heterodimers mGlu2–mGlu4. The chart indicates that the homo- and heterodimers are formed between two separate groups of receptors: mGlu1 and mGlu5 in one group and mGlu2, mGlu3, mGlu4, mGlu7, and mGlu8 in the other

metabotropic receptor for GABA ($GABA_B$), the calcium-sensing receptor (CaSR), the receptor GPCR6A, and the receptors for sweet and umami taste in humans (Kniazeff et al. 2011). These receptors share a large extracellular ligand binding domain (ECD) where the natural ligand binds and a heptahelical transmembrane domain (7TM) that contains the binding site for synthetic allosteric modulators (AMs) and is responsible for G-protein activation (Fig. 7.1a).

7.2 Functional Organization and Structure of the mGluRs

We recently demonstrated that mGluRs must be in a homodimer state to be activated by glutamate (El Moustaine et al. 2012), by purifying, reconstituting, and stabilizing the full-length mGlu2 receptor in nanodiscs that reproduce a native-like environment. Indeed, only the fraction containing the dimeric mGluRs was able to activate a purified G-protein upon stimulation by glutamate. The monomeric receptor bound glutamate, but could not be activated by it, although it could be fully activated by a positive AM (PAM). Furthermore, using a time-resolved Förster resonance energy transfer (trFRET) combined with SNAP-tag technology to label the extracellular N-terminus of the mGlu subunits with fluorophores compatible with FRET, we showed that mGlu homodimers form mostly strict dimers at the cell surface of living cells (Doumazane et al. 2011; Maurel et al. 2008) (Fig. 7.1b). Similar conclusions that mGlu receptors are strict dimers at the surface of cells have been proposed from photobleaching experiments using GFP-tagged mGlu receptors in single-molecule conditions (Levitz et al. 2016). This is in contrast to the $GABA_B$ receptor that forms constitutive, stable, and organized dimers of dimers, and even higher-order oligomers, in similar conditions (Maurel et al. 2008; Calebiro et al. 2013; Comps-Agrar et al. 2011, 2012; Schwenk et al. 2010). Dimers of dimers of mGlu2 receptor were recently detected, but only when their interaction was stabilized by a disulfide bridge, which suggests the oligomeric forms of mGluR are very transient (Xue et al. 2015).

Interestingly, mGlu subunits could also form specific and strict heterodimers at the surface of transfected live cells (Doumazane et al. 2011; Levitz et al. 2016). This was the case for all the groups II (mGlu2 and mGlu3) and III mGlu subunits (mGlu4, mGlu6, mGlu7, and mGlu8) that could form heterodimers together, but not with the two group I mGlu subunits (mGlu1 and mGlu5), while these two latter receptors could form heterodimers with one another (Fig. 7.1c). These heterodimers were revealed in trFRET saturation experiments where the two co-expressed subunits were orthogonally and covalently labeled with a pair of fluorophores attached to the extracellular N-terminal SNAP and CLIP-tagged subunits (Doumazane et al. 2011). Although the existence of such heterodimers in native tissues remains to be established, data indicate they could exist. Indeed, it was shown that different mGlu subunits are co-expressed in the same compartment of neurons (Doumazane et al. 2011; Ferraguti et al. 2005; Ferraguti and Shigemoto 2006). In addition, a recent study has proposed specific pharmacological properties of the mGlu2/mGlu4 heterodimer that was observed in cortical neurons (Yin et al. 2014).

7.3 Crystal Structures and Conformational States of the mGluR

The ECD is composed of both a Venus flytrap (VFT) domain that contains the glutamate binding pocket (also called orthosteric ligand binding site) and a cysteine-rich domain (CRD) that connects the VFT to the 7TM domain. Crystal structures of the isolated VFT dimers have been solved for most mGluRs (mGlu1, mGlu2, mGlu3, mGlu5, and mGlu7)(Kunishima et al. 2000; Monn et al. 2015; Muto et al. 2007; Rondard et al. 2011). They have revealed that the bi-lobal VFT contains the binding site for the orthosteric agonists and competitive antagonists that bind between the upper and lower lobes (Fig. 7.2). In mGluRs, each VFT is connected to

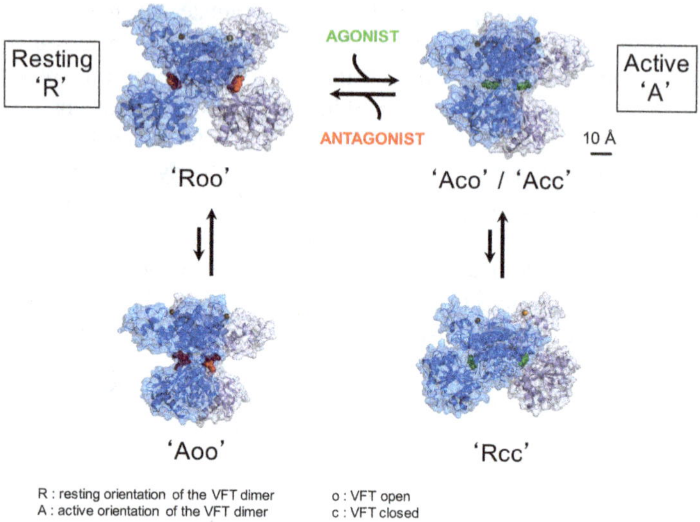

Fig. 7.2 Crystal structures and conformational changes of the extracellular ligand binding domain of the mGluRs (Figure Adapted from Doumazane et al. 2013). Crystal structures of the mGluR VFT dimer were solved in complex with antagonist (*red*) and agonist (*green*), in the resting (R) and active (A) orientation, respectively. In these structures the VFTs are either open (o) or closed (c). Accordingly, the two main conformations, Roo and Aco (or Acc, when both VFT are closed), correspond to the two major states of the VFT dimer. These conformations show the reorientation of the VFT dimer in a scissorlike movement during activation. Whereas the lower VFT lobes are distant in the inactive state they become closed in the active state. Of note, the active reorientation bound to antagonist (Aoo) and the resting conformation bound to agonist (Rcc) have also been observed in the crystal structures (Doumazane et al. 2013). The VFT at the front of the dimer is in *dark blue*, and the one at the back is in *purple*

the 7TM domain through the CRD similarly to other class C GPCRs that are constitutive dimers, such as CaSR, GPRC6A, and the taste receptors. But in contrast, the $GABA_B$ receptor lacks this CRD (Kniazeff et al. 2011; Rondard et al. 2011). It was demonstrated that the CRD of mGluRs is required for the allosteric communication between the VFT and the 7TM domains and is stabilized by disulfide bridges, one of which connects the CRD to the VFT (Muto et al. 2007; Rondard et al. 2006).

From the crystal structures of the isolated ECD dimer of mGluRs (Kunishima et al. 2000), CaSR (Geng et al. 2016; Zhang et al. 2016), and the $GABA_B$ receptor (Geng et al. 2013), it is known that each VFT can adopt two major states: (i) an "open" conformation observed in the absence of ligand or in presence of competitive antagonists and (ii) a "closed" conformation stabilized by agonists. In addition, the ECD dimer can adopt two different orientations, either the "active" (A) state, in which the two ECDs are reoriented by 70° compared to the structure obtained in the presence of the antagonist ("resting" (R) state)(Kunishima et al. 2000; Rondard et al. 2011; Tsuchiya et al. 2002) (Fig. 7.2). To summarize, based on crystal structures, the main conformations that define the inactive and active states of the mGluR are the resting state Roo where both VFTs are opened and the active states Aco or Acc where one or both VFTs are close, respectively.

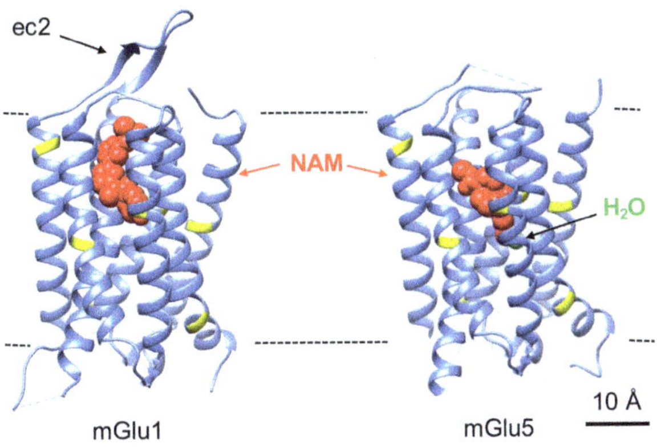

Fig. 7.3 Crystal structures of the 7TM domains of the mGlu1 and mGlu5 receptors (Figure taken with permission from Rondard and Pin 2015). These structures correspond to the inactive state stabilized by a specific NAM (in *red*) of these receptors (Doré et al. 2014; Wu et al. 2014). The most conserved residue of each TM, according to the universal numbering system for the residues in the transmembrane helices of class C GPCRs (Doré et al. 2014), is shown in *yellow*, and the water molecule identified at the bottom of the allosteric binding site in mGlu5 7TM is shown in *green*

The first crystal structures of the isolated 7TM domain of two human mGluRs, mGlu1 and mGlu5, were solved at high resolution (Doré et al. 2014; Wu et al. 2014) (Fig. 7.3). These structures have revealed that despite the lack of sequence conservation (<15% identical residues) between class C and class A GPCRs, the overall fold of the 7TM is conserved between these two classes. These structures enabled the proposal of a universal numbering system for residues in the transmembrane helices of class C GPCRs (Doré et al. 2014; Pin et al. 2003) similar to the nomenclature used for the class A GPCRs (Ballesteros and Weinstein 1995). The most conserved residue in each class C GPCR TM domain was assigned the position .50, and this number is preceded by the TM number (this residue is called X.50, where X is the number of the TM). Although these 7TM structures of mGluRs were only solved in the inactive state, in complex with a negative AM (NAM, which is known to stabilize the receptor ground state), they did reveal for the first time the binding site for AMs in class C GPCRs. The inactive state was further stabilized by the ionic interactions between TM3 and TM6 called the "ionic lock" involving residues $Lys^{3.50}$ and $Glu^{6.35}$ in mGlu1 and mGlu5. In class A GPCRs, this ionic lock is observed between the conserved $Arg^{3.50}$ of the D(E)RY motif and $Asp/Glu^{6.30}$. The ionic lock is known to restrict the activation-related movement of TM6 in class A GPCRs. In the mGlu1 and mGlu5 receptors, this ionic lock is further stabilized by polar interactions with the intracellular loop 1(Doré et al. 2014; Wu et al. 2014).

These structures have also suggested that the mGlu1 7TM domain can form dimers through TM1-TM1 interaction, and this interface is stabilized by cholesterol molecules (Wu et al. 2014). Although this TM1-TM1 interface was found in other GPCR dimers (Cherezov et al. 2007; Ruprecht et al. 2004), a recent study based on cysteine cross-linking revealed that TM4, TM5, and TM6 in the full-length mGlu2 receptor constitute the 7TM dimer interface (Xue et al. 2015) (see details below). TM4-TM6 interfaces are also frequently observed in class A GPCR dimers (Guo et al. 2005; Manglik et al. 2012; Wu et al. 2010). Finally, these 7TM structures did not reveal much information regarding the structural coupling between the ECD and the 7TM in the mGluRs. The extracellular loop 2 of mGlu1 7TM (usually the most important extracellular loop for GPCRs activation (Venkatakrishnan et al. 2013)) adopts a large β hairpin conformation pointing to the extracellular space and potentially to the ECD (Fig. 7.3). A similar conformation of the extracellular loop 2 was previously observed in the crystal structures of class A receptors (Wu et al. 2012). But such a feature was not observed in the mGlu5 7TM structure. In the absence of structure of the full-length mGlu receptors, we are proposing a 3D working model based on the structures of the isolated ECD dimer and 7TMs (Fig. 7.4).

7.4 Structural Dynamics of the mGluR ECD During Activation

Recent structural, biophysical, and functional data have strongly contributed toward elucidating the mechanism of activation and the dynamics of the ECD in the class C GPCRs (Rondard et al. 2011; Rondard and Pin 2015) (Fig. 7.5). The first mandatory step for mGluR activation is the closure of the VFTs upon binding of the agonist between the two lobes of each VFT (Rondard et al. 2011). Bessis et al. have shown that competitive antagonists, that bind into the glutamate binding pocket of the VFT domain, prevent mGlu8 receptor activation by blocking VFT closure by steric hindrance (Bessis et al. 2002). In mGlu8 receptor mutants where the glutamate binding pocket was enlarged to accommodate these competitive antagonists in the closed conformation of the VFT domain, the competitive antagonists became full agonists resulting in a complete activation of the receptor. It was concluded that these antagonists were able to stabilize the VFT of these mGlu8 mutants in a closed state similar to the one stabilized by the agonist. Further evidence of VFT closure being sufficient for mGluR activation came recently from the work of Kiyonaka et al. (Kiyonaka et al. 2016) who described the activation of the mGlu1 receptor by a bipyridine compound that acts like a staple as it is chelated by the upper and lower lobes of the VFT. The resulting VFT that was stabilized in closed conformation produced a fully activated receptor. The binding of this bipyridine reagent was achieved by introducing a histidine residue in each lobe, by site-directed mutagenesis. In addition, it was demonstrated that the closure of one VFT per dimer is sufficient to activate the mGlu receptor, but the closure of both VFTs in the dimer is required for full activity (Kniazeff et al. 2004a). It was also shown that VFT closure is sufficient for the activation of other class C GPCRs such as the $GABA_B$ receptor. The introduction of pair of cysteines that stabilize the closed conformation of the $GABA_{B1}$ subunit VFT responsible for GABA binding, thanks to a disulfide bridge between the upper and lower lobes of the VFT, produced a $GABA_B$ receptor mutant with a strong constitutive activity (Kniazeff et al. 2004b).

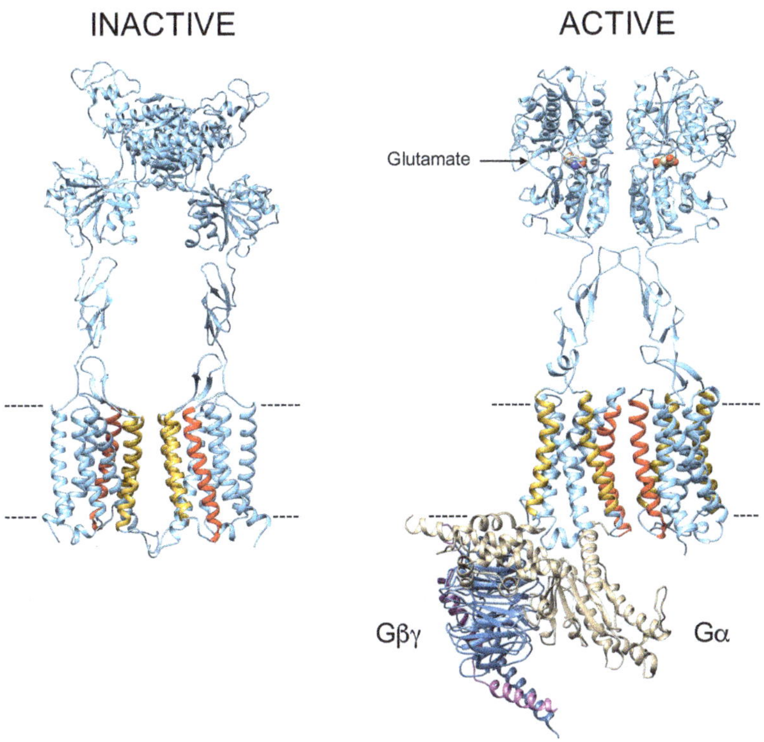

Fig. 7.4 Structural models of the full-length mGlu2 receptor dimer in the resting and active states. To build these models, we used the crystal structures of the mGluR ECD in the absence of ligand or in the presence of agonist, to model the resting and active conformations, respectively, as previously reported (Xue et al. 2015). Model of the 7TM dimers in the resting and active reorientations were previously described (Xue et al. 2015). In the resting state, the lower lobes of the VFT and the CRD are distant, while they become close in the active conformation when bound to glutamate. The 7TM dimer interface is formed by the transmembrane helices TM4 and TM5 (in *orange*) in the resting state and TM6 helices (in *red*) in the active state. In the active state, the trimeric G-protein Gαβγ is shown in the nucleotide-free state and coupled to one 7TM domain (According to the crystal structure of a class A GPCR solved in the active state, in complex with Gαβγ (Rasmussen et al. 2011))

The second critical step for mGluR activation is the reorientation of the VFT in the dimer (Figs. 7.2 and 7.5). It is known that the VFT dimer oscillates between two major conformations, the R state stabilized by antagonists and the A state stabilized by agonists and required for receptor activation. It has been well illustrated for the mGlu1 receptor that the lower lobes of the two VFTs become closer in the active state; meanwhile, they are distant in the resting state (Kunishima et al. 2000; Tsuchiya et al. 2002). Since the VFT is cross-linked to the CRD through a disulfide bridge (Rondard et al. 2006), the reorientation of the VFT dimer during activation brings the two CRDs in contact in the active state. The introduction of a steric hindrance at the CRD dimer interface such as N-glycosylation, which produce a mutant that can no longer be activated by an agonist (Doumazane et al. 2013), is consistent

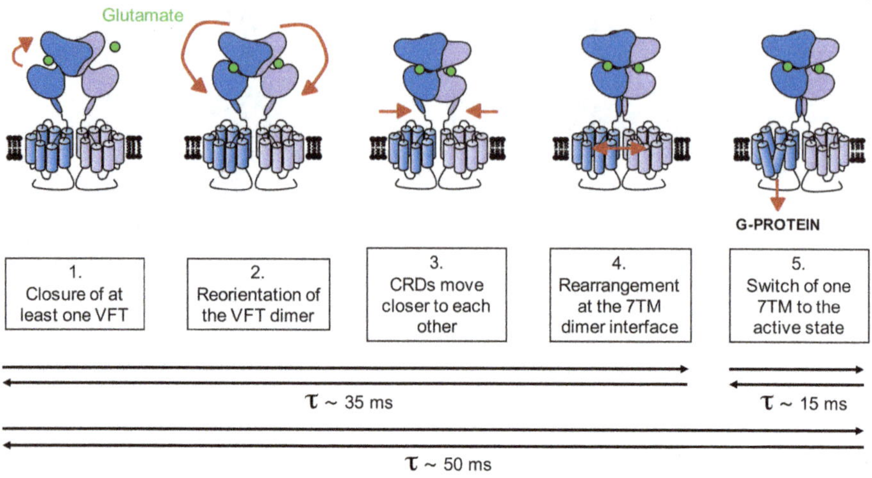

Fig. 7.5 Model of mGluR activation (Figure taken with permission from Rondard and Pin 2015). Schematic illustration of the major steps in mGluR activation that lead to G-protein activation. Glutamate or other agonists bind and induce VFT closure and reorientation (Kunishima et al. 2000), leading to the formation of a specific interface between CRDs (Huang et al. 2011), and finally to the rearrangement at the 7TM dimer interface (Xue et al. 2015) that is followed by the activation of only one of the two 7TMs (Hlavackova et al. 2005). The time course is shown underneath

with the two CRDs being close during activation. Conversely, locking the active orientation of the CRD dimer by a cysteine cross-linking at the CRD dimer interface is sufficient to induce constitutive activity in the mGluRs (Huang et al. 2011). The recent crystal structures of the apo- and agonist-bound conformations of the CaSR (Geng et al. 2016; Zhang et al. 2016) and $GABA_B$ receptors (Geng et al. 2013) show a similar reorientation of the two VFTs during activation. However, these crystal structures suggested that the amplitude of ECD rearrangement of these two receptors is smaller than the one in mGluR ECD. Finally, in the $GABA_B$ receptor, stabilization of the direct contact between the two lower lobes of the VFTs through a disulfide bridge produces a receptor mutant constitutively active (Geng et al. 2013). It is consistent with the two lower lobes of the VFT dimer being in direct interaction in the active state of the $GABA_B$ receptor, as proposed for the mGluRs.

We recently analyzed ECD dynamics by trFRET with a lanthanide donor with a long lifetime and an organic fluorophore as acceptor that were linked to N-terminal extracellular SNAP-tags on the mGluRs (Fig. 7.6a). Also, for kinetic analysis of the ECD reorientation, single-molecule FRET (smFRET) experiments with the same SNAP-tagged isolated ECD dimers (Olofsson et al. 2014) and SNAP-tagged full-length mGlu2 and mGlu3 receptors (Levitz et al. 2016; Vafabakhsh et al. 2015) were used. We thus showed that the rearrangement of the ECD dimer is correlated with receptor activation at the surface of living cells (Doumazane et al. 2013). The ECD dimer is in an equilibrium between the two major conformations described above, R and A, and it oscillates between these states at sub-millisecond timescale

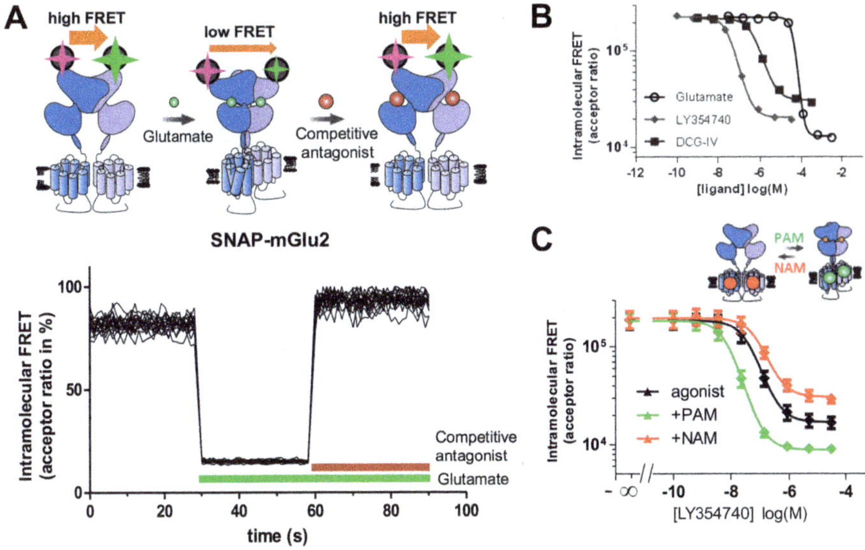

Fig. 7.6 Rearrangement of the mGluR ECD dimer measured by trFRET (Figure Adapted from Rondard and Pin 2015). **(a)** We have developed SNAP-tagged mGlu2 sensors to monitor mGluR activation at the surface of living cells by measuring the trFRET signal between pairs of fluorophores linked to the SNAP-tag, following incubation with glutamate or a competitive antagonist (LY341495). The time course is shown below. **(b, c)** These sensors facilitate the identification of partial agonists such as DCG-IV **(b)**, and of PAMs and NAMs **(c)** in a format compatible with high-throughput drug screening (Charts taken with permission from Doumazane et al. 2013)

between being bound, or not, by an agonist or an antagonist (Olofsson et al. 2014) (Fig. 7.7a). As these two states are energetically similar even in the presence of ligands, this explains why mGluR ECDs could be crystallized in the R or A conformation with both kinds of ligand (Muto et al. 2007; Doumazane et al. 2013) and why there is no difference between the crystal structures of the ECD in complex with partial or full agonists (Muto et al. 2007).

7.5 Dynamics of the mGluR Transmembrane Domain During Activation

How the ECD signal is transmitted to the 7TM domain and then to the G-protein remains an open question. In our current model for receptor activation (Fig. 7.5), agonist binding would cause a scissoring movement between VFTs, resulting in a rearrangement of the CRDs that become stabilized in the active state (Geng et al. 2016; Huang et al. 2011). While these initial steps start to be well understood, the molecular mechanism of the following conformational changes remains unknown. The CRD conformational change, upon agonist binding, is expected to trigger a

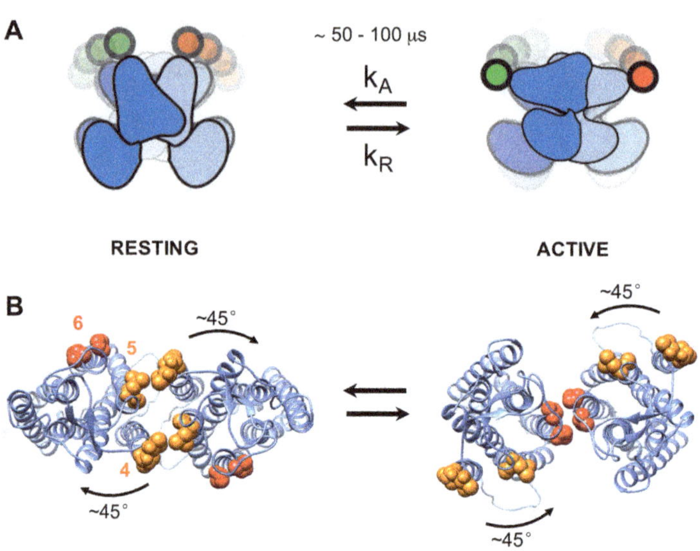

Fig. 7.7 Structural dynamics of the mGluRs (Figure with permission from Rondard and Pin 2015). **(a)** Based on smFRET experiments, we have shown that ECD dimers oscillate between the resting and active states, whether unbound or bound to an agonist or an antagonist (Olofsson et al. 2014). **(b)** Model of the 7TM of the mGlu2 receptor in the resting and active states, illustrating the large rearrangement that occurs at the 7TM dimer interface during receptor activation (Xue et al. 2015). The position of cysteine residues introduced in TM4 and TM5 or in TM6 are indicated in *orange* and *red*, respectively

rearrangement of the two 7TMs (Xue et al. 2015) leading to the activation of only one 7TM at a time in the dimer (Hlavackova et al. 2005). This asymmetry could result from the coupling of a single G-protein to a mGluR 7TM dimer. The G-protein is expected to be coupled to the 7TM which reach the active state (Xue et al. 2015).

Moreover, using a cysteine cross-linking strategy and trFRET experiments, we showed that structural changes at the 7TM interface of mGlu2 dimers cause receptor activation (Xue et al. 2015) (Fig. 7.7b). The main dimer interface is formed by TM4-TM5 in the inactive state and by TM6 in the active state. This major change in the dimer interface is required for receptor activity, because locking the TM4-TM5 interface prevents activation by agonists, while locking the TM6 interface leads to a constitutively active receptor.

Finally, the kinetics of mGlu1 receptor rearrangement was investigated using GFP-based FRET (Hlavackova et al. 2012; Marcaggi et al. 2009; Tateyama and Kubo 2006). The study by Hlavackova et al. revealed a relative movement between interacting 7TM domains during receptor activation (Hlavackova et al. 2012). The relative rearrangement of the two 7TMs was proposed to occur first (time constant of ~35 ms) and then to be followed by activation of a single 7TM within the dimer (time constant of ~50 ms) (Fig. 7.5).

7.6 Molecular Bases of Ligand Efficacy: Partial Agonism

The term « efficacy » was defined as the degree to which different agonists produce varying responses when occupying the same proportion of receptors (Christopoulos et al. 2014). The maximal response that agonists can produce when all their receptors are occupied may vary for different ligands, a partial agonist having a lower efficacy than a full agonist. The molecular bases of such differences can be due either to the stabilization of various active conformations with different coupling efficacies or to the stabilization of only a fraction of the receptors in an active state (Rosenbaum et al. 2009). mGluR partial agonists act by shifting the conformational equilibrium toward the fully active state, to a lower extent compared to full agonists, rather than by stabilizing alternative static conformations (Olofsson et al. 2014; Vafabakhsh et al. 2015). mGluR intramolecular trFRET sensors represent a powerful tool to identify partial agonists (Doumazane et al. 2013) which are otherwise difficult to detect with classical cell signaling assays (Fig. 7.6b).

7.7 Allosteric Modulation of the mGluRs

Reports of the diversity of the ligands that can modulate the activity of mGluR by binding to sites other than the orthosteric site are increasingly numerous. The allosteric ligands acting on the mGluR ECD, called extracellular domain allosteric modulators (Kniazeff et al. 2011), are mostly ions (Tsuchiya et al. 2002; Jiang et al. 2010; Tora et al. 2015) and the monoclonal antibodies (Ullmer et al. 2012). On the contrary, the synthetic compounds developed to act in the mGluR 7TM domain are classically called PAMs and NAMs (Lindsley et al. 2016). Understanding the molecular basis of these different AMs is important for the development of future drugs to modulate mGluR activity in vivo.

The molecular basis for different positive AMs acting in the ECD was proposed for several class C GPCRs including mGluRs. These AMs were shown to stabilize the active conformation of the ECD by facilitating two states: the VFT closure and/or the reorientation of the VFT dimer (Fig. 7.8a). In the umami and sweet taste class C GPCRs, the natural ligands, glutamate, and sweeteners, respectively, bind near the hinge region of the VFT domain and induce initial closure of the two lobes. AM compounds bind near the opening of the VFT binding pocket and further stabilize the closed conformation induced by the natural agonists (Zhang et al. 2008, 2010). However, these kinds of enhancers have not yet been discovered for the mGluRs. Instead, cations and anions were recently shown to play a similar role in stabilizing the ECD dimer active state. Although such ions are not potential drugs, they do highlight the possibility of developing new ligands for mGluR modulation. First, cations such as calcium were shown to potentiate the activation of group I mGluRs (Jiang et al. 2010). The Ca^{2+} binding site is near the entrance of the VFT (Fig. 7.8a).

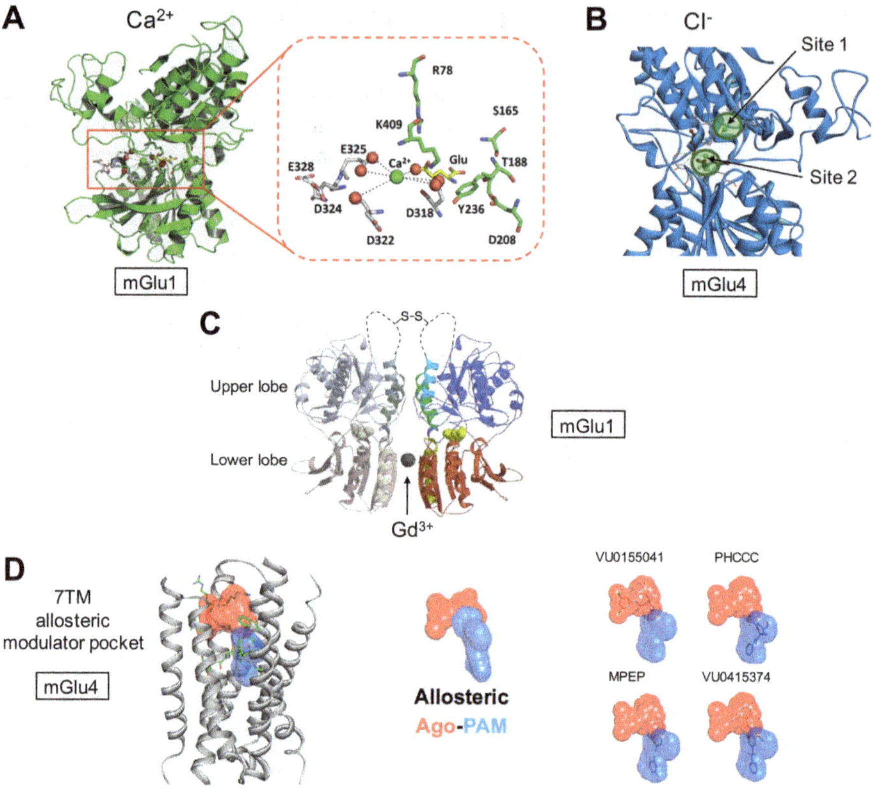

Fig. 7.8 Allosteric modulation of the mGluRs. **(a)** Molecular modeling of the mGlu1 VFT. A close-up is shown on the right with the molecular detail of the Ca^{2+} binding site near the entrance of the VFT binding pocket that was proposed to stabilize the closure of the VFT domain when induced by glutamate (Jiang et al. 2010). **(b)** 3D model of the mGlu4 VFT where the binding of Cl^- has been shown to facilitate the activation of the mGlu4 receptor. While Cl^- binding in the Site 1 is important to stabilize the structure of the VFT, we have proposed that Cl^- binding in the Site 2 (near to the glutamate) would favor VFT closure (Tora et al. 2015). **(c)** Crystal structure of mGlu1 VFT dimer in the active orientation, where the presence of a Gd^{3+} ion has been proposed to stabilize the active state by facilitating the proximity between the two lower lobes (Tsuchiya et al. 2002). **(d)** 3D model of the 7TM AM binding pocket in the mGlu4 receptor. We proposed two overlapping binding pockets, one at the surface (in *red*) and a deeper one (in *blue*). The middle panel highlights the 3D space occupied by these two binding pockets. While the intrinsic agonist activity is proposed to come from the binding in *red* region, cooperativity would mostly depend on the interaction in the deep pocket (Rovira et al. 2015). The right panel shows examples of mGlu4 PAMs that all have agonist activity, except the MPEP, which is predicted to only bind in the deep pocket

More recently, it was shown that chloride ions play an important role in potentiation of all mGluRs and that they confer different degrees of receptor modulation (Tora et al. 2015). Interestingly, for the mGlu4 receptor, in low Cl^- buffer, the receptors had a strong decrease in glutamate potency. A number of binding sites were proposed for Cl^-, including in the VFT, where they facilitate VFT closure (Fig. 7.8b).

Finally, another mechanism to stabilize the active state of the ECD dimer is through the binding of cations at the VFT dimer interface between the two lower lobes. Such cation binding was observed in the crystal structure of the mGlu1 VFT dimer in the active state (Tsuchiya et al. 2002). In this conformation, both VFT are in closed state (conformation Acc) thanks to the presence of one gadolinium (Gd) ion (Fig. 7.8c). It was proposed that the Gd^{3+} acts as an extracellular domain allosteric modulator by facilitating the approach of two negatively charged regions of the lower lobe interface by neutralizing the repulsive charges.

The other AM compounds that bind to the 7TM domain of mGluRs have been well described and extensively developed (Lindsley et al. 2016). 7TM NAMs and PAMs are particularly attractive compounds to generate subtype-selective drug-like mGluR ligands since their binding site are not conserved between mGluR subtypes. In contrast, the development of orthosteric drugs has been hampered by the high conservation of the glutamate binding site within the different mGluRs (Goudet et al. 2012). PAMs and NAMs can modulate the affinity or efficacy of native ligands in positive and negative, respectively (Conn et al. 2009). Some AMs were also reported to be neutral (Christopoulos et al. 2014). The PAMs are of special interest since they can finely tune the activation of the mGluRs by glutamate. In contrast to mGluR agonists, PAMs are expected to enhance receptor activity only when and where needed physiologically (when and where glutamate is produced to act on the mGluRs). The PAMs potentiate the response mediated by orthosteric agonists by stabilizing the ECD active state, a mechanism that originated from the cooperativity between the ECD and 7TM domains (Parmentier et al. 2002). In contrast, the NAMs act as noncompetitive antagonists. We demonstrated this using mGluR intramolecular trFRET sensors, where PAMs further stabilized the active state of the ECD (Doumazane et al. 2013). G-protein coupling to the mGluR 7TM had the same effect (Doumazane et al. 2013) (Fig. 7.6c). Finally, 7TM AMs could modulate just a specific signaling cascade among the several pathways activated by the receptor (Sheffler et al. 2011). This is the concept of functional selectivity or ligand bias signaling (Sheffler et al. 2011; Urban et al. 2007). This stimulus bias was nicely illustrated by a mGlu5 PAM that had an in vivo efficacy in schizophrenia models but that did not potentiate the NMDA receptor function (Rook et al. 2015). This is in contrast to most other mGlu5 PAMs whose NMDA receptor function is potentiated by PAMs.

The crystal structures of mGlu1 and mGlu5 7TMs (Doré et al. 2014; Wu et al. 2014) have provided a 3D framework to understand the molecular recognition of AMs and facilitate structure-based drug design for the mGluRs (Christopher et al. 2015). In these structures, the NAMs bind within a long and narrow pocket in the 7TM domain. This mode of binding is also seen in the orthosteric binding sites of class A GPCRs, although these evolved independently from the class C GPCRs (Nordstrom et al. 2011). Within the mGlu1 and mGlu5 7TM domain, NAMs bind sufficiently deep in the 7TM to make contact with the $Trp^{6.50}$ residue in TM6. This residue, known as the toggle switch residue, is important for 7TM activation and is conserved in all mGluRs as well as in many class A GPCRs (Rosenbaum et al. 2009). PAMs are expected to bind to the same binding pocket as mGluR NAMs

(Lindsley et al. 2016). Indeed, among the AMs that lack selectivity, the same allosteric ligand can have a positive or negative cooperative effect on two distinct mGluRs (Wood et al. 2011). Finally, NAMs and PAMs of a specific mGlu receptor can easily be interconverted by small chemical modifications (Wood et al. 2011; Gomez-Santacana et al. 2014).

Future research is required to compare the physiological benefits of the PAMs with intrinsic agonist activity (called ago-PAMs), versus the compounds that are devoid of this agonist activity (pure PAMs), and to understand why ago-PAMs are agonists. Indeed, while the mGluR PAMs enhance agonist activity (positive cooperativity) by increasing the affinity and/or efficacy of orthosteric agonists, many also display different degrees of intrinsic agonist activity (allosteric agonism). In recent studies, Rook et al. showed that a pure PAM of mGlu5 provides a strong therapeutic benefit compared to a mGlu5 ago-PAM in schizophrenia models (Rook et al. 2012). Such pure mGlu5 PAMs can provide major advantages, avoiding adverse effect observed with mGlu5 agonists. The molecular basis for the difference in effect on the mGluRs between ago-PAM and pure PAM was recently analyzed for the mGlu4 receptor. Rovira et al. proposed that PAMs could bind to two overlapping binding pockets (Rovira et al. 2015) (Fig. 7.8d). The first one, at the entrance to 7TM cleft and analogous to the orthosteric site of class A GPCRs, would be linked to allosteric agonism. The second one, deeper in the 7TM, corresponds to the Na^+ binding pocket of class A GPCRs (Katritch et al. 2014) and would be responsible for the strong positive cooperativity (Rovira et al. 2015). Thus, for mGlu4 receptor the cooperativity is proposed to come from ligand binding in this deep pocket, while the intrinsic efficacy would be mainly due to the interaction in the shallow pocket.

7.8 Innovative Ligands Control the Activity of mGluRs with Light

A major challenge in pharmacology is to enhance drug selectivity to have a therapeutic effect while avoiding side effects. Controlling drug activity with light pulses offers the possibility of enhancing pharmacological selectivity with spatial and temporal resolution. Recent studies have reported the development of innovative ligands that enable the control of mGluRs by light in vivo (Pittolo et al. 2014; Reiner et al. 2015; Rovira et al. 2016). The principle of these studies relies on the use of a ligand containing a photo-switchable chemical moiety that can adopt a high- or low-affinity conformation under the control of light (Fig. 7.9a). Therefore, the activity of such ligands on the mGluRs could be reversibly regulated. Photo-switchable ligands have been developed both for the VFT orthosteric site and for the 7TM allosteric pocket. Two kinds of approaches were developed for the orthosteric site. First, tethered photo-switchable glutamate derivatives were used that are covalently attached to the VFT through an engineered cysteine located on the VFT of the mGluR (Carroll et al. 2015; Levitz et al. 2013). Alternatively, to facilitate their covalent

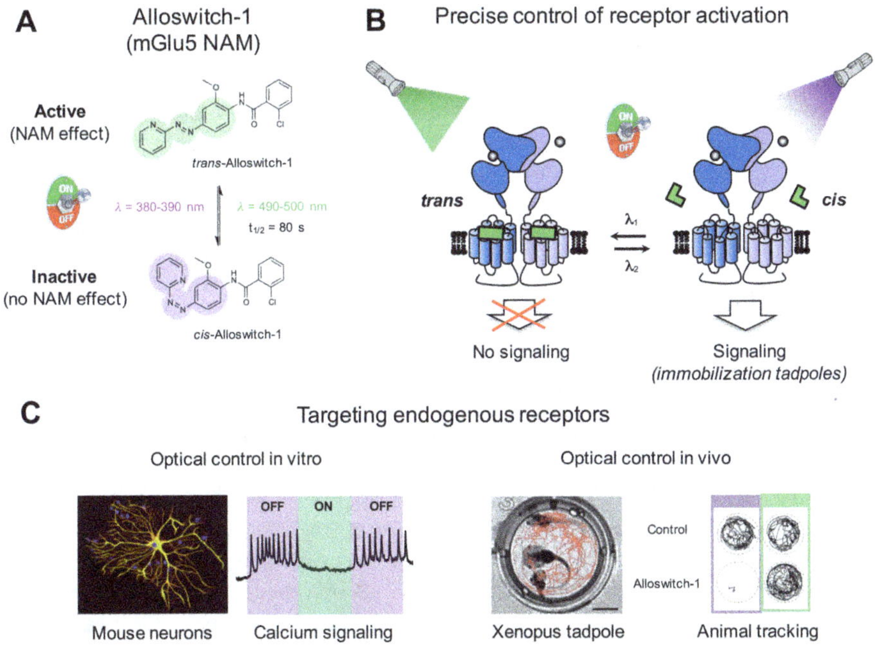

Fig. 7.9 Optopharmacological control of the endogeneous mGluRs. (**a**) Example of a photo-switchable allosteric modulator for the mGlu5 receptor, called Alloswitch-1(Pittolo et al. 2014). Upon ultraviolet light illumination, the compound adopts a *cis* conformation that prevents its NAM effect. The absence of NAM effect of the *cis* conformation allows the basal activity of the mGlu5 receptor to be revealed. Conversely, excitation with *green* light favors the *trans* conformation of the ligand that behaves as a NAM by blocking this basal activity (**b**). (**c**) Alloswitch-1 is efficient both in primary neuronal cultures and in living animals, such as transparent *Xenopus tropicalis* tadpoles

attachment, the photo-switchable glutamate was modified with a SNAP substrate moiety at one end that enabled its covalent binding to the SNAP-tag at the extracellular N-terminal end of mGlu2 (Broichhagen et al. 2015). A long flexible linker was used in this ligand between the photo-switchable glutamate and the SNAP-substrate moieties to enable the attachment of the ligand to the same subunit in the mGluR homodimer.

The second approach relies on the use of an AM that binds to the 7TM of the mGluRs (Fig. 7.9). Pittolo et al. first have established the proof of concept that a photo-switchable NAM of mGluR5 can be used to control mGlu5 receptor activity in vitro and in vivo. This ligand blocked mediated mGlu5 receptor signaling in primary cultures of rat cortical astrocytes, and this blocking can be reversibly controlled by light (Pittolo et al. 2014). Most interestingly, the effect of this photo-switchable drug was also observed in living animals, where light could control the mobility of the transparent *Xenopus tropicalis* tadpoles (Pittolo et al. 2014). Since this first generation of photo-switchable compounds needed ultraviolet light for optical control, which may damage irradiated tissue, the same groups have recently developed a photo-switchable mGluR4 NAMs that work with visible blue light (Rovira et al. 2016). After validating the efficacy of this compound in reversing the analgesic activity of a mGlu4 receptor agonist in a mouse model of chronic pain, they have demonstrated it was possible to control the mobility of zebrafish larvae by blue light illumination (Rovira et al. 2016).

7.9 Conclusion and Future Perspectives

Despite major advances in elucidating the structure of the mGluRs, the structures of the 7TM in the active state and of the full-length dimeric receptor in the inactive and active states remain to be determined. These structures will be important for understanding the functional coupling between the ECD and 7TM domains and the molecular bases of the cooperativity in the 7TM dimer. Further smFRET studies with full-length mGluRs should also help to improve our knowledge of mGluR dynamics and especially the effect of AMs and G-proteins on the ECD. Finally, as well as its potential for drug development, mGluR optopharmacology will be a useful research tool and should open new avenues to elucidate the physiological role of the mGluRs in both normal and pathological conditions.

Acknowledgments We would like to thanks Dr. Francine Acher for providing the mGlu4 receptor figure with chloride. This work was supported by grants from the Agence Nationale de la Recherche (ANR-09-PIRI-0011, ANR-09-BIOT-018 and ANR-13-BSV5-0007), the FRM (Equipe FRM DEQ20130326522) and Fondation Bettencourt Schueller. X.R. received a financial support from the Beatriu de Pinós program of Agència de Gestió d'Ajuts Universitaris i de Recerca (AGAUR).

Conflict of Interest The authors declare no conflicts of interest.

References

Ballesteros J, Weinstein H (1995) Integrated methods for the construction of three-dimensional models of structure-function relations in G protein–coupled receptors. Methods Neurosci 25:366–428

Bessis A-S et al (2002) Closure of the Venus Flytrap module of mGlu8 receptor and the activation process: insights from mutations converting antagonists into agonists. Proc Natl Acad Sci U S A 99:11097–11102

Broichhagen J et al (2015) Orthogonal optical control of a G protein-coupled receptor with a SNAP-tethered photochromic ligand. ACS Cent Sci 1:383–393

Calebiro D et al (2013) Single-molecule analysis of fluorescently labeled G-protein-coupled receptors reveals complexes with distinct dynamics and organization. Proc Natl Acad Sci U S A 110:743–748

Carroll EC et al (2015) Two-photon brightness of azobenzene photoswitches designed for glutamate receptor optogenetics. Proc Natl Acad Sci U S A 112:E776–E785

Cherezov V et al (2007) High-resolution crystal structure of an engineered human beta2-adrenergic G protein-coupled receptor. Science 318:1258–1265
Christopher JA et al (2015) Fragment and structure-based drug discovery for a class C GPCR: discovery of the mGlu5 negative allosteric modulator HTL14242 (3-Chloro-5-[6-(5-fluoropyridin-2-yl)pyrimidin-4-yl]benzonitrile). J Med Chem 58:6653–6664
Christopoulos A et al (2014) International Union of Basic and Clinical Pharmacology. XC. multisite pharmacology: recommendations for the nomenclature of receptor allosterism and allosteric ligands. Pharmacol Rev 66:918–947
Comps-Agrar L et al (2011) The oligomeric state sets GABA(B) receptor signalling efficacy. EMBO J 30:2336–2349
Comps-Agrar L, Kniazeff J, Brock C, Trinquet E, Pin JP (2012) Stability of GABAB receptor oligomers revealed by dual TR-FRET and drug-induced cell surface targeting. FASEB J 26:3430–3439
Conn P, Pin J-P (1997) Pharmacology and functions of metabotropic glutamate receptors. Annu Rev Pharmacol Toxicol 37:205–237
Conn PJ, Christopoulos A, Lindsley CW (2009) Allosteric modulators of GPCRs: a novel approach for the treatment of CNS disorders. Nat Rev Drug Discov 8:41–54
Doré AS et al (2014) Structure of class C GPCR metabotropic glutamate receptor 5 transmembrane domain. Nature 511:557–562
Doumazane E et al (2011) A new approach to analyze cell surface protein complexes reveals specific heterodimeric metabotropic glutamate receptors. FASEB J 25:66–77
Doumazane E et al (2013) Illuminating the activation mechanisms and allosteric properties of metabotropic glutamate receptors. Proc Natl Acad Sci U S A 110:E1416–E1425
El Moustaine D et al (2012) Distinct roles of metabotropic glutamate receptor dimerization in agonist activation and G-protein coupling. Proc Natl Acad Sci U S A 109:16342–16347
Ferraguti F, Shigemoto R (2006) Metabotropic glutamate receptors. Cell Tissue Res 326:483–504
Ferraguti F et al (2005) Metabotropic glutamate receptor 8-expressing nerve terminals target subsets of GABAergic neurons in the hippocampus. J Neurosci 25:10520–10536
Geng Y, Bush M, Mosyak L, Wang F, Fan QR (2013) Structural mechanism of ligand activation in human GABA(B) receptor. Nature 504:254–259
Geng Y et al (2016) Structural mechanism of ligand activation in human calcium-sensing receptor. eLife 5:e13662
Gomez-Santacana X et al (2014) A double effect molecular switch leads to a novel potent negative allosteric modulator of metabotropic glutamate receptor 5. Med Chem Commun 5:1548–1554
Goudet C et al (2012) A novel selective metabotropic glutamate receptor 4 agonist reveals new possibilities for developing subtype selective ligands with therapeutic potential. FASEB J 26:1682–1693
Gregory KJ, Noetzel MJ, Niswender CM (2013) Pharmacology of metabotropic glutamate receptor allosteric modulators: structural basis and therapeutic potential for CNS disorders. Prog Mol Biol Transl Sci 115:61–121
Guo W, Shi L, Filizola M, Weinstein H, Javitch JA (2005) Crosstalk in G protein-coupled receptors: changes at the transmembrane homodimer interface determine activation. Proc Natl Acad Sci U S A 102:17495–17500
Hlavackova V et al (2005) Evidence for a single heptahelical domain being turned on upon activation of a dimeric GPCR. EMBO J 24:499–509
Hlavackova V et al (2012) Sequential inter- and intrasubunit rearrangements during activation of dimeric metabotropic glutamate receptor 1. Sci Signal 5:ra59
Huang S et al (2011) Interdomain movements in metabotropic glutamate receptor activation. Proc Natl Acad Sci U S A 108:15480–15485
Jacquemont S et al (2011) Epigenetic modification of the FMR1 gene in fragile X syndrome is associated with differential response to the mGluR5 antagonist AFQ056. Sci Transl Med 3:64ra1
Jiang Y et al (2010) Elucidation of a novel extracellular calcium-binding site on metabotropic glutamate receptor 1{alpha} (mGluR1{alpha}) that controls receptor activation. J Biol Chem 285:33463–33474

Katritch V et al (2014) Allosteric sodium in class A GPCR signaling. Trends Biochem Sci 39:233–244
Kiyonaka S et al (2016) Allosteric activation of membrane-bound glutamate receptors using coordination chemistry within living cells. Nat Chem 8:958–967
Kniazeff J et al (2004a) Closed state of both binding domains of homodimeric mGlu receptors is required for full activity. Nat Struct Mol Biol 11:706–713
Kniazeff J et al (2004b) Locking the dimeric GABA(B) G-protein-coupled receptor in its active state. J Neurosci 24:370–377
Kniazeff J, Prezeau L, Rondard P, Pin JP, Goudet C (2011) Dimers and beyond: the functional puzzles of class C GPCRs. Pharmacol Ther 130:9–25
Kunishima N et al (2000) Structural basis of glutamate recognition by a dimeric metabotropic glutamate receptor. Nature 407:971–977
Levitz J et al (2013) Optical control of metabotropic glutamate receptors. Nat Neurosci 16:507–516
Levitz J et al (2016) Mechanism of assembly and cooperativity of homomeric and heteromeric metabotropic glutamate receptors. Neuron 92(1):143–159
Lindsley CW et al (2016) Practical strategies and concepts in GPCR allosteric modulator discovery: recent advances with metabotropic glutamate receptors. Chem Rev 116:6707–6741
Manglik A et al (2012) Crystal structure of the micro-opioid receptor bound to a morphinan antagonist. Nature 485:321–326
Marcaggi P, Mutoh H, Dimitrov D, Beato M, Knopfel T (2009) Optical measurement of mGluR1 conformational changes reveals fast activation, slow deactivation, and sensitization. Proc Natl Acad Sci U S A 106:11388–11393
Maurel D et al (2008) Cell-surface protein-protein interaction analysis with time-resolved FRET and snap-tag technologies: application to GPCR oligomerization. Nat Methods 5:561–567
Monn JA et al (2015) Synthesis and pharmacological characterization of C4-disubstituted analogs of 1S,2S,5R,6S-2-aminobicyclo[3.1.0]hexane-2,6-dicarboxylate: identification of a potent, selective metabotropic glutamate receptor agonist and determination of agonist-bound human mGlu2 and mGlu3 amino terminal domain structures. J Med Chem 58:1776–1794
Muto T, Tsuchiya D, Morikawa K, Jingami H (2007) Structures of the extracellular regions of the group II/III metabotropic glutamate receptors. Proc Natl Acad Sci U S A 104:3759–3764
Niswender CM, Conn PJ (2010) Metabotropic glutamate receptors: physiology, pharmacology, and disease. Annu Rev Pharmacol Toxicol 50:295–322
Nordstrom KJ, Sallman Almen M, Edstam MM, Fredriksson R, Schioth HB (2011) Independent HHsearch, Needleman–Wunsch-based, and motif analyses reveal the overall hierarchy for most of the G protein-coupled receptor families. Mol Biol Evol 28:2471–2480
Olofsson L et al (2014) Fine tuning of sub-millisecond conformational dynamics controls metabotropic glutamate receptors agonist efficacy. Nat Commun 5:5206
Parmentier M-L, Prézeau L, Bockaert J, Pin J-P (2002) A model for the functioning of family 3 GPCRs. Trends Pharmacol Sci 23:268–274
Patil ST et al (2007) Activation of mGlu2/3 receptors as a new approach to treat schizophrenia: a randomized Phase 2 clinical trial. Nat Med 13:1102–1107
Pin J-P, Galvez T, Prézeau L (2003) Evolution, structure and activation mechanism of family 3/C G-protein coupled receptors. Pharmacol Ther 98:325–354
Pittolo S et al (2014) An allosteric modulator to control endogenous G protein-coupled receptors with light. Nat Chem Biol 10:813–815
Rasmussen SG et al (2011) Crystal structure of the beta2 adrenergic receptor-Gs protein complex. Nature 477:549–555
Reiner A, Levitz J, Isacoff EY (2015) Controlling ionotropic and metabotropic glutamate receptors with light: principles and potential. Curr Opin Pharmacol 20:135–143
Rondard P, Pin JP (2015) Dynamics and modulation of metabotropic glutamate receptors. Curr Opin Pharmacol 20:95–101
Rondard P et al (2006) Coupling of agonist binding to effector domain activation in metabotropic glutamate-like receptors. J Biol Chem 281:24653–24661

Rondard P, Goudet C, Kniazeff J, Pin JP, Prézeau L (2011) The complexity of their activation mechanism opens new possibilities for the modulation of mGlu and GABAB class C G protein-coupled receptors. Neuropharmacology 60:82–92

Rook JM et al (2012) Unique signaling profiles of positive allosteric modulators of metabotropic glutamate receptor subtype 5 determine differences in in vivo activity. Biol Psychiatry 73(6):501–509

Rook JM et al (2015) Biased mGlu5-positive allosteric modulators provide in vivo efficacy without potentiating mGlu5 modulation of NMDAR currents. Neuron 86:1029–1040

Rosenbaum DM, Rasmussen SG, Kobilka BK (2009) The structure and function of G-protein-coupled receptors. Nature 459:356–363

Rovira X et al (2015) Overlapping binding sites drive allosteric agonism and positive cooperativity in type 4 metabotropic glutamate receptors. FASEB J 29:116–130

Rovira X et al (2016) OptoGluNAM4.1, a photoswitchable allosteric antagonist for real-time control of mGlu4 receptor activity. Cell Chem Biol 23:929–934

Ruprecht JJ, Mielke T, Vogel R, Villa C, Schertler GF (2004) Electron crystallography reveals the structure of metarhodopsin I. EMBO J 23:3609–3620

Schwenk J et al (2010) Native GABA(B) receptors are heteromultimers with a family of auxiliary subunits. Nature 465:231–235

Sheffler DJ, Gregory KJ, Rook JM, Conn PJ (2011) Allosteric modulation of metabotropic glutamate receptors. Adv Pharmacol 62:37–77

Tateyama M, Kubo Y (2006) Dual signaling is differentially activated by different active states of the metabotropic glutamate receptor 1alpha. Proc Natl Acad Sci U S A 103:1124–1128

Tora AS et al (2015) Allosteric modulation of metabotropic glutamate receptors by chloride ions. FASEB J 29:4174–4188

Tsuchiya D, Kunishima N, Kamiya N, Jingami H, Morikawa K (2002) Structural views of the ligand-binding cores of a metabotropic glutamate receptor complexed with an antagonist and both glutamate and Gd^{3+}. Proc Natl Acad Sci U S A 99:2660–2665

Ullmer C et al (2012) Functional monoclonal antibody acts as a biased agonist by inducing internalization of metabotropic glutamate receptor 7. Br J Pharmacol 167:1448–1466

Urban JD et al (2007) Functional selectivity and classical concepts of quantitative pharmacology. J Pharmacol Exp Ther 320:1–13

Vafabakhsh R, Levitz J, Isacoff EY (2015) Conformational dynamics of a class C G-protein-coupled receptor. Nature 524:497–501

Venkatakrishnan AJ et al (2013) Molecular signatures of G-protein-coupled receptors. Nature 494:185–194

Wood MR, Hopkins CR, Brogan JT, Conn PJ, Lindsley CW (2011) "Molecular switches" on mGluR allosteric ligands that modulate modes of pharmacology. Biochemistry 50:2403–2410

Wu B et al (2010) Structures of the CXCR4 chemokine GPCR with small-molecule and cyclic peptide antagonists. Science 330:1066–1071

Wu H et al (2012) Structure of the human kappa-opioid receptor in complex with JDTic. Nature 485:327–332

Wu H et al (2014) Structure of a class C GPCR metabotropic glutamate receptor 1 bound to an allosteric modulator. Science 344:58–64

Xue L et al (2015) Major ligand-induced rearrangement of the heptahelical domain interface in a GPCR dimer. Nat Chem Biol 11:134–140

Yin S et al (2014) Selective actions of novel allosteric modulators reveal functional heteromers of metabotropic glutamate receptors in the CNS. J Neurosci 34:79–94

Zhang F et al (2008) Molecular mechanism for the umami taste synergism. Proc Natl Acad Sci U S A 105:20930–20934

Zhang F et al (2010) Molecular mechanism of the sweet taste enhancers. Proc Natl Acad Sci U S A 107:4752–4757

Zhang C et al (2016) Structural basis for regulation of human calcium-sensing receptor by magnesium ions and an unexpected tryptophan derivative co-agonist. Sci Adv 2:e1600241

Chapter 8
Metabotropic Glutamate Receptor Function in Thalamocortical Circuitry

Thomas E. Salt and Caroline S. Copeland

Abstract The thalamo-corticothalamic network underpins sensory, motor and cognitive processing. All three of the mGlu receptor groups are represented in these pathways. Of the Group I receptors, it seems clear that mGlu1 receptors are located postsynaptically to corticothalamic terminals on dendrites of thalamic relay cells and may function to modulate relay cell transmission under various conditions. Group III mGlu receptors (mGlu4, mGlu7, mGlu8) may also modulate this synapse via a presynaptic mechanism. GABAergic inhibitory processes can be modulated via activation of either Group II (mGlu2 and mGlu3) receptors or Group III receptors, acting either on GABAergic terminals or on thalamic cell bodies, including those in the thalamic reticular nucleus (TRN). There is also evidence that mGlu receptors can modulate astrocyte function in the thalamus. The heterogeneous expression of mGlu receptors at specific sites within the thalamus makes them potential therapeutic targets for a variety of conditions including pain, epilepsy and cognitive disorders.

Keywords Thalamus • Cortex • Thalamic reticular nucleus • GABA inhibition • Pain • Epilepsy • Cognition • Schizophrenia • Psychiatry

8.1 Introduction

The thalamus plays a pivotal role in the integration and processing of sensory and motor information and in cognitive processing. Metabotropic glutamate (mGlu) receptors of all three groups are widespread amongst the different cell types and afferents to

Thomas E. Salt, MA, PhD, FBPhS (✉)
UCL Institute of Ophthalmology, 11-43 Bath Street, London EC1V 9EL, UK
e-mail: t.salt@ucl.ac.uk

Caroline S. Copeland, BSc, MRes, PhD
St George's, University of London, Cranmer Terrace, London SW17 0RE, UK
e-mail: c.copeland@sgul.ac.uk

R.T. Ngomba et al. (eds.), *mGLU Receptors*, The Receptors 31,
DOI 10.1007/978-3-319-56170-7_8

the thalamus and thus are well placed to control and modulate synaptic processing at a local microcircuit level as well as at a more global macro-circuit level. This chapter reviews the location and function of mGlu receptors in the different elements of the thalamus and attempts to provide an integrated overview of their function in health and disease. This is followed by an exploration of mGlu receptors as potential targets for disease treatment and therapy in psychiatric, neurological and other conditions.

8.2 The Thalamus

The thalamus functions as a processing station and filter of information in conjunction with the neocortex. With over 30 distinct nuclear groups, each relaying a characteristic signal to a functionally distinct area of the cortex, the thalamus can be considered as concerned with almost, if not all, functional modalities, with the probable exception of olfaction (Guillery and Sherman 2002; Sherman and Guillery 2011). Excitatory inputs to the thalamus can be categorised as either *driver inputs* or *modulator inputs* on the basis of terminal morphology, synaptic location and synaptic pharmacology (Sherman and Guillery 2011). Driver inputs typically have large round (RL) terminals that contact relay neurones close to the soma, whereas modulator inputs have small round (RS) terminals that contact relay neurones more distally in the dendritic tree (Guillery and Sherman 2002; Sherman and Guillery 2011). Based on this distinction of inputs, thalamic nuclei can therefore be classed as either *first-order* or *higher-order* nuclei based upon the source of their *driver* inputs: first-order nuclei receive driver inputs from the periphery (e.g. auditory, visual, somatosensory), whilst higher-order nuclei receive driver inputs from cortical layer V (Mitchell et al. 2014; Sherman 2007; Sherman and Guillery 2011) (Fig. 8.1). First-order thalamic nuclei can thus be considered as processing information from the peripheral sensory inputs, whereas higher-order thalamic nuclei process information that they receive from layer V of the cortex.

Common to all thalamic nuclei, thalamocortical afferents project to layer IV of the cortex and also receive reciprocal *modulatory* corticothalamic inputs from layer VI, which modulate how information from *driver* inputs is transmitted (Sherman 2007). Both thalamocortical and corticothalamic afferents also innervate the associated thalamic reticular nucleus (TRN), which serves to provide both feedback and feedforward inhibition to thalamic nuclei upon thalamocortical and corticothalamic innervation, respectively (Crabtree 1999; Jones 1985; Pinault and Deschenes 1998a, b; Salt 1989; Shosaku et al. 1989) (Fig. 8.1). Molecular, physiological and pharmacological studies in thalamic nuclei have provided substantial evidence that mGlu receptors are present and functional within thalamocortical circuitry, as described below.

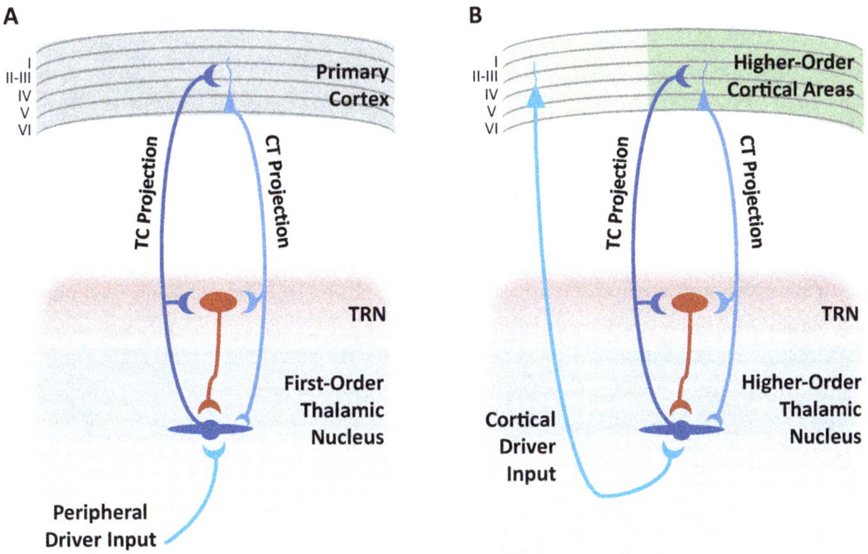

Fig. 8.1 Basic simplified circuitry formed between (**a**) first-order and (**b**) higher-order thalamic nuclei, cortex and thalamic reticular nucleus (TRN). First-order nuclei receive driver inputs from peripheral sources (**a**), with higher-order nuclei receiving driver inputs from pyramidal cells in cortical layer V (**b**). All thalamic nuclei project thalamocortical (TC) afferents to layer IV of the cortex and also receive reciprocal *modulatory* corticothalamic (CT) inputs from layer VI pyramidal cells in the same cortical area. Both TC and CT afferents also innervate the associated TRN, which serves to provide feedback inhibition to thalamic nuclei

8.3 Localisation of mGlu Receptors in Thalamocortical Circuitry

8.3.1 Group I Receptors

Evidence from mRNA and ultrastructural studies indicates that the Group I mGlu receptors are located postsynaptically on the distal dendrites of thalamic neurons beneath terminals of corticothalamic axons in various thalamic nuclei and mammalian species (Godwin et al. 1996; Liu et al. 1998; Martin et al. 1992; Vidnyanszky et al. 1996; Zikopoulos and Barbas 2007) and on the dendrites of local circuit interneurons presynaptic to thalamic neurons (Godwin et al. 1996). mGlu5 receptors may also be expressed on glial cells within the thalamus, as there is evidence to suggest that these receptors may contribute to glia activation upon synaptic input to the somatosensory ventrobasal thalamus (Parri et al. 2010).

8.3.2 Group II Receptors

mRNA analysis in the rat and protein staining in mGlu2-deficient mice have revealed that mGlu3 receptors are heavily localised within GABAergic terminals of the TRN (Neto et al. 2000a; Ohishi et al. 1993a, b; Tamaru et al. 2001). In contrast to this high mGlu3 localisation, mGlu2 has a diffuse staining pattern (Neki et al. 1996): mRNA analysis showed no significant mGlu2 signal in thalamic nuclei or the TRN; however, mGlu2 mRNA was detected in layer VI of S1 cortex (Ohishi et al. 1993b). Ultrastructural studies also provide evidence that Group II receptors are localised on glial processes, some of which appear to surrounding GABAergic TRN terminals in rodent thalamic nuclei (Liu et al. 1998; Mineff and Valtschanoff 1999). Taken together, these results suggest that mGlu3 receptors are localised primarily on TRN axon terminals, that mGlu2 receptors are primarily on glutamatergic thalamocortical axon terminals and that both subtypes may be present on glial processes in thalamic nuclei.

8.3.3 Group III Receptors

The Group III mGlu receptors (except for mGlu6) likely participate in the control of synaptic transmission throughout the thalamocortical network: ultrastructural and electrophysiological studies indicate mGlu4 expression on corticothalamic terminals in both the TRN and thalamic nuclei (Ngomba et al. 2008; Turner and Salt 1999), with mGlu7 and mGlu8 receptors indicated to be expressed on TRN neurons, both postsynaptically to thalamocortical and corticothalamic afferents and also presynaptically on TRN terminals in thalamic nuclei (Bradley et al. 1998, 1999; Corti et al. 1998, 2002; Kinoshita et al. 1998; Kyuyoung and Huguenard 2014; Neto et al. 2000a; Ohishi et al. 1995; Saugstad et al. 1997; Turner and Salt 2003).

8.4 mGlu Function in Thalamocortical Circuitry in Health and Disease

The circuitry formed between the thalamus, cortex and TRN ensures the generation of synchronous activity required for sensory perception and the preparation and execution of distinct motor and/or cognitive tasks. The presence of mGlu receptors throughout synapses within this circuitry is therefore of great functional significance as it is important to understand how afferent inputs to the thalamus are controlled in order to understand how neurophysiological disease states associated with thalamic malfunction precipitate. Therefore, we shall now consider the functional roles of mGlu receptors located at distinct sites within the thalamocortical circuitry (Fig. 8.2).

The Driver Input to Thalamocortical Neurons There is little evidence to suggest that any of the mGlu receptor subtypes are involved in the relay of inputs from driver afferents to thalamocortical neurons; rather, this fast transmission is mediated

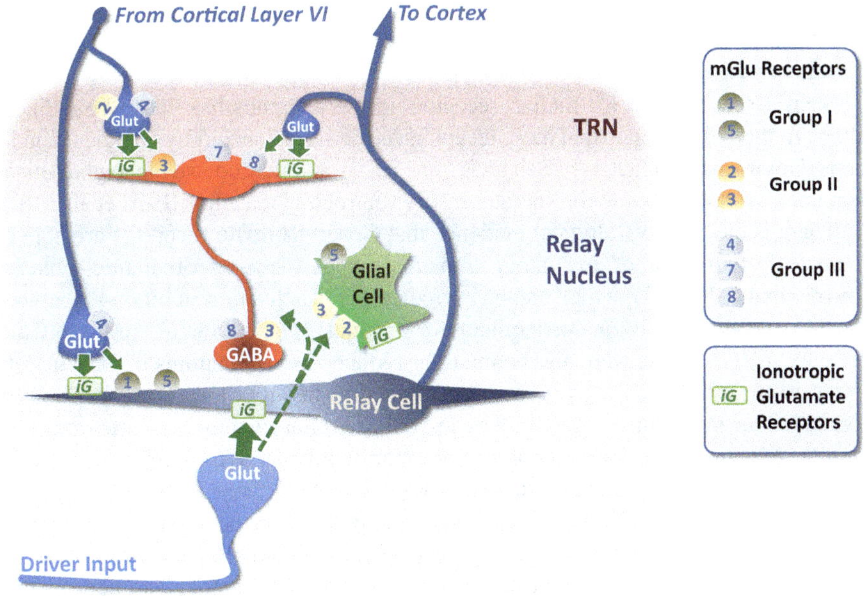

Fig. 8.2 Simplified synaptic circuitry formed within thalamic nuclei, with major inputs and outputs. Locations of ionotropic glutamate receptors and metabotropic glutamate receptor subtypes are indicated. Glutamate release is indicated by green arrows (*Dashed arrows* indicate possible synaptic spillover). For clarity, GABA release is not shown

via the ionotropic glutamate receptors, NMDA and AMPA receptors (Guillery and Sherman 2002; Salt and Eaton 1996). By contrast, in thalamic nuclei where there are intrinsic Golgi II interneurons, Group I mGlu receptors appear to be involved in the mediation of responses of these interneurons to driver afferents (Errington et al. 2011) (see below), and mGlu1 receptor activation may reduce sensory driver input to relay neurons via a presynaptic mechanism (Govindaiah et al. 2012a).

The Modulatory Corticothalamic Input to Thalamic Neurons Extensive study of the corticothalamic projection to the thalamus has revealed that this input is likely relayed by NMDA receptors and modulated by postsynaptic mGlu1/5 receptors and presynaptic mGlu4 receptors.

Activation of the Group I mGlu receptors in thalamic neurons in vitro causes a slow depolarising response associated with an increase in membrane resistance (Govindaiah et al. 2012b; Turner and Salt 2000). Relatively small corticothalamic synaptic potentials can be attributed to mGlu1 receptor activation in vitro (Hughes et al. 2002; Reichova and Sherman 2004; Turner and Salt 2000). However, in vivo studies have indicated that mGlu1 receptor activation can exert a larger influence on the relay of corticothalamic inputs than might be predicted from in vitro studies, as mGlu1 receptor activation has been demonstrated to potentiate thalamic neuron responses mediated via ionotropic glutamate receptors (Salt and Binns 2000; Salt et al. 2012). This facilitation of ionotropic glutamate receptor function upon Group I mGlu receptor activation may be of importance during nociceptive processing in the thalamus (Salt and Binns 2000; Salt et al. 2014). Indeed, thalamic mGlu1 receptor mRNA levels are reduced in experimental arthritis in rats (Neto et al. 2000b), and this may be an adaptation to the increased nociceptive input ascending from the periphery to the thalamus.

Such a specific role for mGlu5 receptors in the thalamus has yet to be defined, although there is evidence that these receptors are activated under physiological conditions (Govindaiah et al. 2012b; Salt and Binns 2000), and may contribute to glia activation upon synaptic input to the somatosensory ventrobasal thalamus (Parri et al. 2010).

There is electrophysiological evidence that Group III mGlu receptors are able to mediate a reduction of excitatory transmission at the corticothalamic synapse (Turner and Salt 1999), which can be attributed to the activation of mGlu4 receptors located presynaptically on corticothalamic terminals (Ngomba et al. 2008). mGlu4 receptor mRNA levels were also found to be reduced in the thalamus in arthritic rats (Neto et al. 2000b). However, whilst a reduction in mGlu1 receptor expression would reduce nociceptive signalling by impeding excitatory glutamatergic transmission, a reduction in mGlu4 receptor expression would serve to increase excitatory transmission of corticothalamic nociceptive inputs. Selective targeting to decrease mGlu1 and increase mGlu4 receptor function may therefore be the most appropriate anti-nociceptive therapeutic strategy. Indeed, Group I mGlu receptor antagonists are of therapeutic potential for the treatment of chronic inflammatory and neuropathic pain states (Lesage 2004; Nicoletti et al. 2011), and agonists targeting the mGlu4 receptors possess analgesic/anti-hyperalgesic qualities (Goudet et al. 2008).

The GABAergic Input to Thalamocortical Neurons The location of Group II and III receptors on TRN terminals and surrounding astrocytic processes indicates they play a pivotal role in modulating inhibition in the thalamus. As TRN terminals are not directly targeted by synaptically released glutamate, these receptors may be activated by endogenous 'glutamate spillover' (Kullmann 2000) from synapses formed between excitatory driver afferents and thalamocortical neurons upon normal sensory processing (Copeland et al. 2012) or under conditions of intense synaptic activation (Alexander and Godwin 2006; Ngomba et al. 2008, 2011).

There is substantial evidence from in vitro and in vivo electrophysiological studies that upon driver afferent stimulation, the Group II mGlu receptors reduce inhibition in thalamic nuclei, likely via a reduction in GABAergic transmission from the TRN (Alexander and Godwin 2005; Copeland et al. 2012, 2015; Salt and Eaton 1995; Salt and Turner 1998; Turner and Salt 2003). In first-order sensory nuclei, this mechanism is likely mediated via both mGlu2 and mGlu3 receptors and could play an important role in discerning relevant information from background activity to enhance sensory perception. In a higher-order nucleus, the mediodorsal thalamus, this mechanism may be mediated solely via the mGlu3 receptor (Copeland et al. 2015), indicating that Group II mGlu receptor distribution across first- and higher-order thalamic nuclei is not uniform. Functionally, this action of the mGlu3 receptor may be of therapeutic importance as increased inhibition in the mediodorsal thalamus has been associated with cognitive deficit onset (Parnaudeau et al. 2013). Indeed, the mGlu3 receptor has been implicated in the aetiological, pathophysiological and pharmacotherapeutic aspects of schizophrenia (Harrison et al. 2008), with polymorphisms in the mGlu3 receptor gene and protein, but not the mGlu2 receptor, detected in patients with schizophrenia (Cherlyn et al. 2010; Ghose et al. 2009; Mounce et al. 2014). The design of future novel therapies targeted to treat deficits in cognitive function may therefore achieve greater success if selectivity and higher efficacy for mGlu3 receptors were achieved.

Group III receptors on TRN presynaptic terminals have also been demonstrated to mediate a presynaptic reduction of TRN-mediated inhibition in thalamic nuclei upon driver afferent stimulation (Turner and Salt 2003) and therefore likely act in concert with the Group II mGlu receptor mechanism to enable relevant information to be discerned from background activity. This is likely mediated via mGlu7 receptors on presynaptic TRN terminals (Kyuyoung and Huguenard 2014). However, unlike the Group II mGlu receptors, their function has not yet been probed in higher-order thalamic nuclei.

In addition to the TRN input onto thalamocortical cells, most thalamic nuclei (apart from the well-described rodent ventrobasal thalamic circuit) also contain intrinsic GABAergic Golgi II interneurons. Their function may be particularly important in the lateral geniculate nucleus (LGN) of the thalamus, the thalamic nucleus that processes visual information. It has been shown that activation of mGlu1 and mGlu5 receptors on these interneurons can increase GABA release onto synaptic and extrasynaptic $GABA_A$ receptors on thalamocortical neurons (Errington et al. 2011; Govindaiah et al. 2012a, b). Furthermore, it appears that these Group I mGlu receptors can be activated by high-frequency activity of the retinal (driver) inputs to the LGN, thus representing a mechanism for local regulation of inhibition during high-frequency input from the retina, which may be important in stimulus detection and visual processing.

The Corticothalamic/Thalamocortical Inputs to TRN Neurons The excitatory inputs from thalamocortical and corticothalamic neurons to the TRN serve to induce feedback and feedforward inhibition to thalamic nuclei, respectively. A number of subtypes from both the Group II and Group III receptor families are located at both pre- and postsynaptic sites, and evidence from a number of studies indicates that they appear to be functioning collectively to reduce TRN neuron activity (Alexander and Godwin 2006; Cox and Sherman 1999; Crabtree et al. 2013; Kyuyoung and Huguenard 2014; Ngomba et al. 2008; Turner and Salt 1999, 2003), which results in reduced GABAergic transmission from the TRN to thalamic nuclei. An important observation may be that Group III receptors appear to presynaptically modulate the cortical input but not the thalamic input to TRN and that this may be via mGlu8 (Kyuyoung and Huguenard 2014). Similarly, mGlu2 receptors appear to tonically regulate glutamate release from cortical afferents in TRN (Crabtree et al. 2013). Furthermore, the GABAergic TRN output onto thalamic relay cells appears to be reduced by presynaptic mGlu7 receptor activation (Kyuyoung and Huguenard 2014).

Within the TRN, postsynaptically localised mGlu3 receptors may act to reduce excitatory transmission from corticothalamic and thalamocortical afferents, therefore reducing activation of TRN neurons and decreasing inhibitory transmission to thalamic nuclei (Cox and Sherman 1999). This mGlu3 receptor function within the TRN therefore complements that of the Group II mGlu receptors located on TRN presynaptic terminals (Copeland et al. 2012; Salt and Turner 1998).

Maladaptation of the Group III receptors within the TRN has been heavily implicated in the manifestation of absence epilepsy. Absence seizures are generated by an abnormal oscillatory activity of a thalamocortical loop comprising the ventrobasal thalamus and its associated cortical area and TRN sector (Blumenfeld 2005). Evidence from a number of studies indicates that compounds antagonising the mGlu4 receptor may possess antiepileptic therapeutic profiles: genetic deletion of the mGlu4 receptor or intra-TRN injection of Group III mGlu receptor antagonists protects mice against pentylenetetrazole-induced absence seizures (Snead et al. 2000) and WAG/Rij rats, which spontaneously develop absence epilepsy at 2–3 months of age and display increased mGlu4 receptor expression in the TRN, which, when challenged with mGlu4 receptor agonists, respond with an increase in the frequency of absence seizures (Ngomba et al. 2008). Interestingly, the mGlu4 receptor gene is localised within a susceptibility focus for juvenile myoclonic epilepsy (Izzi et al. 2003; Wong et al. 2001), which is characterised by absence seizures combined with myoclonic and tonic-clonic seizures. There is also evidence to suggest that mGlu7 receptor activation is protective against paroxysmal activity of the thalamocortical network underlying absence seizures: genetically altered mice with disrupted interactions between mGlu7 receptors and PICK1 develop absence seizures and are more sensitive to pentylenetetrazole (Bertaso et al. 2008). Taken together, the most appropriate therapeutic strategy for modulating Group III mGlu receptor activity with regard to absence epilepsy would therefore be to selectively target to decrease mGlu4 and increase mGlu7 receptor function.

8.5 Summary

Metabotropic glutamate receptor expression is heterogeneic throughout the thalamocortical network. This can be exploited to target discrete network components with pharmacological compounds specific for mGlu receptor subtypes. This could lead to the development of novel therapeutic strategies to ameliorate pathophysiological thalamocortical function, such as that described in chronic pain, epilepsy and schizophrenia.

References

Alexander GM, Godwin DW (2005) Presynaptic inhibition of corticothalamic feedback by metabotropic glutamate receptors. J Neurophysiol 94:163–175

Alexander GM, Godwin DW (2006) Unique presynaptic and postsynaptic roles of Group II metabotropic glutamate receptors in the modulation of thalamic network activity. Neuroscience 141:501–513

Bertaso F et al (2008) PICK1 uncoupling from mGluR7a causes absence-like seizures. Nat Neurosci 11:940–948

Blumenfeld H (2005) Cellular and network mechanisms of spike-wave seizures. Epilepsia 46(Suppl 9):21–33

Bradley SR et al (1998) Distribution and developmental regulation of metabotropic glutamate receptor 7a in rat brain. J Neurochem 71:636–645
Bradley SR et al (1999) Immunohistochemical localization of subtype 4a metabotropic glutamate receptors in the rat and mouse basal ganglia. J Comp Neurol 407:33–46
Cherlyn SY et al (2010) Genetic association studies of glutamate, GABA and related genes in schizophrenia and bipolar disorder: a decade of advance. Neurosci Biobehav Rev 34:958–977
Copeland CS, Neale SA, Salt TE (2012) Positive allosteric modulation reveals a specific role for mGlu2 receptors in sensory processing in the thalamus. J Physiol 590:937–951
Copeland CS, Neale SA, Salt TE (2015) Neuronal activity patterns in the mediodorsal thalamus and related cognitive circuits are modulated by metabotropic glutamate receptors. Neuropharmacology 92:16–24
Corti C et al (1998) Cloning and characterization of alternative mRNA forms for the rat metabotropic glutamate receptors mGluR7 and mGluR8. Eur J Neurosci 10:3629–3641
Corti C et al (2002) Distribution and synaptic localisation of the metabotropic glutamate receptor 4 (mGluR4) in the rodent CNS. Neuroscience 110:403–420
Cox CL, Sherman SM (1999) Glutamate inhibits thalamic reticular neurons. J Neurosci 19:6694–6699
Crabtree JW (1999) Intrathalamic sensory connections mediated by the thalamic reticular nucleus. Cell Mol Life Sci 56:683–700
Crabtree JW et al (2013) GABAA, NMDA and mGlu2 receptors tonically regulate inhibition and excitation in the thalamic reticular nucleus. Eur J Neurosci 37:850–859
Errington AC et al (2011) mGluR control of interneuron output regulates feedforward tonic gabaa inhibition in the visual thalamus. J Neurosci 31:8669–8680
Ghose S et al (2009) Differential expression of metabotropic glutamate receptors 2 and 3 in schizophrenia: a mechanism for antipsychotic drug action? Am J Psychiatr 166:812–820
Godwin DW et al (1996) Ultrastructural localization suggests that retinal and cortical inputs access different metabotropic glutamate receptors in the lateral geniculate nucleus. J Neurosci 16:8181–8192
Goudet C et al (2008) Group III metabotropic glutamate receptors inhibit hyperalgesia in animal models of inflammation and neuropathic pain. Pain 137:112–124
Govindaiah G et al (2012a) Activity-dependent regulation of retinogeniculate signaling by metabotropic glutamate receptors. J Neurosci 32:12820–12831
Govindaiah G et al (2012b) Spatially distinct actions of metabotropic glutamate receptor activation in dorsal lateral geniculate nucleus. J Neurophysiol 107:1157–1163
Guillery RW, Sherman SM (2002) Thalamic relay functions and their role in corticocortical communication: generalizations from the visual system. Neuron 33:163–175
Harrison PJ et al (2008) Review: the group II metabotropic glutamate receptor 3 (mGluR3, mGlu3, GRM3): expression, function and involvement in schizophrenia. J Psychopharmacol 22:308–322
Hughes SW et al (2002) Cellular mechanisms of the slow (< 1 Hz) oscillation in thalamocortical neurons in vitro. Neuron 33:947–958
Izzi C et al (2003) Candidate gene analysis of the human metabotropic glutamate receptor type 4 (GRM4) in patients with juvenile myoclonic epilepsy. Am J Med Genet B Neuropsychiatr Genet 123B:59–63
Jones EG (1985) The thalamus. Plenum Press, New York
Kinoshita A et al (1998) Immunohistochemical localization of metabotropic glutamate receptors, mGluR7a and mGluR7b, in the central nervous system of the adult rat and mouse: a light and electron microscopic study. J Comp Neurol 393:332–352
Kullmann DM (2000) Spillover and synaptic cross talk mediated by glutamate and GABA in the mammalian brain. Prog Brain Res 125:339–351
Kyuyoung CL, Huguenard JR (2014) Modulation of short-term plasticity in the corticothalamic circuit by group III metabotropic glutamate receptors. J Neurosci 34:675–687
Lesage ASJ (2004) Role of Group I metabotropic glutamate receptors mGlu1 and mGlu5 in nociceptive signalling. Curr Neuropharmacol 2:363–393

Liu XB, Munoz A, Jones EG (1998) Changes in subcellular localization of metabotropic glutamate receptor subtypes during postnatal development of mouse thalamus. J Comp Neurol 395:450–465

Martin LJ et al (1992) Cellular localization of a metabotropic glutamate receptor in rat brain. Neuron 9:259–270

Mineff E, Valtschanoff J (1999) Metabotropic glutamate receptors 2 and 3 expressed by astrocytes in rat ventrobasal thalamus. Neurosci Lett 270:95–98

Mitchell AS et al (2014) Advances in understanding mechanisms of thalamic relays in cognition and behavior. J Neurosci 34:15340–15346

Mounce J et al (2014) Association of GRM3 polymorphism with white matter integrity in schizophrenia. Schizophr Res 155:8–14

Neki A et al (1996) Pre- and postsynaptic localization of a metabotropic glutamate receptor, mGluR2, in the rat brain: an immunohistochemical study with a monoclonal antibody. Neurosci Lett 202:197–200

Neto FL et al (2000a) Differential distribution of metabotropic glutamate receptor subtype mRNAs in the thalamus of the rat. Brain Res 854:93–105

Neto FL et al (2000b) Expression of metabotropic glutamate receptors mRNA in the thalamus and brainstem of monoarthritic rats. Mol Brain Res 81:140–154

Ngomba RT et al (2008) Positive allosteric modulation of metabotropic glutamate 4 (mGlu4) receptors enhances spontaneous and evoked absence seizures. Neuropharmacology 54:344–354

Ngomba RT et al (2011) Metabotropic glutamate receptors in the thalamocortical network: strategic targets for the treatment of absence epilepsy. Epilepsia 53:1211–1222

Nicoletti F et al (2011) Metabotropic glutamate receptors: from the workbench to the bedside. Neuropharmacology 60:1017–1041

Ohishi H et al (1993a) Distribution of the mRNA for a metabotropic glutamate receptor (mGluR3) in the rat brain: an in situ hybridization study. J Comp Neurol 335:252–266

Ohishi H et al (1993b) Distribution of the messenger RNA for a metabotropic glutamate receptor, mGluR2, in the central nervous system of the rat. Neuroscience 53:1009–1018

Ohishi H et al (1995) Distributions of the mRNAs for L-2-amino-4-phosphonobutyrate-sensitive metabotropic glutamate receptors, mGluR4 and mGluR7, in the rat brain. J Comp Neurol 360:555–570

Parnaudeau S et al (2013) Inhibition of mediodorsal thalamus disrupts thalamofrontal connectivity and cognition. Neuron 77:1151–1162

Parri HR, Gould TM, Crunelli V (2010) Sensory and cortical activation of distinct glial cell subtypes in the somatosensory thalamus of young rats. Eur J Neurosci 32:29–40

Pinault D, Deschenes M (1998a) Anatomical evidence for a mechanism of lateral inhibition in the rat thalamus. Eur J Neurosci 10:3462–3469

Pinault D, Deschenes M (1998b) Projection and innervation patterns of individual thalamic reticular axons in the thalamus of the adult rat: a three-dimensional, graphic, and morphometric analysis. J Comp Neurol 391:180–203

Reichova I, Sherman SM (2004) Somatosensory corticothalamic projections: distinguishing drivers from modulators. J Neurophysiol 92:2185–2197

Salt TE (1989) Gamma-aminobutyric acid and afferent inhibition in the cat and rat ventrobasal thalamus. Neuroscience 28:17–26

Salt TE, Binns KE (2000) Contributions of mGlu1 and mGlu5 receptors to interactions with N-methyl-D-aspartate receptor-mediated responses and nociceptive sensory responses of rat thalamic neurones. Neuroscience 100:375–380

Salt TE, Eaton SA (1995) Distinct presynaptic metabotropic receptors for L-AP4 and CCG1 on GABAergic terminals: pharmacological evidence using novel a-methyl derivative mGluR antagonists, MAP4 and MCCG, in the rat thalamus in vivo. Neuroscience 65:5–13

Salt TE, Eaton SA (1996) Functions of ionotropic and metabotropic glutamate receptors in sensory transmission in the mammalian thalamus. Prog Neurobiol 48:55–72

Salt TE, Turner JP (1998) Modulation of sensory inhibition in the ventrobasal thalamus via activation of Group II metabotropic glutamate receptors (mGluRs) by (2R,4R)-APDC. Exp Brain Res 121:181–185

Salt TE et al (2012) Potentiation of sensory responses in ventrobasal thalamus in vivo via selective modulation of mGlu1 receptors with a positive allosteric modulator. Neuropharmacology 62:1695–1699

Salt TE et al (2014) Function of mGlu1 receptors in the modulation of nociceptive processing in the thalamus. Neuropharmacology 79:405–411

Saugstad JA et al (1997) Cloning and expression of rat metabotropic glutamate receptor 8 reveals a distinct pharmacological profile. Mol Pharmacol 51:119–125

Sherman SM (2007) The thalamus is more than just a relay. Curr Opin Neurobiol 17:417–422

Sherman SM, Guillery RW (2011) Distinct functions for direct and transthalamic corticocortical connections. J Neurophysiol 106:1068–1077

Shosaku A et al (1989) Analysis of recurrent inhibitory circuit in rat thalamus: neurophysiology of the thalamic reticular nucleus. Prog Neurobiol 32:77–102

Snead OC et al (2000) Modulation of absence seizures by the GABA(A) receptor: a critical role for metabotropic glutamate receptor 4 (mGluR4). J Neurosci 20:6218–6224

Tamaru Y et al (2001) Distribution of metabotropic glutamate receptor mglur3 in the mouse cns: differential location relative to pre- and postsynaptic sites. Neuroscience 106:481–503

Turner JP, Salt TE (1999) Group III metabotropic glutamate receptors control corticothalamic synaptic transmission in the rat thalamus in vitro. J Physiol 519:481–491

Turner JP, Salt TE (2000) Synaptic activation of the Group I metabotropic glutamate receptor mGlu1 on the thalamocortical neurones of the rat dorsal lateral geniculate nucleus in vitro. Neuroscience 100:493–505

Turner JP, Salt TE (2003) Group II and III metabotropic glutamate receptors and the control of the TRN input to rat thalamocortical neurones in vitro. Neuroscience 122:459–469

Vidnyanszky Z et al (1996) Immunohistochemical visualization of the mGluR1a metabotropic glutamate receptor at synapses of corticothalamic terminals originating from area 17 of the rat. Eur J Neurosci 8:1061–1071

Wong CG-T et al (2001) Localization of the human mGluR4 gene within an epilepsy susceptibility locus1. Mol Brain Res 87:109–116

Zikopoulos B, Barbas H (2007) Parallel driving and modulatory pathways link the prefrontal cortex and thalamus. PLoS One 2:e848

Chapter 9
Metabotropic Glutamate Receptors in Cancer

Lumeng J. Yu, Brian A. Wall, Janet Wangari-Talbot, and Suzie Chen

Abstract Metabotropic glutamate receptors (mGluRs) are widely known for their roles in synaptic signaling. However, accumulating evidence suggests roles of mGluRs in human malignancies in addition to synaptic transmission. Somatic cell homeostasis presents intriguing possibilities of mGluRs and glutamate signaling as novel targets for human cancers. More recently, aberrant glutamate signaling has been shown to participate in the transformation and maintenance of various cancer types, including glioma, melanoma skin cancer, breast cancer, and prostate cancer, indicating that genes encoding mGluRs, GRMs, can function as oncogenes. Here, we provide a review on the interactions of mGluRs and their ligand, glutamate, in processes that promote the growth of tumors of neuronal and non-neuronal origins. Further, we discuss the evolution of riluzole, a glutamate release inhibitor approved for amyotrophic lateral sclerosis (ALS), but now fashioned as an mGluR1 inhibitor for melanoma therapy and as a radiosensitizer for tumors that have metastasized to the brain. With the success of riluzole, it is not farfetched to believe that other drugs that may act directly or indirectly on other mGluRs can be beneficial for multiple applications.

Keywords Glutamate • G-protein-coupled receptors • Cell transformation • Therapeutic target

L.J. Yu • J. Wangari-Talbot
Susan Lehman Cullman Laboratory for Cancer Research, Ernest Mario School of Pharmacy, Rutgers University, Piscataway, NJ 08854, USA

B.A. Wall
Susan Lehman Cullman Laboratory for Cancer Research, Ernest Mario School of Pharmacy, Rutgers University, Piscataway, NJ 08854, USA

Global Product Safety, Colgate-Palmolive Company, Piscataway, NJ 08854, USA

S. Chen, Ph.D. (✉)
The Cancer Institute of New Jersey,
195 Little Albany Street, New Brunswick, NJ 08903, USA

Susan Lehman Cullman Laboratory for Cancer Research, Ernest Mario School of Pharmacy, Rutgers University, Piscataway, NJ 08854, USA
e-mail: Suziec@pharmacy.rutgers.edu

R.T. Ngomba et al. (eds.), *mGLU Receptors*, The Receptors 31,
DOI 10.1007/978-3-319-56170-7_9

9.1 Glutamine and Glutamate

Glutamate plays an innate role in the human central nervous system as an excitatory neurotransmitter in processes such as learning and memory formation (Alix and Domingues 2011; Fairman and Amara 1999). Its role in cellular homeostasis is related to both its function in nitrogen metabolism and disposal, as well as its use as a metabolic fuel for energy-producing pathways (Dimski et al. 2008; Kelly and Stanley 2001; Spanaki and Plaitakis 2012). In addition, it has long been established that excess glutamate causes neuronal excitotoxicity; more recent findings have also implicated functional glutamate signaling in transformation and progression of various cancers (Prickett and Samuels 2012; Ribeiro et al. 2010; Willard and Koochekpour 2013). Glutamine is preferred as fuel in tumor cells over glucose because of its properties enabling it to act to both fulfill energy requirements as well as serve as an intermediate for macromolecule synthesis (Deberardinis et al. 2008; Moreadith and Lehninger 1984; Wall et al. 2013). The dual role of glutamine derives from its structure. Reactions such as nucleotide synthesis may directly use its γ-nitrogen, whereas the α-nitrogen and its carbon skeleton can be used indirectly in reactions for energy production and biosynthesis (Gaglio et al. 2009). Cell growth requires these metabolic intermediates and is propagated by the conversion of glutamine to glutamate by phosphate-dependent glutaminase (GLS) in the inner mitochondrial membrane (Fig. 9.1) (Cairns et al. 2011; Marie and Shinjo 2011). This enzyme is overexpressed in many tumor types and is a primary driver of glutamine consumption in cancer cells, leading to large intracellular pools of glutamate and subsequent glutamate release, the implications of which will be discussed later in this chapter (Cairns et al. 2011; Marie and Shinjo 2011; Wall et al. 2013).

The conversion of glutamine to glutamate and ammonia constitutes the first and rate-limiting step of glutaminolysis. The resulting glutamate is used as a primary source of energy for proliferating cells but can then be further metabolized to α–ketoglutarate in the mitochondrial matrix. Subsequent breakdown generates NADPH for use as an electron donor, which is also used to maintain glutathione (GSH), a key antioxidant, in its reduced state, thus keeping oxidative stress in check in rapidly growing cells (DeBerardinis et al. 2007; Estrela et al. 2006; Wall et al. 2013). Recently, several groups identified reductively metabolized glutamine as the major carbon source during hypoxia and limited respiration. Reductive metabolism is known to be preferred within the hypoxic conditions of most tumor environments, the result of a tumor's outpacing the development of an effective blood supply, among other conditions, and is due to the stabilization of a transcription factor, hypoxia-induced factor 1α (HIF-1α). Notably, HIF-1α induction and reductive metabolism occur in tumor cells growing in normoxic conditions as well, pointing to a more general role in sustaining tumor growth (Fendt et al. 2013; Filipp et al. 2012; Gameiro et al. 2013; Gao et al. 2009; Karakas et al. 2015; Niklas and Heinzle 2012; Sun and Denko 2014; Wise et al. 2008, 2011; Zamboni 2011) (see Table 9.1).

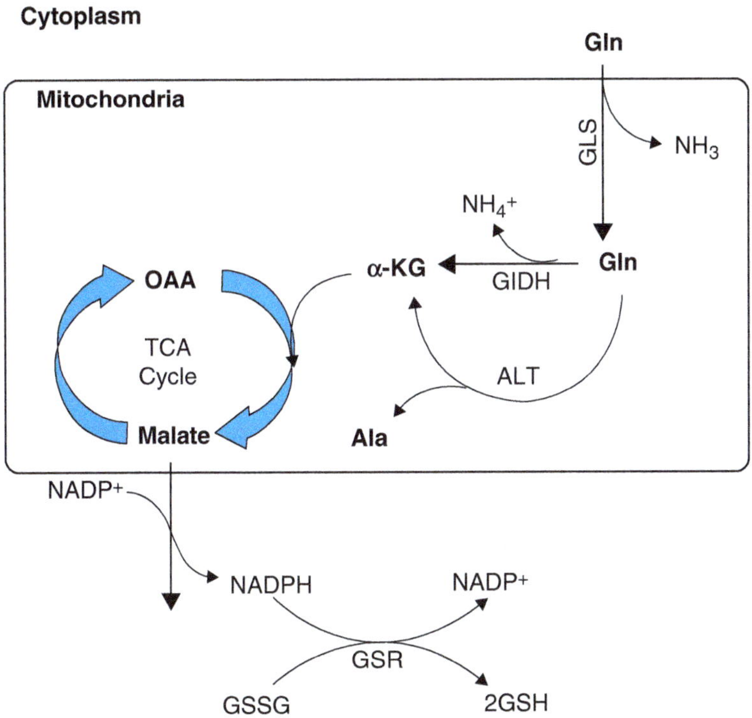

Fig. 9.1 Glutamate metabolism in the mitochondria Glutamine (*Gln*) is converted to glutamate (*Glu*) by the enzyme glutaminase (*GLS*) releasing both Glu and ammonia into the cytosolic compartment of the inner mitochondrial membrane. Glu is then metabolized to α-ketoglutarate (*α-KG*) either by oxidative deamination by glutamate dehydrogenase (*GIDH*) or alanine transaminase (*ALT*). α-KG is metabolized in the tricarboxylic acid (*TCA*) cycle to oxaloacetate through the production of malate. Malate oxidation to pyruvate in the cytosol generates NADPH which is used to maintain glutathione in its reduced state

Table 9.1 Metabotropic glutamate receptors in peripheral tissues

	Group I		Group II		Group III			
	mGluR1	mGluR5	mGluR2	mGluR3	mGluR4	mGluR6	mGluR7	mGluR8
Cartilage	X		X					X
Bone	X				X	X		X
Heart	X	X	X	X				
Esophagus	X				X			
Stomach	X	X	X	X	X	X	X	X
Duodenum	X				X			X
Ileum		X	X	X				X
Colon		X			X		X	X
Gall Bladder					X			X
Pancreas		X	X	X	X			X
Thymus	X	X	X	X	X			
Larynx					X			
Liver		X		X				
Lung				X	X			
Mammary Gland					X			
Ovary			X	X				
Skin	X	X	X	X				
Endothelium	X	X			X			
Dendritic Cells					X			
Testis	X	X	X	X	X	X		X
Adrenal Gland	X	X	X	X	X		X	
Kidney			X	X	X			

Table 9.2 Metabotropic glutamate receptors in cancers

Malignancy	mGluR genes detected	References
Glioma	mGluR1-8	Stepulak et al. (2009) and Takano et al. (2001)
Neuroblastoma	mGluR2-8	Stepulak et al. (2009)
Rhabdomyosarcoma and medulloblastoma	mGluRR2-4, 6-8	Stepulak et al. (2009)
Astrocytoma	mGluR1-8	Stepulak et al. (2009)
Lung carcinoma	mGluR1-2, 4-8	Stepulak et al. (2009)
Colon adenocarcinoma	mGluR1-8	Stepulak et al. (2009)
T-cell leukemia	mGluR1-7	Stepulak et al. (2009)
Multiple myeloma	mGluR2-7	Stepulak et al. (2009)
Breast carcinoma	mGluR1-7	Speyer et al. (2012) and Stepulak et al. (2009)
Thyroid carcinoma	mGluR2, 4, 7	Stepulak et al. (2009)
Prostate carcinoma	mGluR1a, mGluR1-8	Koochekpour (2013) , Miller (1999), Pissimissis et al. (2009) and Stepulak et al. (2009)
Melanoma	mGluR1,5	Namkoong et al. (2007) and Pollock et al. (2003)

9.2 mGluRs in Cancer

Overexpression or aberrant expression of GPCRs has been detected in many cancer cell types and contributes to tumor cell growth by paracrine or autocrine signaling, maintaining an activated state that leads to enhanced cellular proliferation via downstream effector proteins (Table 9.2) (Bhowmick et al. 2004; Burger et al. 1999; Cheng et al. 2008; Takayama et al. 1997). In particular, mGluRs have been shown empirically to be the predominant mediators of glutamatergic signaling in many cancers (Khan et al. 2011; Koochekpour 2013; Martino et al. 2013; Speyer et al. 2014; Teh and Chen 2012; Zhang et al. 2015). The mechanisms by which mGluRs modulate peripheral cell transformation and tumor growth are postulated to be either ectopic expression of wild-type mGluRs, increased proliferative signals arising from receptor overexpression, mutations, or expression of polymorphic variants (Ali et al. 2014; Brocke et al. 2010; Mehta et al. 2013; Namkoong et al. 2007; Wall et al. 2014; Zhang et al. 2015).

9.3 mGluRs in Primary Brain Tumors

The first observations of mGluRs in human malignancies were in neuronal tumors including neuroblastomas, medulloblastomas, and gliomas (Iacovelli et al. 2006; Naarala et al. 1993; Shin et al. 2008; Takano et al. 2001). In normal brain, astrocytes primarily express mGluR3, whereas glial cells express mGluR1 and mGluR5 (Balazs et al. 1997; Petralia et al. 1996; Wroblewska et al. 1998). In a comprehensive study of pediatric brain tumors, mGluR1, mGluR2, and mGluR6 display elevated expression levels in malignant medulloblastomas, ependymomas, and glioblastomas, compared to low-grade astrocytomas (Brocke et al. 2010). Additionally, these overexpression patterns were similar among different tumor cell types, despite distinct histological origins, possibly pointing to common glutamatergic signaling as a key player in these cancers and could be aimed as a therapeutic target in childhood brain tumors (Brocke et al. 2010). Both group I and II mGluRs have been implicated in glioma. Inhibition of mGluR1 significantly decreases cell viability in U87 glioma cells, resulting in apoptosis via mitigated activation of the PI3K/Akt/mTOR pathway (Zhang et al. 2015). These results also translated to a U87 xenograft glioma model, indicating possible clinical relevance for the use of mGluR1 inhibitors in glioma treatment (Zhang et al. 2015). Another group showed that group II mGluRs, mGluR2/3, are expressed in a majority of human glioblastoma specimens. In vitro cultured cell studies showed suppression of cell growth in the presence of an mGluR2/3 antagonist, removal of the antagonist restored cell growth (D'Onofrio et al. 2003). Additionally in U87 glioma cells, use of mGluR2/3 antagonist abrogated cyclin D1/D2 expression and activated both MAPK and PI3K pathways (Arcella et al. 2005). Aronica and colleagues showed that mGluR3 and mGluR5 agonists modulate the expression of the glutamate transport proteins, GLAST, GLT-1, and EAAT. Selective group I agonists downregulated expression of GLAST and GLT-1 in astrocytes, while group II agonist positively induced the expression of the transporters. Similar observations were noted with the EAAT transporter when glioblastoma cells were treated with group I or II agonists. Selective antagonists of group I and II mGluRs prevented the induction of these receptors, and further, inhibition of MAPK and PI3K signaling had a negative effect on the expression of the GLAST and GLT-1 induced by the agonists. Taken together, these results point to complex interactions between various modulators of mGluRs and also suggest the potential use of glutamate receptor agonists to manipulate glutamate levels in brain tumor environments by targeting glutamate transporter expression (Aronica et al. 2003). In malignant gliomas, mGluR3 can regulate chemoresistance to the alkalizing agent, temozolomide, and the level of mGluR3 from human glioblastoma multiforme samples is inversely related to survival following surgery and adjuvant chemotherapy (Ciceroni et al. 2013). On the contrary, activation of the group III mGluR, mGluR4, has been demonstrated to inhibit growth of medulloblastoma. Sixty samples of human medulloblastoma were examined, and approximately 77% expressed mGluR4, which was inversely correlated with tumor severity, spread, and recurrence (Iacovelli et al. 2006). In addition, in medulloblastoma cell lines, mGluR4 activated with a selective enhancer led to inhibition of adenylate cyclase and PI3K pathways, reduced DNA synthesis, and cell proliferation (Iacovelli et al. 2006). Further in vivo studies identified that stimulated mGluR4 inhibits medulloblastoma cell xenograft progression in nude mice, and treatment with an mGluR4 enhancer in the first week of life is able to prevent the development of medulloblastoma in an irradiated, heterozygous Patched-1 mouse model (Iacovelli et al. 2006). These data suggest that disrupted glutamate signaling is involved in primary CNS tumors.

9.4 mGluRs in Non-neuronal Cancers

Among Group I mGluRs, mGluR1 has been shown to induce the neoplastic transformation of immortalized baby mouse kidney epithelial cells (iBMK) in vitro and promote tumor cell progression in vivo (Degenhardt and White 2006; Martino et al. 2013). It was shown that full-length wild-type mGluR1 is oncogenic when exogenously introduced in epithelial cells that led to stimulated MAPK and AKT signaling pathways. In addition, mutations and single-nucleotide polymorphisms (SNPs) of GRM1 have also been described in prostate cancer (Ali et al. 2014). These mutations may serve to modulate mGluR1 by altering gene splicing, ligand binding, and downstream signaling. Eight somatic variations of GRM1 have been identified in cancers, including lung adenocarcinoma where GRM1 mutations have been demonstrated to result in functional downstream signaling with variable modulation of cell proliferation pathways including MAPK/ERK (Esseltine and Ferguson 2013). mGluR5 overexpression has been shown to induce melanoma development in transgenic mice as well (Choi et al. 2011). Increased expression of mGluR5 in the mouse melanocytes was correlated with enhanced levels of MAPK activation, suggesting that glutamatergic signaling plays a significant role not only in the initiation and progression, but also in the maintenance, of tumorigenesis. mGluR5 has also been implicated in oral squamous cell carcinoma as well as laryngeal cancers. Park and colleagues showed that differential levels of mGluR5 are expressed in human oral squamous cell carcinomas with increased mGluR5 immunoreactivity associated with improved overall survival (Park et al. 2007). Additionally, a mGluR5 agonist promoted tumor cell migration, invasion, and adhesion in human tongue cancer cells, which was reversed by an mGluR5 antagonist (Park et al. 2007). Subsequent studies showed that inhibition of mGluR5 with specific antagonist reduces metastasis in CXCR4/stromal-derived factor-1 (SDF-1)-dependent oral cancers (Kuribayashi et al. 2013). In laryngeal cancer, the mGluR5 antagonist, MPEP, inhibits cell proliferation, although its expression is much lower than that in the brain, which may account for the relatively weak response of laryngeal cancer cells to mGluR antagonists (Stepulak et al. 2011). In addition to mGluR5, mGluR4 NMDA- and AMPA-mediated glutamate signaling have also been observed in laryngeal cancers (Chang et al. 2005; Stepulak et al. 2011).

Group II mGluRs have also been implicated in a variety of cancer types, including melanoma. Notably, through exon capture sequencing of GPCRs in malignant melanoma, the Samuels group identified hot spots in GRM3 that are frequently mutated in human melanoma and selectively regulate phosphorylation of MEK, leading to anchorage-independent growth and migratory capabilities in cultured cells (Prickett et al. 2011). The mutations were found throughout the coding region and affected the extracellular domains as well as the seven-transmembrane domain, with two mini-hotspots located proximal to the transmembrane domain. Four somatic mutations (mGluR3^{E767K}, mGluR3^{S610L}, mGluR3^{G561E}, and mGluR3^{E870K}) were found to selectively regulate phosphorylation of MEK1/2 kinase in vitro and induce micrometastasis in vivo. These studies suggest a subset of melanomas in which activating mutations in mGluR3 lead to hypersensitivity to agonist stimulation of the MEK-MAPK pathway without involvement of RAF/RAS genotypes.

Finally, Group III mGluRs have been implicated in colorectal, laryngeal squamous cell, breast cancers, and osteosarcoma, as well as malignant melanoma. Chang and colleagues have reported that mGluR4 is overexpressed in more than 40% of colorectal adenocarcinomas, malignant melanomas, laryngeal squamous cell carcinomas, and breast carcinomas and that overexpression of the receptor was correlated with increased mortality in colorectal carcinoma (Chang et al. 2005). The same group had previously demonstrated that human colon cancer cells resistant to 5-FU overexpress mGluR4 and that mGluR4 agonists enhanced 5-FU resistance while mGluR4 antagonists ablated 5-FU resistance (Yoo et al. 2004). Savage and colleagues also identified two susceptibility loci for osteosarcoma in the gene encoding mGluR4 (Savage et al. 2013). More recently, increased expression of mGluR4 and mGluR8 in a lung carcinoma cell line and human lung adenocarcinoma samples was reported. Further studies revealed that an mGluR8 agonist (S)-3,4-DCPG reduced cell growth and increased apoptosis, indicating that mGluR8 could be a potential target in future lung cancer therapies (Li et al. 2015).

In the next few sections, the role of mGluR1 in melanoma and breast cancer is discussed in detail. Extensive studies on the involvement of mGluR1 in melanoma and triple-negative breasts have been performed and resulted in clinically tested treatments that target the receptor.

9.5 mGluR1 in Melanoma

Our group has previously described an unknown mechanism of melanoma pathogenesis in which the aberrant expression of mGluR1 in melanocytes is sufficient to transform melanocytes in vitro and promote melanoma tumor development in vivo (Chen et al. 1996; Zhu et al. 1998). Chen and colleagues established a transgenic mouse line, TG-3, that harbors a 2-kb genomic fragment, clone B, which had previously been shown to commit fibroblasts to undergo adipocyte differentiation in vitro (Chen et al. 1996; Zhu et al. 1998). Instead of the expected obese phenotype, one founder, TG-3, developed pigmented lesions on the ears, around the eyes, and in the perianal region, identified as melanoma by histological means. Molecular and biochemical analyses revealed that melanoma development in the TG-3 mouse model was a result of a classic case of insertional mutagenesis that led to ectopic expression of mGluR1 in melanocytes. Insertion of clone B resulted in concurrent deletion of approximately 70 kb of host sequence in intron 3 of the gene encoding mGluR1, GRM1 (Pollock et al. 2003; Wall et al. 2013). Further analysis of the pigmented lesions on TG-3 revealed that murine mGluR1 was expressed both on the protein and mRNA levels, but appeared to be low/absent in the normal skin of the animal (Pollock et al. 2003; Wall et al. 2013). A new transgenic mouse line, TG(Grm1)EPv, or "E," was created, using wild-type murine GRM1 cDNA under the control of a melanocyte-specific promoter dopachrome tautomerase (Dct) (Pollock et al. 2003).

This new mouse line displays a melanoma susceptibility profile similar to that of the original TG-3 founder. We concluded that ectopic expression of mGluR1 in melanocytes is sufficient to induce melanoma development in mice in vivo with 100% penetrance in the absence of any additional known exogenous carcinogens (Pollock et al. 2003; Wall et al. 2013).

Further studies demonstrated that constitutive expression of mGluR1 is required to maintain the transformed phenotype in vitro and in vivo. Using an inducible mouse model, Ohtani et al. showed that conditionally induced mGluR1 expression stimulated melanocyte growth leading to melanoma, while silencing mGluR1 expression resulted in tumor abrogation in vivo (Ohtani et al. 2008). In melanomas, stimulation of mGluR1 by glutamate results in near-identical formation of second messengers as previously described for the CNS. Through the second messenger, DAG, PKC is stimulated resulting in activation of the RAS-RAF-MEK-ERK module of the MAPK signaling cascade and the PI3K/AKT pathway, thus upregulating cell proliferation and inhibiting apoptosis (Busca et al. 2000; Choe and Wang 2002; Ferraguti et al. 1999; Marin and Chen 2004; Marin et al. 2006; Thandi et al. 2002).

Furthermore, we demonstrated elevated levels of extracellular glutamate only in mGluR1-expressing melanoma cells, suggesting the existence of autocrine loops, a characteristic of oncogenic GPCRs (Gutkind et al. 1991; Julius et al. 1989; Namkoong et al. 2007). We also demonstrated that MAPK is a downstream target of mGluR1 signaling. Receptor stimulation with the agonist L-quisqualate resulted in increased levels of phosphorylated ERK, indicating active MAPK downstream signaling. Pretreatment of melanoma cells with mGluR1 antagonist, LY367385, followed by agonist treatment did not activate the MAPK pathway suggesting that MAPK activation is dependent upon functional mGluR1 (Namkoong et al. 2007). Other groups have shown that approximately 80% of melanoma tissue samples including 33% of common, blue, and Spitz nevi, 75% of human melanoma cell lines, and 50% of nevus lines are mGluR1 positive, while normal melanocytes are mGluR1 negative (Funasaka et al. 1996). These findings point to mGluR1 involvement in the oncogenesis of a subset of human melanomas (Namkoong et al. 2007; Pollock et al. 2003).

Recent studies from Gelb and colleagues confirm that mGluR1 expression in melanoma cells confers glutamate dependence for cell viability and DNA synthesis in human melanoma cells SK-MEL-2 and SK-MEL-5. Using both shRNA against mGluR1 and a noncompetitive mGluR1 antagonist that had previously never been used in an oncologic setting, JNJ16259685, downregulation of mGluR1 expression decreased cell viability of human melanoma cells in vitro and tumor growth in vivo in a xenograft model (Gelb et al. 2015b). Other new experiments demonstrate that other pathways in addition to MAPK likely mediate mGluR1-mediated cell viability in melanoma cells. They showed that the mGluR1 agonist quisqualate does not always modulate phosphoinositol hydrolysis and promote cell viability as does the native ligand glutamate, indicating that future studies of melanoma cell viability in mGluR1-positive cell lines should not use agonists synonymously with glutamate for stimulation studies. In addition, a novel mGluR1-mediated but G-protein-independent signaling cascade involving dynamin was found to upregulate MAPK signaling with sustained phosphorylation of ERK. A dynamin inhibitor, dynasore, dose-dependently decreased glutamate-dependent cell viability of mGluR1-positive human melanoma cells SK-MEL-2 and SK-MEL-5, but had no effect on viability in mGluR1-negative human melanoma cells UACC930. These results suggest that internalization of mGluR1 receptors via dynamin activity may play an additional role in the cell viability of a subset of mGluR1-positive melanomas (Gelb et al. 2015a).

9.6 mGluR1 as a Therapeutic Target in Treating Melanoma

Results from cultured cell studies suggest that ectopic expression of mGluR1 in cells led to the establishment of autocrine loop with high levels of extracellular glutamate to ensure constitutive activated receptor and upregulated glutamate signaling to participate in neoplastic transformation in vitro and tumorigenesis in vivo. Because decreasing the levels of available glutamate would most likely dampen the glutamatergic signaling and reduce the oncogenic activity of mGluR1. We tested this notion by using riluzole, an inhibitor of glutamate release, in mGluR1-expressing melanoma cells.

Riluzole, marketed by Sanofi-Aventis as Rilutek®, is classified as an anti-excitotoxic, neuroprotective drug that blocks cellular release of glutamate. It is FDA approved for the treatment of amyotrophic lateral sclerosis (ALS). Riluzole has an absolute bioavailability of approximately 60%, which is relatively high, and is also able to cross the blood-brain barrier. Its mechanism of glutamate blockade is thought to be in part due to inactivation of voltage-gated sodium channels on glutamatergic nerve terminals and/or activation of a G-protein-dependent signaling cascade (Doble 1996). Similarly, riluzole has also been shown to block some post-synaptic glutamatergic effects by noncompetitive inhibition of NMDA receptors (Doble 1996).

In vitro riluzole treatment of mGluR1-expressing human melanoma cells reduces cell proliferation and decreases levels of extracellular glutamate, functionally much like the mGluR1 antagonists LY367385 and BAY36-7620 (Fig. 9.2). In vivo human melanoma cell xenograft studies also showed a 50% inhibition of tumor growth compared to controls. Furthermore, cell cycle analysis on riluzole-treated mGluR1-expressing human melanoma cells showed accumulation in the G2-M phase and subsequent buildup in the sub-G1 phase of the cell cycle, consistent with apoptosis, which was confirmed in Westerns with elevated cleaved PARP, an apoptosis marker (Namkoong et al. 2007).

Results from these preclinical studies prompted us to translate into the clinic with a phase 0, first-in-human trial using riluzole in stage III and IV melanoma patients (Yip et al. 2009). The drug was well tolerated and yielded an unexpected significant short-term response rate in 34% of patients. The tumors that shrank in human subjects showed an inhibition of signaling through both the MAPK and PI3K/AKT pathways, correlating with results from preclinical studies (Namkoong et al. 2007; Yip et al. 2009). Additionally,

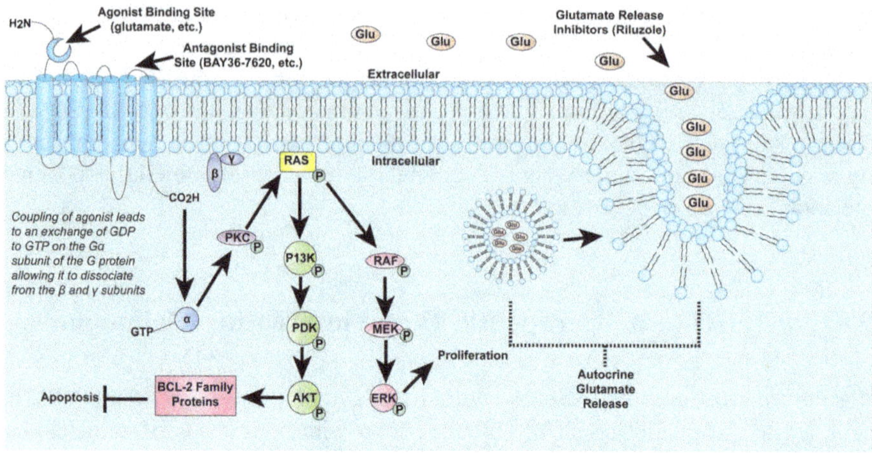

Fig. 9.2 Overall working hypothesis of the mechanisms involved in melanoma development through GRM1 activation

Positron Emission Tomography (PET) documented complete resolution of multiple nodal and cutaneous metastases in several patients. Out of eleven patients that completed the study, only two had disease progression posttreatment. While all patients were positive for mGluR1 expression, the tumor samples also harbored a composite of most common mutations in melanoma, including BRAF V600E as well as NRAS Q61K. From this observation, a possible corollary is that more advanced tumors, as in those found in stage III and IV metastases, have a higher frequency of ectopic mGluR1 expression. A phase 2 trial of riluzole as a single agent for melanoma treatment yielded similar results, with initial stable disease in 30% of patients (Mehnert et al. 2016).

Although radiosensitizers have previously not been shown to provide survival benefit in human melanoma, Khan and colleagues (Khan et al. 2011) hypothesized that riluzole may be different, as cells in the G2/M phase rendered by riluzole treatment would be extremely sensitive to radiation (Ballo and Ang 2004; Ballo et al. 2006; Eichler et al. 2007; Habermalz and Fischer 1976). In vitro studies showed that mGluR1-positive human melanoma cells were significantly less likely to survive following ionizing radiation treatment at the 2 and 4 Gy dose levels when pretreated with riluzole at 25 μM concentration, while there was negligible effect in mGluR1-negative human melanoma cells even at 100μM of riluzole treatment. Furthermore, riluzole with irradiation in mGluR1-positive human melanoma cells induced G2/M synchronization in a sequence-dependent manner, and rendering the receptor non-functional with the noncompetitive antagonist of mGluR1, BAY36-7620, prior to riluzole treatment abrogated these effects. These findings reveal that the mechanism of riluzole is more complex than merely diminish available glutamate. In vivo, mGluR1-positive human melanoma xenografts treated with riluzole plus irradiation resulted in smaller tumor volume and significant increase in apoptotic tumor cells that stained positive for cleaved caspase-3, a marker of apoptosis, and γ-H2AX foci, a marker of DNA double-stranded breaks (Khan et al. 2011). These results suggest that riluzole may be used as a radiosensitizer in a subset of mGluR1-expressing human melanomas.

Further research led to the unexpected discovery that riluzole by itself led to DNA damage and that riluzole in association with irradiation produced additive DNA double-stranded breaks in mGluR1-postive human melanoma cells in vitro and in vivo, as evident by the histological detection of γ-H2AX foci and phosphorylation of histone H2AX on Western immunoblots (Wall et al. 2014). These findings only manifest in mGluR1-positive human melanoma cells but not mGluR1-negative cells. Additional experiments showed co-localization of both γ-H2AX and 53BP1. 53BP1, a checkpoint regulator that binds to p53 and translocates to sites of DNA, breaks in response to DNA damage (Wall et al. 2015). Similar results were observed in isogenic vector control or mGluR1 clones derived from immortalized, non-tumorigenic human melanocytes (Wall et al. 2014).

Using a genetic inducible siRNA (si-GRM1) approach in C8161 human melanoma cells, DNA damage was reduced in riluzole-treated induced si-GRM1 clones, compared to the same clones not subject to the induction of si-GRM1 (Wall et al. 2014). These results support mGluR1-dependent riluzole-mediated DNA damage as the mode of action. In the same report, we proposed that in the presence of riluzole, synthesis of the antioxidant glutathione (GSH), may be interrupted by increased intracellular glutamate that could disrupt the bidirectional transport of glutamate and cysteine required for GSH synthesis (Wall et al. 2014). A decrease in GSH causes increased oxidative stress and reactive oxygen species (ROS) detectible by the ROS-specific marker dihydrorhodamine 123 (DHR123). We confirmed an increase in ROS only in riluzole-treated mGluR1-positive cells but not mGluR1-negative cells (Wall et al. 2014). We also retroactively assessed pre- and posttreatment specimens from the phase 0 and 2 clinical trials of riluzole for DNA damage. A significant number of γ-H2AX staining cells in posttreatment samples of stable disease patients with biologic responses compared to pretreatment samples from the same patients were observed. In comparison, samples from nonresponding patients with disease progression showed low and similar numbers of γ-H2AX foci in pre- and posttreatment specimens (Mehnert et al. 2016; Wall et al. 2014; Yip et al. 2009).

As melanoma is a heterogenous disease and clinical trials utilizing riluzole have shown positive response in only a subset of mGluR1-positive melanomas, we suspect involvement and even crosstalk with other pathways. We showed with mouse melanoma cells that IGF-1R transactivation is involved in mGluR1-mediated melanocyte transformation (Teh et al. 2014). mGluR1 receptor functionality was assessed with the mGluR1 agonist, L-quisqualate [Q], in stable mGluR1 clones derived from a normal mouse melanocytic cell line. AKT activation was observed, but the addition of an IGF-1R inhibitor, PPP, resulted in lack of alterations in AKT activity above baseline, even in the presence of mGluR1-agonist Q, suggesting that functional IGF-1R is necessary for mGluR1-mediated stimulation of the AKT cascade (Shin et al. 2008; Teh et al. 2014). Treatment of human melanoma cells with riluzole and OSI-906, a potent inhibitor of IGF-1R, suppressed cell growth more than either agent alone (Fassnacht et al. 2011; Mulvihill et al. 2009; Teh et al. 2014; Zhao et al. 2012). This synergistic response was reproducible in an in vivo xenograft model, suggesting that IGF signaling is involved in crosstalk with mGluR1 in human melanoma. To validate this hypothesis, we again turned to tumor biopsies from five patients with advanced metastatic melanoma treated with monotherapy riluzole from the completed phase 0 and 2 trials. Notably, responders showed decreased phospho-IGF-1R and phospho-AKT levels, and nonresponders showed increased phospho-IGF-1R and phospho-AKT. While limited by sample size, these results point to a role of the IGF-1R/AKT pathway in single-agent riluzole response in melanoma patients (Teh et al. 2014). Additionally, simultaneous targeting of both MAPK and PI3K signaling pathways in human melanoma brain metastasis-derived cells in vitro can decrease cell survival and invasion (Daphu et al. 2014). As mGluR1 affects both MAPK and PI3K, therapies like riluzole may play a role in overcoming resistance that develops after targeted MAPK and PI3K inhibitor treatment.

Our laboratory has also studied the inhibitors sorafenib or PLX4720/PLX4032 in combination with riluzole. Sorafenib is a small-molecule, multi-kinase inhibitor, and PLX4720/PLX4032 is a small-molecule inhibitor specific to mutated BRAF V600E kinase. We showed that mGluR1-positive cells containing a BRAF V600E mutation were less sensitive to riluzole because both mGluR1 and BRAF V600E stimulate MAPK signaling (Lee et al. 2011a). Furthermore, using a combination of riluzole and sorafenib synergistically reduces growth of wild-type or BRAF V600E human melanoma cells in vitro and in vivo, and the combination of riluzole and PLX4720/PLX4032 also has additive effects on the suppression of melanoma cell growth (Lee et al. 2011b). These results have been formulated into an ongoing phase I clinical trial to assess the safety and efficacy of riluzole with sorafenib in stage III/IV melanoma patients.

9.7 mGluR1 in Brain Metastasis

Excessive glutamate induces excitotoxicity in surrounding neuronal tissue, and in brain tumors this creates space for the expansion of the growing tumor (Van den Bosch 2006; Yu et al. 2015). Melanoma is one of the most common cancers metastasizing to the brain after lung and breast carcinomas, interesting, as melanocytes originate from the neural crest. Recently, our group reviewed brain metastasis from melanoma and current treatments (Yu et al. 2015). In short, these metastases tend to be small, multiple, and hemorrhagic (Chaichana and Chaichana 2011; Ewend et al. 1996; Graus et al. 1985; Wronski and Arbit 2000). Treatment modalities include antiepileptic drugs for symptomatic relief versus definitive treatment with surgical resection, whole brain radiation treatment, stereotactic radiosurgery, targeted therapies, immunotherapies, or a combination of the above (Yu et al. 2015). However, there remains a dearth of effective treatment options for melanoma brain metastasis. Radiation is still one of the mainstays of treatment for primary brain tumors and brain metastases, but efficacy in metastases from melanoma is low compared to that from other primary tumor types (Habermalz and Fischer 1976; McKay and Kefford 1995; Nicholas et al. 2013). Among cancers that commonly metastasize to the brain, lung cancers, especially small-cell lung carcinomas, are most susceptible to radiation, while breast cancers are less sensitive; melanoma and renal cell carcinomas are least sensitive (Barranco et al. 1971).

mGluRs, including those in group I and group II, have been shown to be involved in brain-specific metastasis and proliferation of melanoma. Nygaard and colleagues have shown that a brain-adaptive phenotype of melanoma is found prominently in early metastatic growth phases and occurs via activation of the glutamate receptor-mediated Ca^{2+}-dependent signaling that regulates transcription and neuron-like plasticity (Nygaard et al. 2014). They also reported that elevated levels of extracellular glutamate was detected in the conditioned medium of the MM1 melanoma cells after 72 hours in culture; these cells are not as proliferative but more invasive when compared to the MM5 melanoma cell line, which is rapidly proliferating, but noninvasive (Nygaard et al. 2014). Further studies showed that blockade of upregulated glutamate receptors GRIA1, GRIA2, mGluR3, and mGluR4 by specific antagonists decreased cell viability, establishing a system of functional glutamate signaling in these cells in vitro (Nygaard et al. 2014).

Incidentally, we also speculate that melanoma is one of the most common primary tumors that lead to hemorrhagic intracranial metastases (Lieu et al. 1999; Mandybur et al. 1977). Goydos and colleagues recently showed that melanoma cells with enhanced mGluR1 expression produced not only large tumors in vivo but also more vascular tumors than their lower mGluR1-expressing counterparts. Cultured mGluR1-positive human melanoma cells produced more interleukin-8 (IL-8) and VEGF likely via mGluR1-mediated activation of the AKT–mTOR–HIF1 pathway. Moreover, in clinical posttreatment specimens from patients treated with single-agent riluzole not only showed inhibition of MAPK and PI3K/AKT activities but also decreased density of blood vessels, indicating that suppression of glutamate signaling downstream from mGluR1 could promote anti-angiogenic effects, suggesting inhibition of mGluR1 as a potential target in treatment of melanoma and other cancers with high likelihood of hemorrhagic brain metastases, including renal cell carcinoma (RCC) and hepatocellular carcinoma (HCC) (Lieu et al. 1999; Ti 1977; Wen et al. 2014).

Recently, our laboratory utilized orthotopically implanted mGluR1-expressing human melanoma xenografts in mice as a model for intracranial melanoma metastases. Based on preclinical studies and riluzole's ability to cross the blood-brain barrier, we predicted that riluzole would elicit antiproliferative effects on human melanoma cells in the brain and would act cooperatively with ionizing radiation therapy. We used a bioluminescence enzyme firefly luciferase (Fluc) reporter introduced into the mGluR1-positive human melanoma cell line C8161 to produce several stable clones (C8161-luc), creating a monitoring approach using IVIS, a small animal imaging system for in vivo assessment of tumor burden without having to sacrifice animals at different time intervals. Intraperitoneal (IP) injection of luciferin substrate into animals allowed for oxidation by luciferase, resulting in photon emission/bioluminescence signal intensity acquisition and quantification by IVIS. These signal intensities were first demonstrated to correlate with tumor volume measured by a Vernier caliper in a xenograft tumor experiment using C8161-luc cells injected into the dorsal flanks of athymic nude mice. To model brain metastases, we injected C8161-luc cells intracranially into a new set of mice and monitored for tumor establishment using luminescent imaging. Following tumor detection, the mice were randomized into four treatment groups: vehicle (Veh, DMSO), vehicle and a weekly dose of 4 Gy irradiation (IR), riluzole daily via oral gavage (Ril), and a combination of irradiation and riluzole (Ril+IR). We showed that riluzole or irradiation alone was able to reduce melanoma tumor cell growth in the brain over time as compared to vehicle treatment. However, combination treatment of Ril+IR yielded even smaller tumor burden as demonstrated by the decreased luminescent signal after four weeks of treatment compared with vehicle-treated or either treatment modality alone. From these results, we propose that riluzole could be used as a radiosensitizer for the treatment of intracranial metastasis in combination with irradiation in mGluR1-positive human melanomas (Yu et al. 2015).

9.8 mGluR1 in Breast Cancer

mGluR1 expression has been widely explored in breast cancer. Speyer et al. reported the expression of mGluR1 in five triple-negative breast cancer (TNBC) cell lines, as well as normal mammary epithelial cells with varied expression levels. Silencing of the receptor with GRM1 shRNA resulted in inhibition of cell proliferation. Pharmacologically, treatment of mGluR1-expressing TNBC cell lines with riluzole decreased the available ligand to sustain the activated receptor, therefore functioning as an antagonist. In a series of in vitro and in vivo experiments, inclusion of riluzole in the growing media or administration via oral gavage to xenograft tumor-bearing mice resulted in reduced cell proliferation, tumor progression, and induced apoptosis in vitro and in vivo (Speyer et al. 2012). Additionally, isogenic breast cancer cell lines constructed to mirror the progression of breast cancer demonstrate higher expression of mGluR1 mRNA and protein in premalignant and malignant cells than in normal mammary epithelial cells (Banda et al. 2014). Interestingly, lentiviral-mediated overexpression of mGluR1 in premalignant MCF10A cells showed no significant effects on proliferation, invasion, or soft agar colony formation in vitro, suggesting that the receptor may be involved in later steps during progression. To evaluate this hypothesis, mGluR1 was overexpressed in MCF10AT1 cells, which represent atypical ductal hyperplasia. In this case, a significant increase in the proliferation of the cells was observed. Inhibition of receptor activity using the noncompetitive antagonist, BAY36-7620, or GRM1 shRNA negatively impacted proliferation. With these data, it was hypothesized that mGluR1 may also influence other oncogenic functions that support tumor growth. As such, investigation of the effects of mGluR1 on the migration and invasion of the cell lines was performed. Overexpression of mGluR1 in the MCF10AT1 atypical ductal hyperplasia cells resulted in increased migration and invasion through a reconstructed basement membrane as well as anchorage-independent growth. These alterations were diminished when mGluR1 expression was disrupted using shRNA. In vivo, the mGluR1-overexpressing cells formed multiple invasive carcinomas compared to parental or vector control cells, suggesting a role for mGluR1 in malignant breast cancer phenotypes. Triple-negative breast cancers identified by their lack of expression of progesterone, estrogen, and Her2/Neu receptors remain an important challenge in breast cancer management that impacts on thousands of women because this class of breast cancer does not respond to targeted therapies. For this reason, identification of mGluR1 as a contributor to proliferation, invasion, and malignant progression in TNBC offers a potential therapeutic target.

In addition to the reported roles of mGluR1 in promoting cell proliferation, invasion, anchorage-independent growth, and in vivo carcinoma formation (Banda et al. 2014; Speyer et al. 2012), mGluR1 has also been linked in supporting angiogenesis in breast cancer (Speyer et al. 2014). In vitro studies using primary endothelial cells showed that cells expressing elevated levels of mGluR1 had greater sensitivity to BAY36-7620 and riluzole. Tube formation on matrigel, an ex vivo model of angiogenesis, was suppressed in cells treated with BAY36-7620 or riluzole, suggesting participation of mGluR1 in the angiogenic process. In an in vivo murine mammary tumor model with a strong vascular component, when treated with riluzole or Sunitinib (Roskoski 2007), a similar inhibition in tumor growth was observed as earlier reports (Banda et al. 2014; Speyer et al. 2012), and a reduction in microvessel density was also demonstrated for riluzole at 60% and Sunitinib at 80%, suggesting that mGluR1 may mediate its effect on tumor cell growth and invasion partly by enhancing angiogenesis.

While the contributions of mGluR1 overexpression in breast cancer were explored, a correlation between the polymorphic variants of GRM1 and breast cancer subtypes was demonstrated (Mehta et al. 2013). Earlier, in a genetic association study examining correlation of three single-nucleotide polymorphisms in GRM1 and melanoma susceptibility showed higher frequency of one of the polymorphisms in GRM1-positive melanoma patients compared to controls. It was also reported that the frequency of this polymorphism was greater in subgroup of patients with low level of sun exposure and tumors located on the trunk and extremities (Ortiz et al. 2007). Mehta et al. examined GRM1 polymorphisms in breast cancer. Genotyping of over 1000 breast cancer specimens showed that 2 polymorphisms were associated with age at diagnosis and either estrogen or progesterone receptor-positive disease. In vitro validation of these observations using estrogen- and progesterone-positive (MFC7) or TNBC (MDA-MB-231) cell lines treated with either 17β-estradiol or with the combination of 17β-estradiol and progesterone assessed possible modulation in mGluR1 expression. In MCF7 cells, mGluR1 expression increased 2.7 fold after 17β-estradiol treatment and 2.1-fold after 17β-estradiol plus progesterone treatment. These effects were blocked when the cells were pretreated with tamoxifen. mGluR1 expression was not affected by 17β-estradiol and progesterone in the receptor-negative MDA-MB-231 cells. This study also showed that estrogen receptor-positive tumors were more likely to express mGluR1 in comparison to estrogen-negative tumors. Additionally, it revealed that patients with low mGluR1-expressing tumors were associated with longer periods of distant metastasis-free survival as compared to those with higher mGluR1 expression. These results show that mGluR1 may not only be involved in the disease progression, but that it might also affect response to hormone-based therapy in breast cancer. Given that mGluR1 appears to influence triple-negative breast cancer, estrogen- and progesterone-positive subtypes, riluzole, and other drugs that target this receptor might prove useful in a clinical setting.

9.9 Conclusions

Our knowledge of the functions of metabotropic glutamate receptors is constantly evolving. This evolution has changed the perception that these receptors are mainly involved in neuronal signaling, neurodevelopmental disorders, and neuronal malignancies. A tremendous amount of work has emerged on the role of mGluRs in the biology of melanoma, breast, prostate cancers, and in tumors that originate in the brain or those that metastasize to the brain. The evidence for mGluR1 involvement in melanoma has led to the application of riluzole, a drug used to inhibit glutamate release in ALS, as an anti-melanoma drug and as a radiosensitizer that can be used to treat other tumors that involve mGluR1-mediated glutamate signaling that commonly metastasize to the brain. Further understanding of how other glutamate receptors affect human cancers as well as development of other receptor modulators will result in multiple applications that extend far beyond the traditional roles associated with the glutamate receptors.

References

Ali S, Shourideh M, Koochekpour S (2014) Identification of novel GRM1 mutations and single nucleotide polymorphisms in prostate cancer cell lines and tissues. PloS One 9:e103204

Alix JJ, Domingues AM (2011) White matter synapses: form, function, and dysfunction. Neurology 76:397–404

Arcella A, Carpinelli G, Battaglia G, D'Onofrio M, Santoro F, Ngomba RT, Bruno V, Casolini P, Giangaspero F, Nicoletti F (2005) Pharmacological blockade of group II metabotropic glutamate receptors reduces the growth of glioma cells in vivo. Neuro-Oncology 7:236–245

Aronica E, Gorter JA, Ijlst-Keizers H, Rozemuller AJ, Yankaya B, Leenstra S, Troost D (2003) Expression and functional role of mGluR3 and mGluR5 in human astrocytes and glioma cells: opposite regulation of glutamate transporter proteins. Eur J Neurosci 17:2106–2118

Balazs R, Miller S, Romano C, de Vries A, Chun Y, Cotman CW (1997) Metabotropic glutamate receptor mGluR5 in astrocytes: pharmacological properties and agonist regulation. J Neurochem 69:151–163

Ballo MT, Ang KK (2004). Radiotherapy for cutaneous malignant melanoma: rationale and indications. Oncology (Williston Park) 18, 99–107; discussion 107–110, 113–104

Ballo MT, Ross MI, Cormier JN, Myers JN, Lee JE, Gershenwald JE, Hwu P, Zagars GK (2006) Combined-modality therapy for patients with regional nodal metastases from melanoma. Int J Radiat Oncol Biol Phys 64:106–113

Banda M, Speyer CL, Semma SN, Osuala KO, Kounalakis N, Torres Torres KE, Barnard NJ, Kim HJ, Sloane BF, Miller FR, Goydos JS, Gorski DH (2014) Metabotropic glutamate receptor-1 contributes to progression in triple negative breast cancer. PloS One 9:e81126

Barranco SC, Romsdahl MM, Humphrey RM (1971) The radiation response of human malignant melanoma cells grown in vitro. Cancer Res 31:830–833

Bhowmick NA, Neilson EG, Moses HL (2004) Stromal fibroblasts in cancer initiation and progression. Nature 432:332–337

Brocke KS, Staufner C, Luksch H, Geiger KD, Stepulak A, Marzahn J, Schackert G, Temme A, Ikonomidou C (2010) Glutamate receptors in pediatric tumors of the central nervous system. Cancer Biol Ther 9:455–468

Burger M, Burger JA, Hoch RC, Oades Z, Takamori H, Schraufstatter IU (1999) Point mutation causing constitutive signaling of CXCR2 leads to transforming activity similar to Kaposi's sarcoma herpesvirus-G protein-coupled receptor. J Immunol 163:2017–2022

Busca R, Abbe P, Mantoux F, Aberdam E, Peyssonnaux C, Eychene A, Ortonne JP, Ballotti R (2000) Ras mediates the cAMP-dependent activation of extracellular signal-regulated kinases (ERKs) in melanocytes. EMBO J 19:2900–2910

Cairns RA, Harris IS, Mak TW (2011) Regulation of cancer cell metabolism. Nat Rev Cancer 11:85–95

Chaichana KK, Chaichana KL (2011) Diagnosis and treatment options for brain metastasis of melanoma. In: Morton MR (ed) Treatment of Metastatic Melanoma. InTech, Maastricht

Chang HJ, Yoo BC, Lim SB, Jeong SY, Kim WH, Park JG (2005) Metabotropic glutamate receptor 4 expression in colorectal carcinoma and its prognostic significance. Clin Cancer Res 11:3288–3295

Chen S, Zhu H, Wetzel WJ, Philbert MA (1996) Spontaneous melanocytosis in transgenic mice. J Invest Dermatol 106:1145–1151

Cheng N, Chytil A, Shyr Y, Joly A, Moses HL (2008) Transforming growth factor-beta signaling-deficient fibroblasts enhance hepatocyte growth factor signaling in mammary carcinoma cells to promote scattering and invasion. Mol Cancer Res 6:1521–1533

Choe ES, Wang JQ (2002) Regulation of transcription factor phosphorylation by metabotropic glutamate receptor-associated signaling pathways in rat striatal neurons. Neuroscience 114:557–565

Choi KY, Chang K, Pickel JM, Badger JD 2nd, Roche KW (2011) Expression of the metabotropic glutamate receptor 5 (mGluR5) induces melanoma in transgenic mice. Proc Natl Acad Sci U S A 108:15219–15224

Ciceroni C, Bonelli M, Mastrantoni E, Niccolini C, Laurenza M, Larocca LM, Pallini R, Traficante A, Spinsanti P, Ricci-Vitiani L, Arcella A, De Maria R, Nicoletti F, Battaglia G, Melchiorri D (2013) Type-3 metabotropic glutamate receptors regulate chemoresistance in glioma stem cells, and their levels are inversely related to survival in patients with malignant gliomas. Cell Death Differ 20:396–407

D'Onofrio M, Arcella A, Bruno V, Ngomba RT, Battaglia G, Lombari V, Ragona G, Calogero A, Nicoletti F (2003) Pharmacological blockade of mGlu2/3 metabotropic glutamate receptors reduces cell proliferation in cultured human glioma cells. J Neurochem 84:1288–1295

Daphu I, Horn S, Stieber D, Varughese JK, Spriet E, Dale HA, Skaftnesmo KO, Bjerkvig R, Thorsen F (2014) In vitro treatment of melanoma brain metastasis by simultaneously targeting the MAPK and PI3K signaling pathways. Int J Mol Sci 15:8773–8794

DeBerardinis RJ, Mancuso A, Daikhin E, Nissim I, Yudkoff M, Wehrli S, Thompson CB (2007) Beyond aerobic glycolysis: transformed cells can engage in glutamine metabolism that exceeds the requirement for protein and nucleotide synthesis. Proc Natl Acad Sci U S A 104:19345–19350

Deberardinis RJ, Sayed N, Ditsworth D, Thompson CB (2008) Brick by brick: metabolism and tumor cell growth. Curr Opin Genet Dev 18:54–61

Degenhardt K, White E (2006) A mouse model system to genetically dissect the molecular mechanisms regulating tumorigenesis. Clin Cancer Res 12:5298–5304

Dimski T, Heinen-Lauten J, Krep H (2008) Basal tako-tsubo cardiomyopathy. Induction of a pheochromocytoma after general anesthesia. Anaesthesist 57:143–146

Doble A (1996) The pharmacology and mechanism of action of riluzole. Neurology 47:S233–S241

Eichler AF, Batchelor TT, Henson JW (2007) Diffusion and perfusion imaging in subacute neurotoxicity following high-dose intravenous methotrexate. Neuro-Oncology 9:373–377

Esseltine JL, Ferguson SS (2013) Regulation of G protein-coupled receptor trafficking and signaling by Rab GTPases. Small GTPases 4:132–135

Estrela JM, Ortega A, Obrador E (2006) Glutathione in cancer biology and therapy. Crit Rev Clin Lab Sci 43:143–181

Ewend MG, Carey LA, Brem H (1996) Treatment of melanoma metastases in the brain. Semin Surg Oncol 12:429–435

Fairman WA, Amara SG (1999) Functional diversity of excitatory amino acid transporters: ion channel and transport modes. Am J Physiol 277:F481–F486

Fassnacht M, Libe R, Kroiss M, Allolio B (2011) Adrenocortical carcinoma: a clinician's update. Nat Rev Endocrinol 7:323–335

Fendt SM, Bell EL, Keibler MA, Olenchock BA, Mayers JR, Wasylenko TM, Vokes NI, Guarente L, Vander Heiden MG, Stephanopoulos G (2013) Reductive glutamine metabolism is a function of the alpha-ketoglutarate to citrate ratio in cells. Nat Commun 4:2236

Ferraguti F, Baldani-Guerra B, Corsi M, Nakanishi S, Corti C (1999) Activation of the extracellular signal-regulated kinase 2 by metabotropic glutamate receptors. Eur J Neurosci 11:2073–2082

Filipp FV, Ratnikov B, De Ingeniis J, Smith JW, Osterman AL, Scott DA (2012) Glutamine-fueled mitochondrial metabolism is decoupled from glycolysis in melanoma. Pigment Cell Melanoma Res 25:732–739

Funasaka Y, Harada T, Aiba A, Nishigori C (1996) Expression of metabotropic glutamate receptor and phosphorylated extracellular signal-regulated kinase 1/2 proteins in human melanocytic lesions. Pigment Cell Res 19:656

Gaglio D, Soldati C, Vanoni M, Alberghina L, Chiaradonna F (2009) Glutamine deprivation induces abortive s-phase rescued by deoxyribonucleotides in k-ras transformed fibroblasts. PloS One 4:e4715

Gameiro PA, Yang J, Metelo AM, Perez-Carro R, Baker R, Wang Z, Arreola A, Rathmell WK, Olumi A, Lopez-Larrubia P, Stephanopoulos G, Iliopoulos O (2013) In vivo HIF-mediated reductive carboxylation is regulated by citrate levels and sensitizes VHL-deficient cells to glutamine deprivation. Cell Metab 17:372–385

Gao P, Tchernyshyov I, Chang TC, Lee YS, Kita K, Ochi T, Zeller KI, De Marzo AM, Van Eyk JE, Mendell JT, Dang CV (2009) c-Myc suppression of miR-23a/b enhances mitochondrial glutaminase expression and glutamine metabolism. Nature 458:762–765

Gelb T, Pshenichkin S, Hathaway HA, Grajkowska E, Dalley CB, Wolfe BB, Wroblewski JT (2015a) Atypical signaling of metabotropic glutamate receptor 1 in human melanoma cells. Biochem Pharmacol 98:182–189

Gelb T, Pshenichkin S, Rodriguez OC, Hathaway HA, Grajkowska E, DiRaddo JO, Wroblewska B, Yasuda RP, Albanese C, Wolfe BB, Wroblewski JT (2015b) Metabotropic glutamate receptor 1 acts as a dependence receptor creating a requirement for glutamate to sustain the viability and growth of human melanomas. Oncogene 34:2711–2720

Graus F, Rogers LR, Posner JB (1985) Cerebrovascular complications in patients with cancer. Medicine (Baltimore) 64:16–35

Gutkind JS, Novotny EA, Brann MR, Robbins KC (1991) Muscarinic acetylcholine receptor subtypes as agonist-dependent oncogenes. Proc Natl Acad Sci U S A 88:4703–4707

Habermalz HJ, Fischer JJ (1976) Radiation therapy of malignant melanoma: experience with high individual treatment doses. Cancer 38:2258–2262

Iacovelli L, Arcella A, Battaglia G, Pazzaglia S, Aronica E, Spinsanti P, Caruso A, De Smaele E, Saran A, Gulino A, D'Onofrio M, Giangaspero F, Nicoletti F (2006) Pharmacological activation of mGlu4 metabotropic glutamate receptors inhibits the growth of medulloblastomas. J Neurosci Off J Soc Neurosci 26:8388–8397

Julius D, Livelli TJ, Jessell TM, Axel R (1989) Ectopic expression of the serotonin 1c receptor and the triggering of malignant transformation. Science 244:1057–1062

Karakas E, Regan MC, Furukawa H (2015) Emerging structural insights into the function of ionotropic glutamate receptors. Trends Biochem Sci 40:328–337

Kelly A, Stanley CA (2001) Disorders of glutamate metabolism. Ment Retard Dev Disabil Res Rev 7:287–295

Khan AJ, Wall B, Ahlawat S, Green C, Schiff D, Mehnert JM, Goydos JS, Chen S, Haffty BG (2011) Riluzole enhances ionizing radiation-induced cytotoxicity in human melanoma cells that ectopically express metabotropic glutamate receptor 1 in vitro and in vivo. Clin Cancer Res 17:1807–1814

Koochekpour S (2013) Glutamate, a metabolic biomarker of aggressiveness and a potential therapeutic target for prostate cancer. Asian J Androl 15:212–213

Kuribayashi N, Uchida D, Kinouchi M, Takamaru N, Tamatani T, Nagai H, Miyamoto Y (2013) The role of metabotropic glutamate receptor 5 on the stromal cell-derived factor-1/CXCR4 system in oral cancer. PloS One 8:e80773

Lee HJ, Wall BA, Wangari-Talbot J, Shin SS, Rosenberg S, Chan JL, Namkoong J, Goydos JS, Chen S (2011a) Glutamatergic pathway targeting in melanoma: single-agent and combinatorial therapies. Clin Cancer Res 17:7080–7092

Lee J, Yang EJ, Shin JS, Kim DH, Lee SS, Choi IH (2011b) CKD-712, (S)-1-(alpha-naphthylmethyl)-6,7-dihydroxy-1,2,3,4-tetrahydroisoquinoline, Inhibits the NF-kappaB Activation and Augments Akt Activation during TLR4 Signaling. Immune Netw 11:420–423

Li L, Homan KT, Vishnivetskiy SA, Manglik A, Tesmer JJ, Gurevich VV, Gurevich EV (2015) G Protein-coupled Receptor Kinases of the GRK4 Protein Subfamily Phosphorylate Inactive G Protein-coupled Receptors (GPCRs). J Biol Chem 290:10775–10790

Lieu AS, Hwang SL, Howng SL, Chai CY (1999) Brain tumors with hemorrhage. J Formos Med Assoc 98:365–367

Mandybur TI, Nagpaul AS, Pappas Z, Niklowitz WJ (1977) Alzheimer neurofibrillary change in subacute sclerosing panencephalitis. Ann Neurol 1:103–107

Marie SK, Shinjo SM (2011) Metabolism and brain cancer. Clinics (Sao Paulo) 66(Suppl 1):33–43

Marin YE, Chen S (2004) Involvement of metabotropic glutamate receptor 1, a G protein coupled receptor, in melanoma development. J Mol Med (Berl) 82:735–749

Marin YE, Namkoong J, Cohen-Solal K, Shin SS, Martino JJ, Oka M, Chen S (2006) Stimulation of oncogenic metabotropic glutamate receptor 1 in melanoma cells activates ERK1/2 via PKCepsilon. Cell Signal 18:1279–1286

Martino JJ, Wall BA, Mastrantoni E, Wilimczyk BJ, La Cava SN, Degenhardt K, White E, Chen S (2013) Metabotropic glutamate receptor 1 (Grm1) is an oncogene in epithelial cells. Oncogene 32:4366–4376

McKay MJ, Kefford RF (1995) The spectrum of in vitro radiosensitivity in four human melanoma cell lines is not accounted for by differential induction or rejoining of DNA double strand breaks. Int J Radiat Oncol Biol Phys 31:345–352

Mehnert JM, Wen Y, Lee JH, Jeong BS, Li J, Dudek Pruski-Clark LL, Kane MM, Lin H, Shih W, Chen S, Goydos JS (2016) A phase II trial of riluzole, an antagonist of metabotropic glutamate receptor 1 (GRM1) signaling, in patients with advanced melanoma. J Clin Oncol 29:8557

Mehta MS, Dolfi SC, Bronfenbrener R, Bilal E, Chen C, Moore D, Lin Y, Rahim H, Aisner S, Kersellius RD, Teh J, Chen S, Toppmeyer DL, Medina DJ, Ganesan S, Vazquez A, Hirshfield KM (2013) Metabotropic glutamate receptor 1 expression and its polymorphic variants associate with breast cancer phenotypes. PloS One 8:e69851

Miller RG (1999) Carrell-Krusen Symposium invited lecture. Clinical trials in motor neuron diseases. J Child Neurol 14:173–179

Moreadith RW, Lehninger AL (1984) The pathways of glutamate and glutamine oxidation by tumor cell mitochondria. Role of mitochondrial NAD(P)+-dependent malic enzyme. J Biol Chem 259:6215–6221

Mulvihill MJ, Cooke A, Rosenfeld-Franklin M, Buck E, Foreman K, Landfair D, O'Connor M, Pirritt C, Sun Y, Yao Y, Arnold LD, Gibson NW, Ji QS (2009) Discovery of OSI-906: a selective and orally efficacious dual inhibitor of the IGF-1 receptor and insulin receptor. Future Med Chem 1:1153–1171

Naarala J, Nykvist P, Tuomala M, Savolainen K (1993) Excitatory amino acid-induced slow biphasic responses of free intracellular calcium in human neuroblastoma cells. FEBS Lett 330:222–226

Namkoong J, Shin SS, Lee HJ, Marin YE, Wall BA, Goydos JS, Chen S (2007) Metabotropic glutamate receptor 1 and glutamate signaling in human melanoma. Cancer Res 67:2298–2305

Nicholas S, Mathios D, Jackson C, Lim M (2013) Metastatic melanoma to the brain: surgery and radiation is still the standard of care. Curr Treat Options Oncol 14:264–279

Niklas J, Heinzle E (2012) Metabolic flux analysis in systems biology of mammalian cells. Adv Biochem Eng Biotechnol 127:109–132

Nygaard V, Prasmickaite L, Vasiliauskaite K, Clancy T, Hovig E (2014) Melanoma brain colonization involves the emergence of a brain-adaptive phenotype. Oncoscience 1:82–94

Ohtani Y, Harada T, Funasaka Y, Nakao K, Takahara C, Abdel-Daim M, Sakai N, Saito N, Nishigori C, Aiba A (2008) Metabotropic glutamate receptor subtype-1 is essential for in vivo growth of melanoma. Oncogene 27:7162–7170

Ortiz P, Vanaclocha F, Lopez-Bran E, Esquivias JI, Lopez-Estebaranz JL, Martin-Gonzalez M, Arrue I, Garcia-Romero D, Ochoa C, Gonzalez-Perez A, Ruiz A, Real LM (2007) Genetic analysis of the GRM1 gene in human melanoma susceptibility. Eur J Hum Genet 15:1176–1182

Park SY, Lee SA, Han IH, Yoo BC, Lee SH, Park JY, Cha IH, Kim J, Choi SW (2007) Clinical significance of metabotropic glutamate receptor 5 expression in oral squamous cell carcinoma. Oncol Rep 17:81–87

Petralia RS, Wang YX, Niedzielski AS, Wenthold RJ (1996) The metabotropic glutamate receptors, mGluR2 and mGluR3, show unique postsynaptic, presynaptic and glial localizations. Neurosci 71:949–976

Pissimissis N, Papageorgiou E, Lembessis P, Armakolas A, Koutsilieris M (2009) The glutamatergic system expression in human PC-3 and LNCaP prostate cancer cells. Anticancer Res 29:371–377

Pollock PM, Cohen-Solal K, Sood R, Namkoong J, Martino JJ, Koganti A, Zhu H, Robbins C, Makalowska I, Shin SS, Marin Y, Roberts KG, Yudt LM, Chen A, Cheng J, Incao A, Pinkett HW, Graham CL, Dunn K, Crespo-Carbone SM, Mackason KR, Ryan KB, Sinsimer D, Goydos J, Reuhl KR, Eckhaus M, Meltzer PS, Pavan WJ, Trent JM, Chen S (2003) Melanoma mouse model implicates metabotropic glutamate signaling in melanocytic neoplasia. Nat Genet 34:108–112

Prickett TD, Samuels Y (2012) Molecular pathways: dysregulated glutamatergic signaling pathways in cancer. Clin Cancer Res 18:4240–4246

Prickett TD, Wei X, Cardenas-Navia I, Teer JK, Lin JC, Walia V, Gartner J, Jiang J, Cherukuri PF, Molinolo A, Davies MA, Gershenwald JE, Stemke-Hale K, Rosenberg SA, Margulies EH, Samuels Y (2011) Exon capture analysis of G protein-coupled receptors identifies activating mutations in GRM3 in melanoma. Nat Genet 43:1119–1126

Ribeiro FM, Paquet M, Cregan SP, Ferguson SS (2010) Group I metabotropic glutamate receptor signalling and its implication in neurological disease. CNS Neurol Disord Drug Targets 9:574–595

Roskoski R Jr (2007) Sunitinib: a VEGF and PDGF receptor protein kinase and angiogenesis inhibitor. Biochem Biophys Res Commun 356:323–328

Savage SA, Mirabello L, Wang Z, Gastier-Foster JM, Gorlick R, Khanna C, Flanagan AM, Tirabosco R, Andrulis IL, Wunder JS, Gokgoz N, Patino-Garcia A, Sierrasesumaga L, Lecanda F, Kurucu N, Ilhan IE, Sari N, Serra M, Hattinger C, Picci P, Spector LG, Barkauskas DA, Marina N, de Toledo SR, Petrilli AS, Amary MF, Halai D, Thomas DM, Douglass C, Meltzer PS, Jacobs K, Chung CC, Berndt SI, Purdue MP, Caporaso NE, Tucker M, Rothman N, Landi MT, Silverman DT, Kraft P, Hunter DJ, Malats N, Kogevinas M, Wacholder S, Troisi R, Helman L, Fraumeni JF Jr, Yeager M, Hoover RN, Chanock SJ (2013) Genome-wide association study identifies two susceptibility loci for osteosarcoma. Nat Genet 45:799–803

Shin SS, Namkoong J, Wall BA, Gleason R, Lee HJ, Chen S (2008) Oncogenic activities of metabotropic glutamate receptor 1 (Grm1) in melanocyte transformation. Pigment Cell Melanoma Res 21:368–378

Spanaki C, Plaitakis A (2012) The role of glutamate dehydrogenase in mammalian ammonia metabolism. Neurotox Res 21:117–127

Speyer CL, Smith JS, Banda M, DeVries JA, Mekani T, Gorski DH (2012) Metabotropic glutamate receptor-1: a potential therapeutic target for the treatment of breast cancer. Breast Cancer Res Treat 132:565–573

Speyer CL, Hachem AH, Assi AA, Johnson JS, DeVries JA, Gorski DH (2014) Metabotropic glutamate receptor-1 as a novel target for the antiangiogenic treatment of breast cancer. PloS One 9:e88830

Stepulak A, Luksch H, Gebhardt C, Uckermann O, Marzahn J, Sifringer M, Rzeski W, Staufner C, Brocke KS, Turski L, Ikonomidou C (2009) Expression of glutamate receptor subunits in human cancers. Histochem Cell Biol 132:435–445

Stepulak A, Luksch H, Uckermann O, Sifringer M, Rzeski W, Polberg K, Kupisz K, Klatka J, Kielbus M, Grabarska A, Marzahn J, Turski L, Ikonomidou C (2011) Glutamate receptors in laryngeal cancer cells. Anticancer Res 31:565–573

Sun RC, Denko NC (2014) Hypoxic regulation of glutamine metabolism through HIF1 and SIAH2 supports lipid synthesis that is necessary for tumor growth. Cell Metab 19:285–292

Takano T, Lin JH, Arcuino G, Gao Q, Yang J, Nedergaard M (2001) Glutamate release promotes growth of malignant gliomas. Nat Med 7:1010–1015

Takayama H, LaRochelle WJ, Sharp R, Otsuka T, Kriebel P, Anver M, Aaronson SA, Merlino G (1997) Diverse tumorigenesis associated with aberrant development in mice overexpressing hepatocyte growth factor/scatter factor. Proc Natl Acad Sci U S A 94:701–706

Teh JL, Chen S (2012) Glutamatergic signaling in cellular transformation. Pigment Cell Melanoma Res 25:331–342

Teh JL, Shah R, Shin SS, Wen Y, Mehnert JM, Goydos J, Chen S (2014) Metabotropic glutamate receptor 1 mediates melanocyte transformation via transactivation of insulin-like growth factor 1 receptor. Pigment Cell Melanoma Res 27:621–629

Thandi S, Blank JL, Challiss RA (2002) Group-I metabotropic glutamate receptors, mGlu1a and mGlu5a, couple to extracellular signal-regulated kinase (ERK) activation via distinct, but overlapping, signalling pathways. J Neurochem 83:1139–1153

Ti TK (1977) Advances in the surgical treatment of peptic ulcer. Med J Malaysia 32:186–188

Van den Bosch L (2006) The causes and mechanism of selective motor neuron death in amyotrophic lateral sclerosis. Verh K Acad Geneeskd Belg 68:249–269

Wall BA, Shin SS, Chen S (2013) Glutamate signaling in human cancers. In: Davids LM (ed) Recent Advances in the Biology, Therapy and Management of Melanoma. InTech, Maastricht

Wall BA, Wangari-Talbot J, Shin SS, Schiff D, Sierra J, Yu LJ, Khan A, Haffty B, Goydos JS, Chen S (2014) Disruption of GRM1-mediated signalling using riluzole results in DNA damage in melanoma cells. Pigment Cell Melanoma Res 27:263–274

Wall BA, Yu LJ, Khan A, Haffty B, Goydos JS, Chen S (2015) Riluzole is a radio-sensitizing agent in an in vivo model of brain metastasis derived from GRM1 expressing human melanoma cells. Pigment Cell Melanoma Res 28:105–109

Wen Y, Li J, Koo J, Shin SS, Lin Y, Jeong BS, Mehnert JM, Chen S, Cohen-Sola KA, Goydos JS (2014) Activation of the glutamate receptor GRM1 enhances angiogenic signaling to drive melanoma progression. Cancer Res 74:2499–2509

Willard SS, Koochekpour S (2013) Glutamate, glutamate receptors, and downstream signaling pathways. Int J Biol Sci 9:948–959

Wise DR, DeBerardinis RJ, Mancuso A, Sayed N, Zhang XY, Pfeiffer HK, Nissim I, Daikhin E, Yudkoff M, McMahon SB, Thompson CB (2008) Myc regulates a transcriptional program that stimulates mitochondrial glutaminolysis and leads to glutamine addiction. Proc Natl Acad Sci U S A 105:18782–18787

Wise DR, Ward PS, Shay JE, Cross JR, Gruber JJ, Sachdeva UM, Platt JM, DeMatteo RG, Simon MC, Thompson CB (2011) Hypoxia promotes isocitrate dehydrogenase-dependent carboxylation of alpha-ketoglutarate to citrate to support cell growth and viability. Proc Natl Acad Sci U S A 108:19611–19616

Wroblewska B, Santi MR, Neale JH (1998) N-acetylaspartylglutamate activates cyclic AMP-coupled metabotropic glutamate receptors in cerebellar astrocytes. Glia 24:172–179

Wronski M, Arbit E (2000) Surgical treatment of brain metastases from melanoma: a retrospective study of 91 patients. J Neurosurg 93:9–18

Yip D, Le MN, Chan JL, Lee JH, Mehnert JA, Yudd A, Kempf J, Shih WJ, Chen S, Goydos JS (2009) A phase 0 trial of riluzole in patients with resectable stage III and IV melanoma. Clin Cancer Res 15:3896–3902

Yoo BC, Jeon E, Hong SH, Shin YK, Chang HJ, Park JG (2004) Metabotropic glutamate receptor 4-mediated 5-Fluorouracil resistance in a human colon cancer cell line. Clin Cancer Res 10:4176–4184

Yu LJ, Wall BA, Chen S (2015) The current management of brain metastasis in melanoma: a focus on riluzole. Expert Rev Neurother 15:779–792

Zamboni N (2011) 13C metabolic flux analysis in complex systems. Curr Opin Biotechnol 22:103–108

Zhang C, Yuan XR, Li HY, Zhao ZJ, Liao YW, Wang XY, Su J, Sang SS, Liu Q (2015) Anti-cancer effect of metabotropic glutamate receptor 1 inhibition in human glioma U87 cells: involvement of PI3K/Akt/mTOR pathway. Cell Physiol Biochem 35:419–432

Zhao H, Desai V, Wang J, Epstein DM, Miglarese M, Buck E (2012) Epithelial-mesenchymal transition predicts sensitivity to the dual IGF-1R/IR inhibitor OSI-906 in hepatocellular carcinoma cell lines. Mol Cancer Ther 11:503–513

Zhu H, Reuhl K, Zhang X, Botha R, Ryan K, Wei J, Chen S (1998) Development of heritable melanoma in transgenic mice. J Invest Dermatol 110:247–252

Chapter 10
mGlu5 Receptors in Parkinson's Disease and MPTP-Lesioned Monkeys: Behavior and Brain Molecular Correlates

Nicolas Morin and Thérèse Di Paolo

Abstract Glutamate overactivity is well documented in Parkinson's disease (PD) and dyskinesias induced by L-3,4-dihydroxyphenylalanine (L-DOPA), the gold-standard treatment for this disease. The contribution of metabotropic glutamate receptors type 5 (mGlu5 receptors) in PD and L-DOPA-induced dyskinesias (LID) was the topic of investigations in human PD patients and in 1-methyl-4-phenyl-1,2,3,6-tetrahydropyridine (MPTP) monkeys.

Behaviorally, it has been shown that the prototypical mGlu5 antagonist 2-methyl-6-(phenylethynyl)pyridine (MPEP) as well as the mGlu5 receptor antagonists 3-[(2-methyl-1,3-thiazol-4-yl)ethynyl]pyridine (MTEP) and mavoglurant (AFQ056) acutely attenuated LID in MPTP monkeys. Moreover, a chronic 1-month administration of MPEP to previously drug-naïve MPTP monkeys (de novo treatment) attenuated the development of LID. Acute and chronic MPEP treatments of MPTP monkeys maintained the antiparkinsonian effect of L-DOPA. Mavoglurant was also shown in some clinical studies to reduce LID in PD patients.

Using the selective mGlu5 receptors ligand [^{3}H]ABP688, mGlu5 receptor-specific binding was measured by autoradiography in brains slices of normal and PD patients in relation to motor complications associated with an L-DOPA treatment. PD patients with motor complications (either LID or wearing-off) had higher [^{3}H]ABP688-specific binding compared to those without motor complications and controls in putamen, external and internal globus pallidus. In monkeys with a MPTP lesion and controls, [^{3}H]ABP688- and [^{3}H]MPEP-specific bindings were elevated in the striatum of dyskinetic L-DOPA-treated MPTP monkeys but not in MPTP monkeys without LID compared to controls.

The brain molecular correlates of the long-term effect of a 1-month administration of MPEP with L-DOPA that attenuated the development of LID were shown to

N. Morin, Ph.D. • T. Di Paolo, Ph.D. (✉)
Faculty of Pharmacy, Université Laval, Quebec City, Canada, G1K 7P4

Neuroscience Research Unit, Centre de recherche du CHU de Québec,
2705 Laurier Boulevard, Quebec City, Canada, G1V 4G2
e-mail: nicolas.morin.4@ulaval.ca; therese.dipaolo@crchudequebec.ulaval.ca

R.T. Ngomba et al. (eds.), *mGLU Receptors*, The Receptors 31,
DOI 10.1007/978-3-319-56170-7_10

extend beyond mGlu5 receptors. In the basal ganglia, it has been showed that the L-DOPA-induced changes of NMDA and AMPA ionotropic glutamate receptors as well as mGlu2/mGlu3 receptors were prevented with the addition of MPEP. Moreover, MPEP normalized the L-DOPA-induced changes of dopamine D2 receptors, their associated signaling (ERK and Akt) and neuropeptides (preproenkephalin, preprodynorphin), as well as the serotonin receptors 5-HT_{2A} and 5-HT_{1B}.

In conclusion, these results have shown in humans and in a nonhuman primate model of PD reduction of LID with mGlu5 antagonism. In the basal ganglia, LID were associated with changes of various glutamatergic, dopaminergic, and serotoninergic markers that were normalized with adjunct treatment with an mGlu5 antagonist supporting the mGlu5 receptor as a good target for the treatment of LID.

Keywords Parkinson's disease • L-DOPA-induced dyskinesia • Motor complications • Glutamate receptor • Basal ganglia • Direct pathway • Indirect pathway • Receptor interaction

Abbreviations

5-HT	serotonin
6-OHDA	6-hydroxydopamine
Akt	protein kinase B
AMPA	α-amino-3-hydroxy-5-methyl-4-isoxazolepropionic acid
cAMP	cyclic adenosine monophosphate
DA	dopamine
DHA	docosahexaenoic acid
dipraglurant	ADX-48621
ERK	extracellular signal-regulated kinase
GABA	γ-aminobutyric acid
GAD	glutamic acid decarboxylase
GSK3β	glycogen synthase kinase-3 β
GP	globus pallidus;
iGlu	ionotropic glutamate
KA	kainate
L-DOPA	levodopa (L-3,4-dihydroxyphenylalanine)
LID	L-DOPA-induced dyskinesias
MAPK	mitogen-activated protein kinase
mavoglurant	AFQ056
mGlu	metabotropic glutamate
MPEP	2-methyl-6-(phenylethynyl)pyridine
MPTP	1-methyl-4-phenyl-1,2,3,6-tetrahydropyridine
MTEP	3-[(2-methyl-1,3-thiazol-4-yl)ethynyl]pyridine
NAM	negative allosteric modulator

NMDA	*N*-methyl-D-aspartate
PD	Parkinson's disease
PI	polyphosphoinositide
SERT	serotonin transporter
STN	subthalamic nucleus

10.1 Introduction

Parkinson's disease (PD) is the most common neurodegenerative movement disorder and is likely to increase due to the aging population (Siderowf and Stern 2003). The cardinal motor manifestations of PD, tremor, bradykinesia, and rigidity are secondary to a loss of dopamine in the striatum (Olanow et al. 2009; Toulouse and Sullivan 2008). PD is principally attributed to the death of dopamine (DA) neurons in the substantia nigra, but other neurotransmitters and neuromodulators, such as glutamate, serotonin (5-HT), and adenosine, are also affected (Toulouse and Sullivan 2008). Gene mutations in familial PD are reported, but the cause for the majority of PD cases remains unknown (Olanow et al. 2009). There is currently no cure for PD. Neuroprotection or disease modification defined as an intervention that would protect or rescue vulnerable neurons, thereby slowing, stopping, or reversing disease progression, is not yet available for PD, but laboratory studies are finding promising agents (Olanow et al. 2009).

Restoring lost DA with its precursor, L-3,4-dihydroxyphenylalanine (L-DOPA), introduced 50 years ago still remains a very effective and commonly used treatment for PD (Mercuri and Bernardi 2005). However, up to 80% of PD patients will develop motor complications after 10 years of treatment with L-DOPA (Olanow and Koller 1998). These motor complications include motor fluctuations and abnormal involuntary movements, such as L-DOPA-induced dyskinesias (LID), and contribute to limit the quality of life in PD patients and can be very difficult to manage (Fabbrini et al. 2007). Motor fluctuations such as "wearing-off" are also common. Wearing-off is defined as a reduced duration of benefit from an individual L-DOPA dose and a recurrence of parkinsonian symptoms before the next normal dose of L-DOPA (Fahn et al. 2004). Hence, most PD patients initiated with DA agonist monotherapy will eventually require L-DOPA as disease progresses, and after 10 years their motor complications appear similar as they would have if started initially on L-DOPA therapy (Katzenschlager et al. 2008; Parkinson Study Group 2009). This suggests that disease progression plays the major role in the onset of dyskinesia rather than the type of dopaminergic drug treatment used. Involuntary movements such as LID are paralleled by aberrant forms of plasticity characterized by changes in various neurotransmitter systems and their intracellular signaling pathways (Jenner 2008).

Aside from some benefit of amantadine, that has anti-glutamatergic properties, no drug is yet available for LID (Meissner et al. 2011). Glutamate neurotransmission is increased in the basal ganglia in PD and LID (Klockgether and Turski 1993;

Chase and Oh 2000; Calon et al. 2003a). The pathophysiology and the mechanisms involved in the development of LID are still not fully understood. However, altered dopaminergic and nondopaminergic neurotransmission in the basal ganglia is observed in LID (Blandini and Armentero 2012). Treating LID with adjunct drugs targeting nondopaminergic neurotransmitter systems is proposed such as glutamate to indirectly modulate basal ganglia DA neurotransmission (Brotchie 1998; Morin et al. 2014a). Much emphasis has therefore been placed on finding alternative non-dopaminergic drugs that could circumvent some or all these problems. The design of novel agents to prevent dyskinesias requires elucidation of the adaptive changes produced in the parkinsonian brain by repeated administration of L-DOPA. The use of adjunct drugs to modulate basal ganglia DA neurotransmission is an important strategy to treat LID (Brotchie 1998, 2003; Henry et al. 2001; Calon and Di Paolo 2002; Blanchet et al. 1999; Grondin et al. 1999).

LID are typically observed at the peak of the effect of L-DOPA in PD patients. There is also diphasic dyskinesia at the beginning and at the end of the L-DOPA dosing cycle appearing with the rise and fall of DA levels in the brain (Luquin et al. 1992) and off-dystonia (Marsden et al. 1982). LID occur in 30–80% of PD patients treated with L-DOPA (Barbeau 1980; Nutt 1990). Two conditions are necessary for their appearance: (1) the loss of DA in the nigrostriatal pathway and (2) treatment with L-DOPA or DA agonists. The development of dyskinesias in man usually requires daily treatment for 3–5 years in idiopathic PD (Klawans et al. 1977), and for parkinsonism induced in man by 1-methyl-4-phenyl-1,2,3,6-tetrahydropyridine (MPTP), it occurs after only weeks or months of treatment (Ballard et al. 1985). The same applies to MPTP monkeys where only weeks of L-DOPA therapy are enough before dyskinesias appear (Falardeau et al. 1988; Bedard et al. 1986). MPTP primates respond to DA therapies in a similar manner than idiopathic PD patients (Jenner 2003a, b) and are currently the best model for studying LID.

Glutamate is involved in many physiological functions through its interactions with ionotropic glutamate (iGlu), ligand-gated channel, and metabotropic G-protein-coupled glutamate (mGlu) receptors. iGlu receptor drugs suppressing glutamate excitatory transmission often create undesirable side effects (Johnson et al. 2009), whereas acting on mGlu receptors could lead to a subtler and/or circuit-selective modulation of excitatory transmission (Conn et al. 2005). Pharmacologic characterization of mGlu5 receptors and its selective negative allosteric modulators (NAMs) shows therapeutic potential in animal models of PD (Grégoire et al. 2011; Johnston et al. 2010; Morin et al. 2010) and recently efficacy in human PD (Berg et al. 2011; Stocchi et al. 2013). mGlu5 receptors have been shown to play a crucial role in regulating L-DOPA-induced motor behavior, but the mechanisms involved remain unclear (Gasparini et al. 2013).

This review will cover relevant studies investigating mGlu5 receptor subtypes in the pathophysiology of PD and LID. Brain biochemical correlates of motor complications in PD patients and MPTP-lesioned monkeys will be reviewed (Morin et al. 2014a).

10.2 Nonhuman Primate Models of L-DOPA-Induced Dyskinesias

Similar etiology and functions of a particular human disease are two important characteristics to model a disease. The neurotoxin MPTP closely mimics both behavioral changes and cellular loss as seen in PD (Albanese et al. 1993). The administration of MPTP in primates promotes the development of rigidity, bradykinesia, and tremor, which are the primary motor features of PD. MPTP induces neuronal death of dopaminergic cells through a cascade of intracellular reactions by targeting specifically cells expressing the DA transporter (Smeyne and Jackson-Lewis 2005). MPTP-treated primates remain after nearly 30 years the gold standard for the study of PD and LID, as well as to evaluate the efficacy of novel compounds (Morin et al. 2014a). Acute and de novo experimental approaches are used to test new antiparkinsonian and antidyskinetic pharmacological agents. In acute approaches, animals are rendered parkinsonian with the administration of MPTP and then chronically treated with L-DOPA for several weeks, until they express well-established, constant, and stable LID. Then, the acute effects of compounds are tested with the co-administration of L-DOPA, and the motor behavior of the animals is evaluated (Grégoire et al. 2009, 2011; Bezard et al. 2004). This approach allows rapid testing of new compounds and its tolerability, and animals may be used for several studies (Morin et al. 2014a). These animals can also be used in a chronic treatment paradigm to measure possible development of tolerance to the investigated drug. In the de novo approach, two or more groups of never treated animals are rendered parkinsonian and then treated with L-DOPA alone or in combination with the new agent. This approach allows verifying if the compounds tested can reduce or prevent the development of LID and evaluating if the duration of the L-DOPA effect decreases over time, a motor complication also called "wearing-off" (Morin et al. 2012; Samadi et al. 2006; Rylander et al. 2010; Grégoire et al. 2008; Hadj Tahar et al. 2004). Moreover, measurements of biochemical changes in the brains are made possible if the animals are euthanized at the end of the protocol (Morin et al. 2012; Samadi et al. 2008a; Ouattara et al. 2010).

As observed in PD patients, the macaque monkey model will display its own pattern of parkinsonian symptoms (Rajput et al. 2009). Indeed, LID involves one or more parts of the body, and each of them should therefore be quantified separately (Hadj Tahar et al. 2000). Several scales are currently available to measure and quantify dyskinesia and were recently reviewed (Fox et al. 2012). Objective measures of bradykinesia with specific motor tasks are also important to separate the antidyskinetic from antiparkinsonian activity of compounds tested (Jourdain et al. 2013). Finally, monkeys will tend to exhibit a worsening of motor symptoms before and after an acute L-DOPA challenge (Kuoppamäki et al. 2002), as seen in some PD patients (Evans et al. 2012). Thus monkey model can replicate PD conditions and underlines their usefulness in the study of new treatments (Morin et al. 2014a).

Individual titration of L-DOPA is often needed to elicit the same amount of LID among the animals (Grégoire et al. 2011; Johnston et al. 2010) even if these animals are equally denervated (Guigoni et al. 2005). However, positive correlations between L-DOPA dose and the duration and the severity of LID were observed (Kuoppamäki et al. 2007). Moreover, in dyskinetic macaques, the administration of L-DOPA after drug holiday lasting few weeks will trigger the same LID as measured before. The same observation was made in PD patients in whom L-DOPA was stopped (Goetz et al. 1982; Mayeux et al. 1985). This supports the feasibility of acute studies with the same animals therefore keeping the number needed and consequently reducing the costs.

The reappearance of LID after a withdrawal indicates that permanent or at least long-term changes are occurring in the brain basal ganglia and these changes may be studied in postmortem brain tissues. Wearing-off, described as shortening in the duration of response to L-DOPA with gradual reappearance of parkinsonian symptoms, is another important motor side effect of chronic L-DOPA administration (Fahn et al. 2004). Parkinsonian patients usually experience such end-of-dose deterioration after several months or years of treatment (Pahwa and Lyons 2009). Wearing-off can be also replicated in de novo MPTP monkeys with a shortening of the antiparkinsonian effect of L-DOPA as reported after 2 weeks of treatment (Morin et al. 2013a). Thus, the MPTP-lesioned monkey model is probably one of the best models for studying LID and PD in humans, and this model has brought significant advances for the treatment of LID. For example, docosahexaenoic acid (DHA) and cabergoline were shown to reduce the severity or delay the development of LID in MPTP-lesioned monkey in chronic and de novo treatments (Samadi et al. 2006; Belanger et al. 2003).

10.3 Glutamate Neurotransmission in PD and L-DOPA-Induced Dyskinesias

Glutamate is the brain most abundant excitatory neurotransmitter mediating as much as 70% of synaptic transmission in the central nervous system, and its overactivity is well documented in PD and LID (Klockgether and Turski 1993; Morin and Di Paolo 2014; Samadi et al. 2007). Amantadine, a noncompetitive antagonist at N-methyl-D-aspartate (NMDA) receptors, is currently the only drug used in the clinic shown to reduce the severity of LID in some PD patients without worsening parkinsonian symptoms (Meissner et al. 2011; Verhagen Metman et al. 1998a; Sawada et al. 2010). Amantadine also improves akinesia, rigidity, and tremor (Olanow et al. 2009). However, the antidyskinetic effect of amantadine may be transient and is often lost within the first year of treatment (Stocchi et al. 2008). In addition, many PD patients cannot tolerate high doses of amantadine because of cognitive impairment, which significantly limits its use (Stocchi et al. 2008).

Riluzole, a nonselective inhibitor of glutamate neurotransmission, was shown to reduce the severity of L-DOPA-induced motor complications in 6-hydroxydopamine (6-OHDA)-lesioned rat model of PD (Papa et al. 1995; Marin et al. 1996;

Engber et al. 1994; Marin et al. 2000). Moreover, in the same animal model, the glutamate transporter GLT1 was reported to be increased (Oueslati et al. 2007; Robelet et al. 2004). However, riluzole was not effective in humans to reduce the expression of LID (Braz et al. 2004; Bara-Jimenez et al. 2006). Despite intensive search, no drug other than amantadine has demonstrated in the clinic an antidykinesic effect that is not associated with a worsening of parkinsonism (Olanow et al. 2009) underlying the complexity of brain changes associated with dyskinesias. mGlu receptors are topics of more recent interest in PD (Samadi et al. 2007).

Inhibiting glutamate neurotransmission in PD can also be neuroprotective since glutamate overactivity is excitotoxic (Planells-Cases et al. 2006; Gardoni and Di Luca 2006). Motor complications are in part L-DOPA dose related; protecting DA neurons could delay motor complications by using less L-DOPA. Moreover, nonmotor features of PD such as depression and anxiety are common (Olanow et al. 2009) and could benefit from mGlu receptor drugs such as mGlu5 receptor antagonists that show anxiolytic and antidepressant activity (Palucha and Pilc 2007; Witkin et al. 2007).

10.4 Ionotropic Glutamate Receptors and L-DOPA-Induced Dyskinesias

iGlu receptors are classified into NMDA, α-amino-3-hydroxy-5-methyl-4-isoxazolepropionic acid (AMPA), and kainate (KA) receptors (Finlay and Duty 2014; Michaelis 1998; Dingledine et al. 1999), and they mediate fast excitatory neurotransmission, whereas mGlu receptors mediate slower modulatory neurotransmission (Samadi et al. 2007). The blockade of NMDA and AMPA receptors with specific antagonists was shown to reduce the development of L-DOPA-induced motor complications in 6-OHDA-lesioned rat model (Marin et al. 2000). Moreover, in both parkinsonian patients with LID (Calon et al. 2003a) and dyskinetic MPTP monkeys (Calon et al. 2002; Huot et al. 2013), important increases in striatal NMDA and AMPA receptor-binding levels were observed. The NMDA antagonist, CI-1041, is reported to prevent the development of LID in parkinsonian monkeys (Hadj Tahar et al. 2004) and associated brain molecular changes (Morissette et al. 2006). Interestingly, in these monkeys, CI-1041 also prevented the increase of striatal mGlu5 receptor levels (Ouattara et al. 2011). Clinical trials also show the antidyskinetic profile of dextrorphan, dextromethorphan, and amantadine, known to block NMDA receptors (Meissner et al. 2011; Sawada et al. 2010; Blanchet et al. 1996, 1997; Verhagen Metman et al. 1998b, 1999; Snow et al. 2000; Rajput et al. 1998; Ruzicka et al. 2000; Luginger et al. 2000).

Kynurenic acid antagonizes glycine B site of AMPA, NMDA, and KA receptors (Schwarcz and Pellicciari 2002; Hilmas et al. 2001; Stone 1993, 2000; Moroni 1999) and inhibits glutamate release (Carpenedo et al. 2001; Nemeth et al. 2006). RO 61-8048 an inhibitor of kynurenine hydroxylase activity increases kynurenic acid levels (Stone 2001). In MPTP monkeys, an acute treatment with RO 61-8048 reduced the severity of LID (Samadi et al. 2005); chronically in de novo treated MPTP monkeys, it reduced their development (Grégoire et al. 2008).

Recent studies have explored the role and the implication of NMDA and AMPA receptor subunits in rodent and nonhuman primate models of PD in LID including the glycine site, NMDA GluN2D subunits, AMPA receptor subunit composition, and NMDA/AMPA receptor ratio (Finlay and Duty 2014; Bagetta et al. 2012; Kobylecki et al. 2010; Zhang et al. 2014; Errico et al. 2011; Heresco-Levy et al. 2013). Nevertheless, significant adverse effects such as cognitive impairment in many patients can occur which limit the use of an iGlu receptor antagonist (Stocchi et al. 2008b; Stayte and Vissel 2014). Recent efforts were devoted to pharmacologically manipulate glutamate transmission with selective mGlu receptor ligands.

10.5 Metabotropic Glutamate Receptors and L-DOPA-Induced Dyskinesias

The mGlu receptors constitute a family of G-protein-coupled receptors comprising eight subtypes that are classified into three groups based on the signal transduction pathway, homology of the amino acid sequence, and receptor pharmacology (Pisani et al. 2003; Conn and Pin 1997). Group I (mGlu1, 5) couples to Gq and stimulates polyphosphoinositide (PI) hydrolysis, while groups II (mGlu2, 3) and III (mGluR4, mGluR6, mGluR7, mGluR8) couple to Gi/Go and inhibit increase in cyclic adenosine monophosphate (cAMP) (Conn and Pin 1997). All mGlu receptor subtypes are mainly located in the brain basal ganglia, except mGlu6 receptor found primarily in the retina (Niswender and Conn 2010) (Fig. 10.1). Presynaptically localized group II and group III mGlu receptors are thought to represent the classical inhibitory autoreceptor mechanism suppressing excess glutamate release from presynaptic terminals (Schoepp 2001). The majority of group I mGlu receptor, including mGlu5, is located postsynaptically on the perisynaptic annulus of dendritic spines, which lead to enhanced neuronal excitation (Lujan et al. 1997).

While the term "antagonist" is widely used when applied to ligands that inhibit the function of specific mGlu receptors, the more appropriate term is "negative allosteric modulator" (NAM), since they inhibit the function of the receptor at a site distal to the actual orthosteric ligand-binding domain of the receptor and only in the presence of glutamate (Olive 2009). Group I NAM may also function as inverse agonists; in cell-based assays they can inhibit the basal (constitutive) activity of group I mGlu receptors in absence of any orthosteric agonist or even when the glutamate-binding domain is removed or mutated (Olive 2009). Moreover, mGlu receptors have a higher affinity for glutamate than iGlu receptors (Conn et al. 2005; Marino et al. 2002). On the basis of these considerations, combined with the rich distribution and diverse physiological roles of mGlu receptors within the basal ganglia, recent attention has been placed on these receptors as alternative targets to

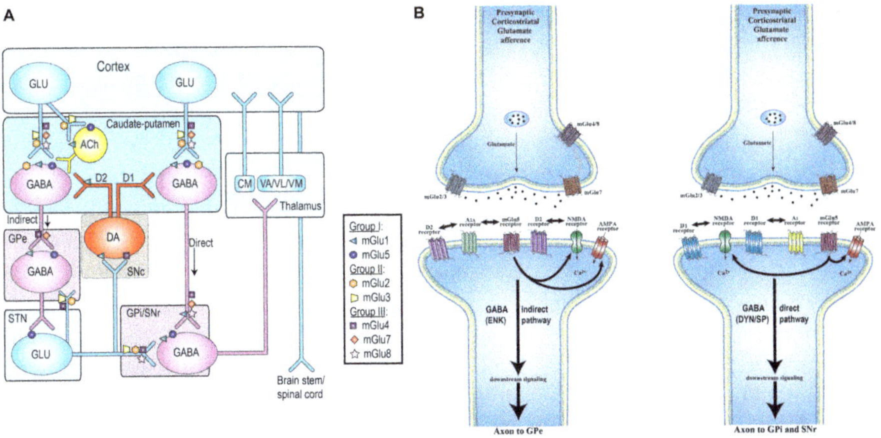

Fig. 10.1 **(a)** Schematic representation of the localization of metabotropic glutamate receptor subtypes (groups I, II, and II) in the basal ganglia motor circuit. mGlu6 receptor subtype (group III) is absent from the figure since it is present only in the retina (Conn et al. 2005; Rascol et al. 2014; Amalric 2015). In the classical and simplified pathophysiological model (not mentioning the collateral projections from the direct to indirect pathway) of the basal ganglia in PD, reduced dopamine input by the loss of dopaminergic neurons in the substantia nigra pars compacta (SNc) has a dual effect on the striatal efferents that project to the GPe (indirect pathway) and the GPi (direct pathway). Increased or facilitated neuronal activity of the indirect pathway, via decrease of the D2 receptor inhibition, increased activity of striatal GABA neurons projecting to the GPe. Over-inhibition of GABA neurons in the GPe disinhibits the STN, which, in turn, overdrives inhibitory output neurons in the GPi and SNr projecting to the thalamus, overall resulting in a decrease in thalamocortical input. **(b)** Postsynaptic receptors on striatal GABA efferences susceptible to be antagonized to counteract excessive corticostriatal glutamate transmission (absence of glutamate release regulation by D2 autoreceptors on corticostriatal neurons) seen in PD state in order to normalize activities of the direct and indirect pathway. Among these, mGlu5 seems to be a relevant target since this receptor acts primarily as modulator of synaptic activity. The *double arrow* indicates that these receptors can form complexes (heteroreceptor) providing a high degree of complexity and plasticity at different levels of basal ganglia circuitry (Fuxe et al. 2015)

modulate glutamate hyperactivity in PD and LID (Conn et al. 2005). Studies in animal models and PD patients indicate that antagonists of group I mGlu receptor, especially mGlu5 receptor, could be considered as a suitable therapeutic approach in PD and LID.

mGlu5 receptor levels were shown to be increased in the putamen of dyskinetic compared to non-dyskinetic MPTP monkeys (Samadi et al. 2008) and parkinsonian patients with motor complications (LID or wearing-off) compared to those without motor complications (Ouattara et al. 2011). Moreover, the prototypal mGlu5 receptor antagonist, 2-methyl-6-(phenylethynyl)pyridine (MPEP), and a more selective analog 3-[(2-methyl-1,3-thiazol-4-yl)ethynyl]pyridine (MTEP) (Carroll 2008) improved motor performance (Breysse et al. 2002) and showed antidyskinetic activity in 6-OHDA rat model (Levandis et al. 2008; Mela et al. 2007). However, the other group I mGlu receptor drugs were not effective (Dekundy et al. 2006; Rylander et al. 2009). In acute treatment in already dyskinetic MPTP-lesioned monkeys, the mGlu5 receptor antagonists MPEP, MTEP, fenobam, and AFQ056 (mavoglurant) were found effective to reduce the severity of LID (Morin et al. 2010; Rylander et al. 2010; Gregoire et al. 2011; Johnston et al. 2010). In a de novo treatment, MPEP reduced the development of L-DOPA-induced motor complications (~70%) in a chronic administration of 1 month (Morin et al. 2013a). Similarly, chronic administration of fenobam to drug-naïve monkeys attenuated the development of dyskinesia without compromising the antiparkinsonian effect of L-DOPA (Rylander et al. 2010). Moreover, the mGlu5 receptor antagonists, mavoglurant and ADX-48621 (dipraglurant), were shown to reduce LID in parkinsonian patients and were well tolerated without worsening motor symptoms (Addex Therapeutics 2012; Stocchi et al. 2013; Berg et al. 2011). For more information the reader is referred to a recent review on the use of mGlu5 antagonists for treatment of L-DOPA-induced dyskinesias including a comparison the clinical trials with mavoglurant and dipraglurant. (Rascol et al. 2014).

Group II mGlu receptor agonists have also proven effective in animal models of PD (Pisani et al. 2003). Moreover, an important decrease in mGlu2/mGlu3 receptor density in dyskinetic compared to non-dyskinetic MPTP-lesioned monkeys was observed (Morin et al. 2013b). Furthermore, changes in mGlu2/mGlu3 receptors were only observed in relation to wearing-off in postmortem brains of parkinsonian patients (Samadi et al. 2009).

More recently, agonists of group III receptors were also shown effective in rodent models of PD (Niswender and Conn 2010). In 6-OHDA-lesioned rat model, the administration of mGlu4 receptor agonist Lu AF21934 combined with L-DOPA treatment reduced the effective dose of L-DOPA and reduced the development of LID (Bennouar et al. 2013). mGlu4 receptor agonists can reduce γ-aminobutyric acid (GABA)ergic neurotransmission at striato-pallidal synapse that is overactive in PD (Macinnes and Duty 2008; Matsui and Kita 2003).

mGlu8 receptor is expressed at lower levels than mGlu4 and mGlu7 receptors but widely distributed in the brain; mGlu7 receptor has low affinity for glutamate only becoming active when glutamate levels are high thus serving as a brake for glutamate overstimulation (Niswender and Conn 2010). AMN082, an mGlu7 receptor agonist, was shown to reverse motor dysfunction associated with reduced DA activity in rodent models (Greco et al. 2010). However, the contribution of mGlu7 and mGlu8 receptors in LID is not yet reported.

10.6 The Effect of a Chronic Treatment with MPEP on L-DOPA-Induced Dyskinesias and Brain Biochemical Borrelates

The mechanisms underlying the development of LID are still unknown, but evidence suggests that LID is the result of maladaptive plasticity at striatal synapses (Cenci and Lundblad 2006; Jenner 2008; Calabresi et al. 2010; Iravani et al. 2012). Altered dopaminergic and nondopaminergic neurotransmission in the basal ganglia is observed in LID (Blandini and Armentero 2012). Glutamate neurotransmission plays a crucial role in the modulation of corticostriatal inputs and striatal output to downstream nuclei of the basal ganglia circuit (Blandini and Armentero 2012). Increased glutamate transmission involves presynaptic changes, such as increased striatal concentration of extracellular glutamate (Dupre et al. 2011) or changes in mGlu2/mGlu3 receptors (Samadi et al. 2008; Gregoire et al. 2009). Postsynaptic changes in AMPA, NMDA, and mGlu5 receptors also seem to play a major role in the development of LID (Duty 2012). mGlu5 receptors are highly expressed in the striatum and other basal ganglia nuclei including subthalamic nucleus (STN), substantia nigra, and globus pallidus (GP) (Ferraguti and Shigemoto 2006). Numerous interactions between mGlu5 receptor and NMDA, D_1 DA, D_2 DA, and A_{2A} adenosine receptors suggest that these receptors may function together as closely associated signaling partners in the appearance of LID (Samadi et al. 2007). mGlu5 receptor activation of NMDA receptors (Pisani et al. 2001) as well as NMDA and mGlu5 receptor co-localization is described (Perroy et al. 2008). The blockade of mGlu5 receptors leads to antidyskinetic actions, which is associated with normalization of firing of striatal signals (Duty 2012). Moreover, chronic mGluR5 NAM treatments are reported to protect DA neurons from MPTP toxicity in mice (Battaglia et al. 2004) and monkeys (Masilamoni 2009). Moreover, Norbin, a neuron-specific protein, can physically interact in vivo with mGlu5 receptors, increases the cell surface localization of the receptor, and positively regulates mGlu5 signaling while maintaining the total amount of this receptor unchanged (Wang et al. 2009). Intracellular mGlu5 receptor was reported to activate signaling cascades distinct from cell surface counterparts (Jong et al. 2009). Thus there is still much to be learned on mGuR5 receptors and their regulation.

We reported that development of LID over a month of treatment was lower by overall ~70% with addition of MPEP to the L-DOPA treatment in de novo MPTP monkeys (Morin et al. 2013a), and this was associated with a normalization of glutamate (Morin et al. 2013b), DA (Morin et al. 2014b), and 5-HT neurotransmission (Morin et al. 2015).

More precisely, in the brain basal ganglia of these monkeys (Morin et al. 2013a), the addition of MPEP to L-DOPA treatment prevented the increase of postsynaptic mGlu5, NMDA NR1/NR2B, and AMPA glutamate receptors, while this treatment prevented the decrease of mGlu2/mGlu3 presynaptic autoreceptors (Morin et al. 2013b). The mGlu5 receptor subtype is highly expressed in striatal medium spiny neurons (Conn et al. 2005; Testa et al. 1994; Paquet and Smith 2003) and plays a key role in modulating the responses mediated by NMDA receptors and L-type calcium channels (Gubellini et al. 2004). In addition, an antagonistic interaction between the D_2 DA receptor and mGlu5 receptors is reported (Fuxe et al. 2008). In rodent models of PD, striatal molecular changes relevant to LID are reported to be reversed by MPEP or MTEP, including delta FosB protein (Jimenez et al. 2009), prodynorphin mRNA (Mela et al. 2007), glutamic acid decarboxylase (GAD65 and GAD67) mRNA (Yamamoto and Soghomonian 2009), and phosphorylated extracellular signal-regulated kinases 1 and 2 (ERK1/ERK2) protein levels (Rylander et al. 2009). Hence, mGlu5 receptor antagonists reverse hyperactive GABAergic transmission in the basal ganglia in rodent models of PD and its downstream molecular changes associated with LID. Hence, mGlu5 receptor antagonist reversed hyperactive glutamate transmission in the basal ganglia of a primate model of PD thus showing the widespread normalization activity of mGlu5 receptor antagonists in the basal ganglia in PD animal models (Morin et al. 2013a).

Moreover, a chronic treatment with MPEP prevented the decrease of D_2 DA receptors, but it did not affect the D_1 DA receptors (Morin et al. 2014b). MPEP also prevented the increase of striatal preproenkephalin/preprodynorphin mRNA levels and phosphorylated proteins ERK1/ERK2 as well as protein kinase B (Akt) and glycogen synthase kinase-3 (GSK3β) (Morin et al. 2014b). Denervation-induced supersensitivity of D_1 and D_2 receptors was initially recognized as a plausible mechanism of LID (Creese et al. 1977; Lee et al. 1978). Numerous studies measured the density of D_1 and D_2 receptors in the brain of human and animal models, but no general consensus emerged. Postmortem studies have shown that striatal DA receptors particularly the D_2 subtype were increased in PD patients (Lee et al. 1978; Guttman et al. 1986) or unchanged (Quik et al. 1979; Rinne et al. 1981), while both D_1 and D_2 receptor subtypes were increased in MPTP monkeys (Falardeau et al. 1988; Bedard et al. 1986; Gagnon et al. 1990). Administration of L-DOPA was shown to reverse these increases in PD patients (Lee et al. 1978; Guttman et al. 1986) and primates in many studies (Falardeau et al. 1988; Gagnon et al. 1990; Berretta et al. 1997). No general consensus also emerged for DA receptor mRNA (Goulet et al. 1997, 2000; Morissette et al. 1996; Aubert et al. 2005; Herrero et al. 1996). These reports support that LID are more complex than hypersensitivity due to a simple increase in the density of striatal DA receptors and its mRNA. MPEP did not affect D_1 receptor levels and its mRNA but was associated with an increased in D_2 receptors levels, its mRNA, and its associated signaling proteins (Morin et al. 2014b). Furthermore, mGlu receptors also have the potential to regulate the mitogen-activated protein kinase (MAPK) pathway. It has been shown that intracaudate injection of a group I mGlu receptor agonist upregulates ERK1/ERK2 phosphorylation (Choe and Wang 2001). mGlu5 receptor stimulation was also reported to lead to activation of other signaling pathways important for cell survival/proliferation, such as ERK and Akt (Mao et al. 2005).

The antidyskinetic effect of MPEP was associated with lower levels of 5-HT_{2A} (Morin et al. 2015) and 5-HT_{1B} (Morin et al. 2015) serotonin receptors, while 5-HT_{1A} receptors and brain serotonin transporter (SERT) remained unaffected (Morin et al. 2015). LID are probably the result of an abnormal adaptation in the striatum due to faulty interaction between glutamate/5-HT and DA inputs in the nigrostriatal pathway. 5-HT neuron terminals in the striatum are suggested to participate in the mechanisms of action of L-DOPA (Navailles and De Deurwaerdere 2012). 5-HT axon terminals can release DA, which is considered as a false neurotransmitter and is probably one of the main presynaptic determinants of LID (Carta et al. 2007). Multiple changes in the basal ganglia glutamate and serotoninergic systems and their specific receptors have been observed (Huot et al. 2013). 5-HT receptor drugs are reported to reduce LID in PD patients (Fox et al. 2009), but changes in 5-HT activity and 5-HT receptors in the brain caused by PD and its pharmacological treatment are not completely understood, and further studies are needed.

Stimulation of 5-HT_{1B} receptors was also shown to reduce dyskinesias induced by D_1 DA receptor agonists (Jaunarajs et al. 2009). This observation suggests that in order to prevent activation of D_1-expressing neurons, an increase of the inhibitory 5-HT_{1B} heteroreceptors might represent another compensatory phenomenon to prevent activation of the direct pathway neurons, thought to be involved in the expression of LID (Greengard 2001). Moreover, an increase of 5-HT_{1B} autoreceptors might also play a crucial role to regulate or inhibit the release of DA by 5-HT axon terminals. 5-HT axon terminals have been suggested to release DA in a nonphysiological manner following L-DOPA treatment, leading to the development of LID (Carta et al. 2007). Hence, 5-HT neurons can take up L-DOPA and convert it into DA where it is sequestered and stored for subsequent release (Maeda et al. 2005). However, 5-HT neurons lack a DA transporter and D_2 DA autoreceptors, thus leading to excessive nonphysiological DA efflux that promotes LID (Navailles and De Deurwaerdere 2012). Hence, multiple changes in the basal ganglia dopaminergic, glutamate and serotoninergic systems, and their specific receptors have been observed such as modulation in the expression and the activity of subtypes of receptors, G proteins, effectors, transcription factors, and protein kinases (Samadi et al. 2007; Huot et al. 2013).

Moreover, functional interactions between 5-HT_{1B} receptors and GABA should also be considered in the molecular mechanisms of LID. It was reported that L-DOPA administration or stimulation of D_1 receptors in 6-OHDA-lesioned animals increases GABA levels in the substantia nigra pars reticulata (Aceves et al. 1992; Ochi et al. 2004; Yamamoto et al. 2006). GABA concentrations were reported to be high in the striatum of PD patients compared to controls (Kish et al. 1986), and changes in $GABA_A$ and $GABA_B$ receptor levels were also previously observed in the same PD patients, as presented in the present study, with motor complications as compared to those without and controls (Calon et al. 2003b). Hence, 5-HT_{1B} inhibitory receptors located on GABAergic axons of striatal neurons are probably implicated in the alterations of GABA receptors and GABA concentrations measured in these brain areas (Castro et al. 1998). Hence, activation of 5-HT_{1B} receptors could reduce GABA release (Morikawa et al. 2000; Stanford and Lacey 1996). Thus, this upregulation of 5-HT_{1B} receptors in the basal ganglia of dyskinetic monkeys and PD patients with motor complications could be a compensatory mechanism in order to normalize GABA concentrations.

Thus, these reported results suggest that the prevention of the development of motor complications with an mGlu5 receptor antagonist was associated with the normalization of important markers and receptors of glutamate, DA, and 5-HT neurotransmission, supporting the therapeutic use of an mGlu5 receptor antagonist to treat LID.

10.7 Conclusion

Overall, these studies suggest that glutamate receptor stimulation is involved in the pathogenesis of L-DOPA-induced motor complications in PD, and some receptor subtypes, such as mGlu5 receptor subtypes, are potential selective targets for treatment of these adverse effects (Blanchet et al. 1997; Chase and Oh 2000;

Verhagen Metman et al. 2000; Montastruc et al. 1997). mGlu5 receptor antagonists are very promising drugs for the management of various brain disorders, and one of its potential applications is LID therapy. Therefore, it is essential to know the long-term behavioral effects of this class of compounds, and ensuing possible biochemical adaptations of the brain, it is critical to learn as much as possible from MPTP primate models, which closely mimic human PD conditions (Morin et al. 2014a). The mGlu5 antagonist treatment in parkinsonian primates and rats affects not only glutamate receptors but also DA and 5-HT receptors. This may be indirect by restoring glutamate neurotransmission but could also involve the direct interactions of the trio mGlu5-D_2-A_{2A} receptor cross talk (Fiorentini et al. 2013; Fuxe et al. 2007). Moreover, supporting a close mGlu5-A_{2A} receptor interaction, MTEP treatment was reported to decrease mice brain A_{2A} receptor-specific binding and regulate the conditioned effects of cocaine (Brown et al. 2012). The implication of these receptor interactions in mental and neurodegenerative diseases and more specifically in the development and expression of PD symptoms and LID needs further investigation to find novel targets and, ultimately, novel pharmacological treatments.

References

Aceves J, Floran B, Martinez-Fong D, Benitez J, Sierra A, Flores G (1992) Activation of D1 receptors stimulates accumulation of gamma-aminobutyric acid in slices of the pars reticulata of 6-hydroxydopamine-lesioned rats. Neurosci Lett 145(1):40–42

Addex Therapeutics (2012) Geneva, Switzerland [media release]. 2012 Mar 21 [online]. Available from URL: http://www.addextherapeutics.com/investors/press-releases/news-details/article/addex-reports-positive-top-line-phase-iia-data-for-dipraglurant-in-parkinsons-disease-levodopa-indu/. Accessed 11 Feb 2013

Albanese A, Granata R, Gregori B, Piccardi MP, Colosimo C, Tonali P (1993) Chronic administration of 1-methyl-4-phenyl-1,2,3,6-tetrahydropyridine to monkeys: behavioural, morphological and biochemical correlates. Neuroscience 55(3):823–832

Amalric M (2015) Targeting metabotropic glutamate receptors (mGluRs) in Parkinson's disease. Curr Opin Pharmacol 20:29–34

Aubert I, Guigoni C, Hakansson K, Li Q, Dovero S, Barthe N et al (2005) Increased D1 dopamine receptor signaling in levodopa-induced dyskinesia. Ann Neurol 57(1):17–26

Bagetta V, Sgobio C, Pendolino V, Del Papa G, Tozzi A, Ghiglieri V et al (2012) Rebalance of striatal NMDA/AMPA receptor ratio underlies the reduced emergence of dyskinesia during D2-like dopamine agonist treatment in experimental Parkinson's disease. J Neurosci 32(49):17921–17931

Ballard PA, Tetrud JW, Langston JW (1985) Permanent human parkinsonism due to 1-methyl-4-phenyl-1,2,3,6-tetrahydropyridine (MPTP): seven cases. Neurology 35(7):949–956

Bara-Jimenez W, Dimitrova TD, Sherzai A, Aksu M, Chase TN (2006) Glutamate release inhibition ineffective in levodopa-induced motor complications. Mov Disord Off J Mov Disord Soc 21(9):1380–1383

Barbeau A (1980) High level levodopa therapy in several akinetic parkinsonian patients: twelve years later. In: Rinne JK, Klinger H, Stamm G (eds) Parkinson's disease: current progress, problems and management. Elsevier, Amsterdam, pp 229–239

Battaglia G, Busceti CL, Molinaro G, Biagioni F, Storto M, Fornai F et al (2004) Endogenous activation of mGlu5 metabotropic glutamate receptors contributes to the development of nigrostriatal damage induced by 1-methyl-4-phenyl-1,2,3,6-tetrahydropyridine in mice. J Neurosci Off J Soc Neurosci 24(4):828–835

Bedard PJ, Di Paolo T, Falardeau P, Boucher R (1986) Chronic treatment with L-DOPA, but not bromocriptine induces dyskinesia in MPTP-parkinsonian monkeys. Correlation with [3H]spiperone binding. Brain Res 379(2):294–299

Belanger N, Gregoire L, Hadj Tahar A, Bedard PJ (2003) Chronic treatment with small doses of cabergoline prevents dopa-induced dyskinesias in parkinsonian monkeys. Mov Disord Off J Mov Disord Soc 18(12):1436–1441

Bennouar KE, Uberti MA, Melon C, Bacolod MD, Jimenez HN, Cajina M et al (2013) Synergy between L-DOPA and a novel positive allosteric modulator of metabotropic glutamate receptor 4: implications for Parkinson's disease treatment and dyskinesia. Neuropharmacology 66:158–169

Berg D, Godau J, Trenkwalder C, Eggert K, Csoti I, Storch A et al (2011) AFQ056 treatment of levodopa-induced dyskinesias: results of 2 randomized controlled trials. Mov Disord 26(7):1243–1250

Berretta S, Parthasarathy HB, Graybiel AM (1997) Local release of GABAergic inhibition in the motor cortex induces immediate-early gene expression in indirect pathway neurons of the striatum. J Neurosci Off J Soc Neurosci 17(12):4752–4763

Bezard E, Hill MP, Crossman AR, Brotchie JM, Michel A, Grimée R et al (2004) Levetiracetam improves choreic levodopa-induced dyskinesia in the MPTP-treated macaque. Eur J Pharmacol 485(1-3):159–164

Blanchet PJ, Metman LV, Mouradian MM, Chase TN (1996) Acute pharmacologic blockade of dyskinesias in Parkinson's disease. Mov Disord Off J Mov Disord Socy 11(5):580–581

Blanchet PJ, Papa SM, Metman LV, Mouradian MM, Chase TN (1997) Modulation of levodopa-induced motor response complications by NMDA antagonists in Parkinson's disease. Neurosci Biobehav Rev 21(4):447–453

Blanchet PJ, Konitsiotis S, Whittemore ER, Zhou ZL, Woodward RM, Chase TN (1999) Differing effects of N-methyl-D-aspartate receptor subtype selective antagonists on dyskinesias in levodopa-treated 1-methyl-4-phenyl- tetrahydropyridine monkeys. J Pharmacol Exp Ther 290(3):1034–1040

Blandini F, Armentero MT (2012) New pharmacological avenues for the treatment of L-DOPA-induced dyskinesias in Parkinson's disease: targeting glutamate and adenosine receptors. Expert Opin Investig Drugs 21(2):153–168

Braz CA, Borges V, Ferraz HB (2004) Effect of riluzole on dyskinesia and duration of the on state in Parkinson disease patients: a double-blind, placebo-controlled pilot study. Clin Neuropharmacol 27(1):25–29

Breysse N, Baunez C, Spooren W, Gasparini F, Amalric M (2002) Chronic but not acute treatment with a metabotropic glutamate 5 receptor antagonist reverses the akinetic deficits in a rat model of parkinsonism. J Neurosci Off J Soc Neurosci 22(13):5669–5678

Brotchie JM (1998) Adjuncts to dopamine replacement: a pragmatic approach to reducing the problem of dyskinesia in Parkinson's disease. Mov Disord 13(6):871–876

Brotchie JM (2003) CB1 cannabinoid receptor signalling in Parkinson's disease. Curr Opin Pharmacol 3(1):54–61

Brown RM, Duncan JR, Stagnitti MR, Ledent C, Lawrence AJ (2012) mGlu5 and adenosine A2A receptor interactions regulate the conditioned effects of cocaine. Int J Neuropsychopharmacol/Off Sci J Coll Int Neuropsychopharmacol 15(7):995–1001

Calabresi P, Di Filippo M, Ghiglieri V, Tambasco N, Picconi B (2010) Levodopa-induced dyskinesias in patients with Parkinson's disease: filling the bench-to-bedside gap. Lancet Neurol 9(11):1106–1117

Calon F, Di Paolo T (2002) Levodopa response motor complications -GABA receptors and preproenkephalin expression in human brain. Parkinsonism Relat Disord 8(6):449–454

Calon F, Morissette M, Ghribi O, Goulet M, Grondin R, Blanchet PJ et al (2002) Alteration of glutamate receptors in the striatum of dyskinetic 1-methyl-4-phenyl-1,2,3,6-tetrahydropyridine-treated monkeys following dopamine agonist treatment. Prog Neuro-Psychopharmacol Biol Psychiatry 26(1):127–138

Calon F, Rajput AH, Hornykiewicz O, Bedard PJ, Di Paolo T (2003a) Levodopa-induced motor complications are associated with alterations of glutamate receptors in Parkinson's disease. Neurobiol Dis 14(3):404–416

Calon F, Morissette M, Rajput AH, Hornykiewicz O, Bedard PJ, Di Paolo T (2003b) Changes of GABA receptors and dopamine turnover in the postmortem brains of parkinsonians with levodopa-induced motor complications. Mov Disord Off J Mov Disord Soc 18(3):241–253

Carpenedo R, Pittaluga A, Cozzi A, Attucci S, Galli A, Raiteri M et al (2001) Presynaptic kynurenate-sensitive receptors inhibit glutamate release. Eur J Neurosci 13(11):2141–2147

Carroll FI (2008) Antagonists at metabotropic glutamate receptor subtype 5: structure activity relationships and therapeutic potential for addiction. Ann N Y Acad Sci 1141:221–232

Carta M, Carlsson T, Kirik D, Bjorklund A (2007) Dopamine released from 5-HT terminals is the cause of L-DOPA-induced dyskinesia in parkinsonian rats. Brain J Neurol 130(Pt 7):1819–1833

Castro ME, Pascual J, Romon T, Berciano J, Figols J, Pazos A (1998) 5-HT1B receptor binding in degenerative movement disorders. Brain Res 790(1-2):323–328

Cenci MA, Lundblad M (2006) Post- versus presynaptic plasticity in L-DOPA-induced dyskinesia. J Neurochem 99(2):381–392

Chase TN, Oh JD (2000) Striatal mechanisms and pathogenesis of parkinsonian signs and motor complications. Ann Neurol 47(4 Suppl 1):S122–S129. discussion S9-30

Choe ES, Wang JQ (2001) Group I metabotropic glutamate receptor activation increases phosphorylation of cAMP response element-binding protein, Elk-1, and extracellular signal-regulated kinases in rat dorsal striatum. Brain Res Mol Brain Res 94(1-2):75–84

Conn PJ, Pin JP (1997) Pharmacology and functions of metabotropic glutamate receptors. Annu Rev Pharmacol Toxicol 37:205–237

Conn PJ, Battaglia G, Marino MJ, Nicoletti F (2005) Metabotropic glutamate receptors in the basal ganglia motor circuit. Nat Rev Neurosci 6(10):787–798

Creese I, Burt DR, Snyder SH (1977) Dopamine receptor binding enhancement accompanies lesion-induced behavioral supersensitivity. Science 197(4303):596–598

Dekundy A, Pietraszek M, Schaefer D, Cenci MA, Danysz W (2006) Effects of group I metabotropic glutamate receptors blockade in experimental models of Parkinson's disease. Brain Res Bull 69(3):318–326

Dingledine R, Borges K, Bowie D, Traynelis SF (1999) The glutamate receptor ion channels. Pharmacol Rev 51(1):7–61

Dupre KB, Ostock CY, Eskow Jaunarajs KL, Button T, Savage LM, Wolf W et al (2011) Local modulation of striatal glutamate efflux by serotonin 1A receptor stimulation in dyskinetic, hemiparkinsonian rats. Exp Neurol 229(2):288–299

Duty S (2012) Targeting glutamate receptors to tackle the pathogenesis, clinical symptoms and levodopa-induced dyskinesia associated with Parkinson's disease. CNS drugs 26(12):1017–1032

Engber TM, Papa SM, Boldry RC, Chase TN (1994) NMDA receptor blockade reverses motor response alterations induced by levodopa. Neuroreport 5(18):2586–2588

Errico F, Bonito-Oliva A, Bagetta V, Vitucci D, Romano R, Zianni E et al (2011) Higher free D-aspartate and N-methyl-D-aspartate levels prevent striatal depotentiation and anticipate L-DOPA-induced dyskinesia. Exp Neurol 232(2):240–250

Evans AH, Farrell MJ, Gibson SJ, Helme RD, Lim SY (2012) Dyskinetic patients show rebound worsening of affect after an acute L-dopa challenge. Parkinsonism Relat Disord 18(5):514–519

Fabbrini G, Brotchie JM, Grandas F, Nomoto M, Goetz CG (2007) Levodopa-induced dyskinesias. Mov Disord 22(10):1379–1389. quiz 523

Fahn S, Oakes D, Shoulson I, Kieburtz K, Rudolph A, Lang A et al (2004) Levodopa and the progression of Parkinson's disease. N Engl J Med 351(24):2498–2508

Falardeau P, Bouchard S, Bedard PJ, Boucher R, Di Paolo T (1988) Behavioral and biochemical effect of chronic treatment with D-1 and/or D-2 dopamine agonists in MPTP monkeys. Eur J Pharmacol 150(1-2):59–66

Ferraguti F, Shigemoto R (2006) Metabotropic glutamate receptors. Cell Tissue Res 326(2):483–504
Finlay C, Duty S (2014) Therapeutic potential of targeting glutamate receptors in Parkinson's disease. J Neural Transm 121(8):861–880
Fiorentini C, Savoia P, Savoldi D, Missale C (2013) Receptor heteromers in Parkinson's disease and L-DOPA-induced dyskinesia. CNS Neurol Disord Drug Targets 12(8):1101–1113
Fox SH, Chuang R, Brotchie JM (2009) Serotonin and Parkinson's disease: on movement, mood, and madness. Mov Disord Off J Mov Disord Soc 24(9):1255–1266
Fox SH, Johnston TH, Li Q, Brotchie J, Bezard E (2012) A critique of available scales and presentation of the non-human primate dyskinesia rating scale. Mov Disord Off J Mov Disord Soc 27(11):1373–1378
Fuxe K, Ferre S, Genedani S, Franco R, Agnati LF (2007) Adenosine receptor-dopamine receptor interactions in the basal ganglia and their relevance for brain function. Physiol Behav 92(1-2):210–217
Fuxe K, Marcellino D, Rivera A, Diaz-Cabiale Z, Filip M, Gago B et al (2008) Receptor-receptor interactions within receptor mosaics. Impact on neuropsychopharmacology. Brain Res Rev 58(2):415–452
Fuxe K, Guidolin D, Agnati LF, Borroto-Escuela DO (2015) Dopamine heteroreceptor complexes as therapeutic targets in Parkinson's disease. Expert Opin Ther Targets 19(3):377–398
Gagnon C, Bedard PJ, Di Paolo T (1990) Effect of chronic treatment of MPTP monkeys with dopamine D-1 and/or D-2 receptor agonists. Eur J Pharmacol 178(1):115–120
Gardoni F, Di Luca M (2006) New targets for pharmacological intervention in the glutamatergic synapse. Eur J Pharmacol 545(1):2–10
Gasparini F, Di Paolo T, Gomez-Mancilla B (2013) Metabotropic glutamate receptors for Parkinson's disease therapy. Park Dis 2013:196028
Goetz CG, Tanner CM, Klawans HL (1982) Drug holiday in the management of Parkinson disease. Clin Neuropharmacol 5(4):351–364
Goulet M, Morissette M, Calon F, Blanchet PJ, Falardeau P, Bedard PJ et al (1997) Continuous or pulsatile chronic D2 dopamine receptor agonist (U91356A) treatment of drug-naive 4-phenyl-1,2,3,6-tetrahydropyridine monkeys differentially regulates brain D1 and D2 receptor expression: in situ hybridization histochemical analysis. Neuroscience 79(2):497–507
Goulet M, Grondin R, Morissette M, Maltais S, Falardeau P, Bedard PJ et al (2000) Regulation by chronic treatment with cabergoline of dopamine D1 and D2 receptor levels and their expression in the striatum of Parkinsonian-monkeys. Prog Neuro-Psychopharmacol Biol Psychiatry 24(4):607–617
Greco B, Lopez S, van der Putten H, Flor PJ, Amalric M (2010) Metabotropic glutamate 7 receptor subtype modulates motor symptoms in rodent models of Parkinson's disease. J Pharmacol Exp Ther 332(3):1064–1071
Greengard P (2001) The neurobiology of slow synaptic transmission. Science 294(5544):1024–1030
Grégoire L, Rassoulpour A, Guidetti P, Samadi P, Bédard PJ, Izzo E et al (2008) Prolonged kynurenine 3-hydroxylase inhibition reduces development of levodopa-induced dyskinesias in parkinsonian monkeys. Behav Brain Res 186(2):161–167
Grégoire L, Samadi P, Graham J, Bédard PJ, Bartoszyk GD, Di Paolo T (2009) Low doses of sarizotan reduce dyskinesias and maintain antiparkinsonian efficacy of L-Dopa in parkinsonian monkeys. Parkinsonism Relat Disord 15(6):445–452
Gregoire L, Morin N, Ouattara B, Gasparini F, Bilbe G, Johns D et al (2011) The acute antiparkinsonian and antidyskinetic effect of AFQ056, a novel metabotropic glutamate receptor type 5 antagonist, in L-Dopa-treated parkinsonian monkeys. Parkinsonism Relat Disord 17(4):270–276
Grondin R, Bedard PJ, Hadj Tahar A, Gregoire L, Mori A, Kase H (1999) Antiparkinsonian effect of a new selective adenosine A2A receptor antagonist in MPTP-treated monkeys. Neurology 52(8):1673–1677

Gubellini P, Pisani A, Centonze D, Bernardi G, Calabresi P (2004) Metabotropic glutamate receptors and striatal synaptic plasticity: implications for neurological diseases. Prog Neurobiol 74(5):271–300

Guigoni C, Dovero S, Aubert I, Li Q, Bioulac BH, Bloch B et al (2005) Levodopa-induced dyskinesia in MPTP-treated macaques is not dependent on the extent and pattern of nigrostrial lesioning. Eur J Neurosci 22(1):283–287

Guttman M, Seeman P, Reynolds GP, Riederer P, Jellinger K, Tourtellotte WW (1986) Dopamine D2 receptor density remains constant in treated Parkinson's disease. Ann Neurol 19(5):487–492

Hadj Tahar A, Grégoire L, Bangassoro E, Bédard PJ (2000) Sustained cabergoline treatment reverses levodopa-induced dyskinesias in parkinsonian monkeys. Clin Neuropharmacol 23(4):195–202

Hadj Tahar A, Grégoire L, Darré A, Bélanger N, Meltzer L, Bédard PJ (2004) Effect of a selective glutamate antagonist on L-dopa-induced dyskinesias in drug-naive parkinsonian monkeys. Neurobiol Dis 15(2):171–176

Henry B, Fox SH, Crossman AR, Brotchie JM (2001) Mu- and delta-opioid receptor antagonists reduce levodopa-induced dyskinesia in the MPTP-lesioned primate model of Parkinson's disease. Exp Neurol 171(1):139–146

Heresco-Levy U, Shoham S, Javitt DC (2013) Glycine site agonists of the N-methyl-D-aspartate receptor and Parkinson's disease: a hypothesis. Mov Disord 28(4):419–424

Herrero MT, Augood SJ, Asensi H, Hirsch EC, Agid Y, Obeso JA et al (1996) Effects of L-DOPA-therapy on dopamine D2 receptor mRNA expression in the striatum of MPTP-intoxicated parkinsonian monkeys. Brain Res Mol Brain Res 42(1):149–155

Hilmas C, Pereira EF, Alkondon M, Rassoulpour A, Schwarcz R, Albuquerque EX (2001) The brain metabolite kynurenic acid inhibits alpha7 nicotinic receptor activity and increases non-alpha7 nicotinic receptor expression: physiopathological implications. J Neurosci Off J Soc Neurosci 21(19):7463–7473

Huot P, Johnston TH, Koprich JB, Fox SH, Brotchie JM (2013) The pharmacology of L-DOPA-induced dyskinesia in Parkinson's disease. Pharmacol Rev 65(1):171–222

Iravani MM, McCreary AC, Jenner P (2012) Striatal plasticity in Parkinson's disease and L-dopa induced dyskinesia. Parkinsonism Relat Disord 18(Suppl 1):S123–S125

Jaunarajs KL, Dupre KB, Steiniger A, Klioueva A, Moore A, Kelly C et al (2009) Serotonin 1B receptor stimulation reduces D1 receptor agonist-induced dyskinesia. Neuroreport 20(14):1265–1269

Jenner P (2003a) The contribution of the MPTP-treated primate model to the development of new treatment strategies for Parkinson's disease. Parkinsonism Relat Disord 9(3):131–137

Jenner P (2003b) The MPTP-treated primate as a model of motor complications in PD: primate model of motor complications. Neurology 61(6 Suppl 3):S4–11

Jenner P (2008) Molecular mechanisms of L-DOPA-induced dyskinesia. Nat Rev Neurosci 9(9):665–677

Jimenez A, Bonastre M, Aguilar E, Marin C (2009) Effect of the metabotropic glutamate antagonist MPEP on striatal expression of the Homer family proteins in levodopa-treated hemiparkinsonian rats. Psychopharmacology 206(2):233–242

Johnson KA, Conn PJ, Niswender CM (2009) Glutamate receptors as therapeutic targets for Parkinson's disease. CNS Neurol Disord Drug Targets 8(6):475–491

Johnston TH, Fox SH, McIldowie MJ, Piggott MJ, Brotchie JM (2010) Reduction of L-DOPA-induced dyskinesia by the selective metabotropic glutamate receptor 5 antagonist 3-[(2-methyl-1,3-thiazol-4-yl)ethynyl]pyridine in the 1-methyl-4-phenyl-1,2,3,6-tetrahydropyridine-lesioned macaque model of Parkinson's disease. J Pharmacol Exp Ther 333(3):865–873

Jong YJ, Kumar V, O'Malley KL (2009) Intracellular metabotropic glutamate receptor 5 (mGluR5) activates signaling cascades distinct from cell surface counterparts. J Biol Chem 284(51):35827–35838

Jourdain VA, Gregoire L, Morissette M, Morin N, Parent M, Di Paolo T (2013) Potentiation of response to low doses of levodopa in MPTP-injected monkeys by chemical unilateral subthalamotomy. J Neurosurg 118(1):180–191

Katzenschlager R, Head J, Schrag A, Ben-Shlomo Y, Evans A, Lees AJ et al (2008) Fourteen-year final report of the randomized PDRG-UK trial comparing three initial treatments in PD. Neurology 71(7):474–480

Kish SJ, Rajput A, Gilbert J, Rozdilsky B, Chang LJ, Shannak K et al (1986) Elevated gamma-aminobutyric acid level in striatal but not extrastriatal brain regions in Parkinson's disease: correlation with striatal dopamine loss. Ann Neurol 20(1):26–31

Klawans HL, Goetz C, Nausieda PA, Weiner WJ (1977) Levodopa-induced dopamine receptor hypersensitivity. Trans Am Neurol Assoc 102:80–83

Klockgether T, Turski L (1993) Toward an understanding of the role of glutamate in experimental parkinsonism: agonist-sensitive sites in the basal ganglia. Ann Neurol 34(4):585–593

Kobylecki C, Cenci MA, Crossman AR, Ravenscroft P (2010) Calcium-permeable AMPA receptors are involved in the induction and expression of l-DOPA-induced dyskinesia in Parkinson's disease. J Neurochem 114(2):499–511

Kuoppamäki M, Al-Barghouthy G, Jackson M, Smith L, Zeng B-Y, Quinn N et al (2002) Beginning-of-dose and rebound worsening in MPTP-treated common marmosets treated with levodopa. Mov Disord Off J Mov Disord Soc 17(6):1312–1317

Kuoppamäki M, Al-Barghouthy G, Jackson MJ, Smith LA, Quinn N, Jenner P (2007) L-dopa dose and the duration and severity of dyskinesia in primed MPTP-treated primates. J Neural Transm 114(9):1147–1153

Lee T, Seeman P, Rajput A, Farley IJ, Hornykiewicz O (1978) Receptor basis for dopaminergic supersensitivity in Parkinson's disease. Nature 273(5657):59–61

Levandis G, Bazzini E, Armentero MT, Nappi G, Blandini F (2008) Systemic administration of an mGluR5 antagonist, but not unilateral subthalamic lesion, counteracts l-DOPA-induced dyskinesias in a rodent model of Parkinson's disease. Neurobiol Dis 29(1):161–168

Luginger E, Wenning GK, Bosch S, Poewe W (2000) Beneficial effects of amantadine on L-dopa-induced dyskinesias in Parkinson's disease. Mov Disord Off J Mov Disord Soc 15(5):873–878

Lujan R, Roberts JD, Shigemoto R, Ohishi H, Somogyi P (1997) Differential plasma membrane distribution of metabotropic glutamate receptors mGluR1 alpha, mGluR2 and mGluR5, relative to neurotransmitter release sites. J Chem Neuroanat 13(4):219–241

Luquin MR, Scipioni O, Vaamonde J, Gershanik O, Obeso JA (1992) Levodopa-induced dyskinesias in Parkinson's disease: clinical and pharmacological classification. Mov Disord Off J Mov Disord Soc 7(2):117–124

Macinnes N, Duty S (2008) Group III metabotropic glutamate receptors act as hetero-receptors modulating evoked GABA release in the globus pallidus in vivo. Eur J Pharmacol 580(1-2):95–99

Maeda T, Nagata K, Yoshida Y, Kannari K (2005) Serotonergic hyperinnervation into the dopaminergic denervated striatum compensates for dopamine conversion from exogenously administered l-DOPA. Brain Res 1046(1-2):230–233

Mao L, Yang L, Tang Q, Samdani S, Zhang G, Wang JQ (2005) The scaffold protein Homer1b/c links metabotropic glutamate receptor 5 to extracellular signal-regulated protein kinase cascades in neurons. J Neurosci Off J Soc Neurosci 25(10):2741–2752

Marin C, Papa S, Engber TM, Bonastre M, Tolosa E, Chase TN (1996) MK-801 prevents levodopa-induced motor response alterations in parkinsonian rats. Brain Res 736(1-2):202–205

Marin C, Jimenez A, Bonastre M, Chase TN, Tolosa E (2000) Non-NMDA receptor-mediated mechanisms are involved in levodopa-induced motor response alterations in parkinsonian rats. Synapse 36(4):267–274

Marino MJ, Awad H, Poisik O, Wittmann M, Conn PJ (2002) Localization and physiological roles of metabotropic glutamate receptors in the direct and indirect pathways of the basal ganglia. Amino Acids 23(1-3):185–191

Marsden C, Parkes J, Quinn N (1982) Fluctuations of disability in Parkinson's disease – clinical aspects. In: Marsden CD, Fahn S (eds) Movement disorders. Butterworths Scientific, London, pp 96–122

Masilamoni GJ, Bogenpohl JW, Alagille D, Delevich K, Tamagnan G, Votaw JR, Wichmann T, Smith Y (2011) Metabotropic glutamate receptor 5 antagonist protects dopaminergic and noradrenergic neurons from degeneration in MPTP-treated monkeys. Brain 134:2057–2073

Matsui T, Kita H (2003) Activation of group III metabotropic glutamate receptors presynaptically reduces both GABAergic and glutamatergic transmission in the rat globus pallidus. Neuroscience 122(3):727–737

Mayeux R, Stern Y, Mulvey K, Cote L (1985) Reappraisal of temporary levodopa withdrawal ("drug holiday") in Parkinson's disease. N Engl J Med 313(12):724–728

Meissner WG, Frasier M, Gasser T, Goetz CG, Lozano A, Piccini P et al (2011) Priorities in Parkinson's disease research. Nat Rev Drug Discov 10(5):377–393

Mela F, Marti M, Dekundy A, Danysz W, Morari M, Cenci MA (2007) Antagonism of metabotropic glutamate receptor type 5 attenuates l-DOPA-induced dyskinesia and its molecular and neurochemical correlates in a rat model of Parkinson's disease. J Neurochem 101(2):483–497

Mercuri NB, Bernardi G (2005) The 'magic' of L-dopa: why is it the gold standard Parkinson's disease therapy? Trends Pharmacol Sci 26(7):341–344

Michaelis EK (1998) Molecular biology of glutamate receptors in the central nervous system and their role in excitotoxicity, oxidative stress and aging. Prog Neurobiol 54(4):369–415

Montastruc JL, Rascol O, Senard JM (1997) Glutamate antagonists and Parkinson's disease: a review of clinical data. Neurosci Biobehav Rev 21(4):477–480

Morikawa H, Manzoni OJ, Crabbe JC, Williams JT (2000) Regulation of central synaptic transmission by 5-HT(1B) auto- and heteroreceptors. Mol Pharmacol 58(6):1271–1278

Morin N, Di Paolo T (2014) Pharmacological treatments inhibiting levodopa-induced dyskinesias in MPTP-lesioned monkeys: brain glutamate biochemical correlates. Front Neurol 5:144

Morin N, Gregoire L, Gomez-Mancilla B, Gasparini F, Di Paolo T (2010) Effect of the metabotropic glutamate receptor type 5 antagonists MPEP and MTEP in parkinsonian monkeys. Neuropharmacology 58(7):981–986

Morin N, Grégoire L, Morissette M, Desrayaud S, Gomez-Mancilla B, Gasparini F et al (2012) MPEP, an mGlu5 receptor antagonist, reduces the development of l-DOPA-induced motor complications in de novo parkinsonian monkeys: biochemical correlates. Neuropharmacology 66:355–364

Morin N, Gregoire L, Morissette M, Desrayaud S, Gomez-Mancilla B, Gasparini F et al (2013a) MPEP, an mGlu5 receptor antagonist, reduces the development of L-DOPA-induced motor complications in de novo parkinsonian monkeys: biochemical correlates. Neuropharmacology 66:355–364

Morin N, Morissette M, Gregoire L, Gomez-Mancilla B, Gasparini F, Di Paolo T (2013b) Chronic treatment with MPEP, an mGlu5 receptor antagonist, normalizes basal ganglia glutamate neurotransmission in L-DOPA-treated parkinsonian monkeys. Neuropharmacology 73:216–231

Morin N, Jourdain VA, Di Paolo T (2014a) Modeling dyskinesia in animal models of Parkinson disease. Exp Neurol 256:105–116

Morin N, Jourdain VA, Morissette M, Gregoire L, Di Paolo T (2014b) Long-term treatment with l-DOPA and an mGlu5 receptor antagonist prevents changes in brain basal ganglia dopamine receptors, their associated signaling proteins and neuropeptides in parkinsonian monkeys. Neuropharmacology 79:688–706

Morin N, Morissette M, Gregoire L, Di Paolo T (2015) Effect of a chronic treatment with an mGlu5 receptor antagonist on brain serotonin markers in parkinsonian monkeys. Prog Neuropsychopharmacol Biol Psychiatry 56:27–38

Morissette M, Goulet M, Calon F, Falardeau P, Blanchet PJ, Bedard PJ et al (1996) Changes of D1 and D2 dopamine receptor mRNA in the brains of monkeys lesioned with 1-methyl-4-phenyl-1,2,3,6-tetrahydropyridine: correction with chronic administration of L-3,4-dihydroxyphenylalanine. Mol Pharmacol 50(5):1073–1079

Morissette M, Dridi M, Calon F, Hadj Tahar A, Meltzer LT, Bedard PJ et al (2006) Prevention of levodopa-induced dyskinesias by a selective NR1A/2B N-methyl-D-aspartate receptor antagonist in parkinsonian monkeys: implication of preproenkephalin. Mov Disord 21(1):9–17

Moroni F (1999) Tryptophan metabolism and brain function: focus on kynurenine and other indole metabolites. Eur J Pharmacol 375(1-3):87–100

Navailles S, De Deurwaerdere P (2012) Imbalanced Dopaminergic Transmission Mediated by Serotonergic Neurons in L-DOPA-Induced Dyskinesia. Park Dis 2012:1–16

Nemeth H, Toldi J, Vecsei L (2006) Kynurenines, Parkinson's disease and other neurodegenerative disorders: preclinical and clinical studies. J Neural Transm Suppl 70:285–304

Niswender CM, Conn PJ (2010) Metabotropic glutamate receptors: physiology, pharmacology, and disease. Annu Rev Pharmacol Toxicol 50:295–322

Nutt JG (1990) Levodopa-induced dyskinesia: review, observations, and speculations. Neurology 40(2):340–345

Ochi M, Shiozaki S, Kase H (2004) L-DOPA-induced modulation of GABA and glutamate release in substantia nigra pars reticulata in a rodent model of Parkinson's disease. Synapse 52(2):163–165

Olanow CW, Koller WC (1998) An algorithm (decision tree) for the management of Parkinson's disease: treatment guidelines. American Academy of Neurology. Neurology 50(3 Suppl 3):S1–57

Olanow CW, Stern MB, Sethi K (2009) The scientific and clinical basis for the treatment of Parkinson disease (2009). Neurology 72(21 Suppl 4):S1–136

Olive MF (2009) Metabotropic glutamate receptor ligands as potential therapeutics for addiction. Curr Drug Abuse Rev 2(1):83–989

Ouattara B, Gasparini F, Morissette M, Grégoire L, Samadi P, Gomez-Mancilla B et al (2010) Effect of L-Dopa on metabotropic glutamate receptor 5 in the brain of parkinsonian monkeys. J Neurochem 113(3):715–724

Ouattara B, Gregoire L, Morissette M, Gasparini F, Vranesic I, Bilbe G et al (2011) Metabotropic glutamate receptor type 5 in levodopa-induced motor complications. Neurobiol Aging 32(7):1286–1295

Oueslati A, Sgambato-Faure V, Melon C, Kachidian P, Gubellini P, Amri M et al (2007) High-frequency stimulation of the subthalamic nucleus potentiates L-DOPA-induced neurochemical changes in the striatum in a rat model of Parkinson's disease. J Neurosci Off J Soc Neurosci 27(9):2377–2386

Pahwa R, Lyons KE (2009) Levodopa-related wearing-off in Parkinson's disease: identification and management. Curr Med Res Opin 25(4):841–849

Palucha A, Pilc A (2007) Metabotropic glutamate receptor ligands as possible anxiolytic and anti-depressant drugs. Pharmacol Ther 115(1):116–147

Papa SM, Boldry RC, Engber TM, Kask AM, Chase TN (1995) Reversal of levodopa-induced motor fluctuations in experimental parkinsonism by NMDA receptor blockade. Brain Res 701(1-2):13–18

Paquet M, Smith Y (2003) Group I metabotropic glutamate receptors in the monkey striatum: subsynaptic association with glutamatergic and dopaminergic afferents. J Neurosci Off J Soc Neurosci 23(20):7659–7669

Parkinson Study Group (2009) Long-term effect of initiating pramipexole vs levodopa in early Parkinson disease. Arch Neurol 66(5):563–570

Perroy J, Raynaud F, Homburger V, Rousset MC, Telley L, Bockaert J et al (2008) Direct interaction enables cross-talk between ionotropic and group I metabotropic glutamate receptors. J Biol Chem 283(11):6799–6805

Pisani A, Gubellini P, Bonsi P, Conquet F, Picconi B, Centonze D et al (2001) Metabotropic glutamate receptor 5 mediates the potentiation of N-methyl-D-aspartate responses in medium spiny striatal neurons. Neuroscience 106(3):579–587

Pisani A, Bonsi P, Centonze D, Gubellini P, Bernardi G, Calabresi P (2003) Targeting striatal cholinergic interneurons in Parkinson's disease: focus on metabotropic glutamate receptors. Neuropharmacology 45(1):45–56

Planells-Cases R, Lerma J, Ferrer-Montiel A (2006) Pharmacological intervention at ionotropic glutamate receptor complexes. Curr Pharm Des 12(28):3583–3596

Quik M, Spokes EG, Mackay AV, Bannister R (1979) Alterations in [3H]spiperone binding in human caudate nucleus, substantia nigra and frontal cortex in the Shy-Drager syndrome and Parkinson's disease. J Neurol Sci 43(3):429–437

Rajput AH, Rajput A, Lang AE, Kumar R, Uitti RJ, Galvez-Jimenez N (1998) New use for an old drug: amantadine benefits levodopa-induced dyskinesia. Mov Disord Off J Mov Disord Soc 13(5):851

Rajput AH, Voll A, Rajput ML, Robinson CA, Rajput A (2009) Course in Parkinson disease subtypes: a 39-year clinicopathologic study. Neurology 73(3):206–212

Rascol O, Fox S, Gasparini F, Kenney C, Di Paolo T, Gomez-Mancilla B (2014) Use of metabotropic glutamate 5-receptor antagonists for treatment of levodopa-induced dyskinesias. Parkinsonism Relat Disord 20(9):947–956

Rinne UK, Lonnberg P, Koskinen V (1981) Dopamine receptors in the Parkinsonian brain. J Neural Transm 51(1-2):97–106

Robelet S, Melon C, Guillet B, Salin P, Kerkerian-Le GL (2004) Chronic L-DOPA treatment increases extracellular glutamate levels and GLT1 expression in the basal ganglia in a rat model of Parkinson's disease. Eur J Neurosci 20(5):1255–1266

Ruzicka E, Streitova H, Jech R, Kanovsky P, Roth J, Rektorova I et al (2000) Amantadine infusion in treatment of motor fluctuations and dyskinesias in Parkinson's disease. J Neural Transm 107(11):1297–1306

Rylander D, Recchia A, Mela F, Dekundy A, Danysz W, Cenci MA (2009) Pharmacological modulation of glutamate transmission in a rat model of L-DOPA-induced dyskinesia: effects on motor behavior and striatal nuclear signaling. J Pharmacol Exp Ther 330(1):227–235

Rylander D, Iderberg H, Li Q, Dekundy A, Zhang J, Li H et al (2010) A mGluR5 antagonist under clinical development improves L-DOPA-induced dyskinesia in parkinsonian rats and monkeys. Neurobiol Dis 39(3):352–361

Samadi P, Gregoire L, Rassoulpour A, Guidetti P, Izzo E, Schwarcz R et al (2005) Effect of kynurenine 3-hydroxylase inhibition on the dyskinetic and antiparkinsonian responses to levodopa in Parkinsonian monkeys. Mov Disord 20(7):792–802

Samadi P, Grégoire L, Rouillard C, Bédard PJ, Di Paolo T, Lévesque D (2006) Docosahexaenoic acid reduces levodopa-induced dyskinesias in 1-methyl-4-phenyl-1,2,3,6-tetrahydropyridine monkeys. Ann Neurol 59(2):282–288

Samadi P, Rouillard C, Bedard PJ, Di Paolo T (2007) Functional neurochemistry of the basal ganglia. Handb Clin Neurol 83:19–66

Samadi P, Grégoire L, Morissette M, Calon F, Hadj Tahar A, Dridi M et al (2008a) mGluR5 metabotropic glutamate receptors and dyskinesias in MPTP monkeys. Neurobiol Aging 29(7):1040–1051

Samadi P, Gregoire L, Morissette M, Calon F, Hadj Tahar A, Belanger N et al (2008b) Basal ganglia group II metabotropic glutamate receptors specific binding in non-human primate model of L-Dopa-induced dyskinesias. Neuropharmacology 54(2):258–268

Samadi P, Rajput A, Calon F, Gregoire L, Hornykiewicz O, Rajput AH et al (2009) Metabotropic glutamate receptor II in the brains of Parkinsonian patients. J Neuropathol Exp Neurol 68(4):374–382

Sawada H, Oeda T, Kuno S, Nomoto M, Yamamoto K, Yamamoto M et al (2010) Amantadine for dyskinesias in Parkinson's disease: a randomized controlled trial. PLoS One 5(12):e15298

Schoepp DD (2001) Unveiling the functions of presynaptic metabotropic glutamate receptors in the central nervous system. J Pharmacol Exp Ther 299(1):12–20

Schwarcz R, Pellicciari R (2002) Manipulation of brain kynurenines: glial targets, neuronal effects, and clinical opportunities. J Pharmacol Exp Ther 303(1):1–10

Siderowf A, Stern M (2003) Update on Parkinson disease. Ann Intern Med 138(8):651–658

Smeyne RJ, Jackson-Lewis V (2005) The MPTP model of Parkinson's disease. Brain Res Mol Brain Res 134(1):57–66

Snow BJ, Macdonald L, McAuley D, Wallis W (2000) The effect of amantadine on levodopa-induced dyskinesias in Parkinson's disease: a double-blind, placebo-controlled study. Clin Neuropharmacol 23(2):82–85

Stanford IM, Lacey MG (1996) Differential actions of serotonin, mediated by 5-HT1B and 5-HT2C receptors, on GABA-mediated synaptic input to rat substantia nigra pars reticulata neurons in vitro. J Neurosci Off J Soc Neurosci 16(23):7566–7573

Stayte S, Vissel B (2014) Advances in non-dopaminergic treatments for Parkinson's disease. Front Neurosci 8:113

Stocchi F, Tagliati M, Olanow CW (2008) Treatment of levodopa-induced motor complications. Mov Disord Off J Mov Disord Soc 23(Suppl 3):S599–S612

Stocchi F, Rascol O, Destee A, Hattori N, Hauser RA, Lang AE et al (2013) AFQ056 in Parkinson patients with levodopa-induced dyskinesia: 13-week, randomized, dose-finding study. Mov Disord 28(13):1838–1846

Stone TW (1993) Neuropharmacology of quinolinic and kynurenic acids. Pharmacol Rev 45(3):309–379

Stone TW (2000) Development and therapeutic potential of kynurenic acid and kynurenine derivatives for neuroprotection. Trends Pharmacol Sci 21(4):149–154

Stone TW (2001) Kynurenines in the CNS: from endogenous obscurity to therapeutic importance. Prog Neurobiol 64(2):185–218

Testa CM, Standaert DG, Young AB, Penney JB Jr (1994) Metabotropic glutamate receptor mRNA expression in the basal ganglia of the rat. J Neurosci Off J Soc Neurosci 14(5 Pt 2):3005–3018

Toulouse A, Sullivan AM (2008) Progress in Parkinson's disease-where do we stand? Prog Neurobiol 85(4):376–392

Verhagen Metman L, Del Dotto P, van den Munckhof P, Fang J, Mouradian MM, Chase TN (1998a) Amantadine as treatment for dyskinesias and motor fluctuations in Parkinson's disease. Neurology 50(5):1323–1326

Verhagen Metman L, Del Dotto P, Natte R, van den Munckhof P, Chase TN (1998b) Dextromethorphan improves levodopa-induced dyskinesias in Parkinson's disease. Neurology 51(1):203–206

Verhagen Metman L, Del Dotto P, LePoole K, Konitsiotis S, Fang J, Chase TN (1999) Amantadine for levodopa-induced dyskinesias: a 1-year follow-up study. Arch Neurol 56(11):1383–1386

Verhagen Metman L, Konitsiotis S, Chase TN (2000) Pathophysiology of motor response complications in Parkinson's disease: hypotheses on the why, where, and what. Mov Disord Off J Mov Disord Soc 15(1):3–8

Wang H, Westin L, Nong Y, Birnbaum S, Bendor J, Brismar H et al (2009) Norbin is an endogenous regulator of metabotropic glutamate receptor 5 signaling. Science 326(5959):1554–1557

Witkin JM, Marek GJ, Johnson BG, Schoepp DD (2007) Metabotropic glutamate receptors in the control of mood disorders. CNS Neurol Disord Drug Targets 6(2):87–100

Yamamoto N, Soghomonian JJ (2009) Metabotropic glutamate mGluR5 receptor blockade opposes abnormal involuntary movements and the increases in glutamic acid decarboxylase mRNA levels induced by l-DOPA in striatal neurons of 6-hydroxydopamine-lesioned rats. Neuroscience 163(4):1171–1180

Yamamoto N, Pierce RC, Soghomonian JJ (2006) Subchronic administration of L-DOPA to adult rats with a unilateral 6-hydroxydopamine lesion of dopamine neurons results in a sensitization of enhanced GABA release in the substantia nigra, pars reticulata. Brain Res 1123(1):196–200

Zhang X, Feng ZJ, Chergui K (2014) GluN2D-containing NMDA receptors inhibit neurotransmission in the mouse striatum through a cholinergic mechanism: implication for Parkinson's disease. J Neurochem 129(4):581–590

Chapter 11
Is There a Future for mGlu5-Positive Allosteric Modulators in Absence Epilepsy? A Comparison with Ethosuximide

Gilles van Luijtelaar, Valerio D'Amore, Ines Santolini, and Richard T. Ngomba

Abstract Ethosuximide is the drug of choice in the treatment of various types of absence seizures. However, there is plenty of room for other anti-absence drugs, considering that not all subjects (57–74%) become seizure-free and about 47% of ethosuximide therapy fails. New anti-absence drugs may target or modulate glutamatergic and or GABAergic neurotransmission, the key players in the circuitry involved in the cortico-thalamo-cortical oscillations responsible for the highly stereotyped spike–wave discharges (SWDs). Cortical highly excitable cells in the focal region form the trigger for the occurrence of SWDs. In contrast, enhanced tonic inhibition is dominant in the thalamus. Biochemical studies have shown that symptomatic WAG/Rij rats differ from age-matched controls in metabotropic glutamate (mGlu) receptor expression and function: mGlu5 receptor expression and function are increased in the somatosensory cortex, and mGlu1 receptor expression is decreased in the thalamus. The two group I mGlu receptor-positive allosteric

G. van Luijtelaar, Ph.D. (✉)
Donders Centre of Cognition, Radboud University, Nijmegen, The Netherlands
e-mail: g.vanLuijtelaar@donders.ru.nl

V. D'Amore, Ph.D., Pharm.D.
Donders Centre of Cognition, Radboud University, Nijmegen, The Netherlands

I.R.C.C.S. NEUROMED, Neuropharmacology Unit, Parco Tecnologico, Località Camerelle, 86077 Pozzilli, IS, Italy
e-mail: damorevalerio@gmail.com

I. Santolini, M.S., Pharm.D
I.R.C.C.S. NEUROMED, Neuropharmacology Unit, Parco Tecnologico, Località Camerelle, 86077 Pozzilli, IS, Italy
e-mail: ines.santolini@gmail.com

R.T. Ngomba, Pharm.D., M.R.Pharms., Ph.D.
I.R.C.C.S. NEUROMED, Neuropharmacology Unit, Parco Tecnologico, Località Camerelle, 86077 Pozzilli, IS, Italy

School of Pharmacy in College of Science, University of Lincoln, Lincoln LN6 7TS, UK
e-mail: richardngomba@hotmail.com

R.T. Ngomba et al. (eds.), *mGLU Receptors*, The Receptors 31,
DOI 10.1007/978-3-319-56170-7_11

modulators (PAMs) VU0360172 and RO0711401 have an interesting profile in acute and (sub)chronic pharmacological studies and produce a dose-dependent decrease of SWDs. Moreover, both compounds are effective in reducing SWDs in the cortex and thalamus. Interestingly, the GABA reuptake blocker tiagabine reduces SWDs in the cortex and not in the thalamus, while the efficacy of ethosuximide is higher in the cortex than in the thalamus. It is thought that VU0360172 stimulates cortex GABA interneurons, which inhibit highly excitable cortical neurons in the focal area. In the thalamus, VU0360172 most likely reduces tonic inhibition. Thus, group I mGlu receptor PAMs might be further developed as anti-absence drugs, with putative disease-modifying effects on epileptogenesis. The preclinical profile of group I mGlu receptor PAMS deserves to be further explored in models of generalized epilepsy and focal types of epilepsy.

Keywords VU0360172 • Positive allosteric modulator • Group I mGlu receptors • Absence epilepsy • Absence seizures • WAG/Rij • Genetic absence model • Antiepileptic drug development

11.1 Introduction

Childhood absence epilepsy (CAE) is the most common form of pediatric epilepsy, accounting for 10–17% of all cases. It typically begins between 4 and 8 years of age and is characterized by brief staring spells, often with the eyes fluttering and the child being unresponsive to external stimuli. Children may experience dozens to hundreds of these episodes per day. Ethosuximide (Etx), dating back to the 1950s, and valproate acid (VPA) have been for many decades and still are the drugs of first choice for absence seizures (Wallace 1986), while the relatively new antiepileptic drug lamotrigine was proposed and approved only in the mid-1990s (Frank et al. 1999; Posner et al. 2005). It was only recently that the efficacy, safety, and tolerability of the three compounds were compared in a large multicentered study. In 2010 a double-blind, randomized, comparative controlled study, including 446 patients, examined the efficacy and tolerability of Etx, VPA, and lamotrigine. At the week 16–20 visit, subjects on Etx (53%) and VPA (58%) had significantly higher freedom from failure rates than subjects on lamotrigine. Subjects on Etx also had significantly less attentional dysfunction compared to subjects on VPA (Glauser et al. 2010; Tenney and Glauser 2013). This combination of findings implied Etx as the optimal initial monotherapy for CAE. Despite these rather favorable outcomes and Etx being the "winner" at week 16–20, Etx therapy failed in 47% of subjects: 14% due to seizures and 24% due to intolerable side effects, while 13% withdrew from the study. Epidemiologic cohort studies showed only seizure freedom in the range from 21 to 74%. In five prospective cohort studies, the proportion of seizure-free subjects was 57–74%. Although labeled as a "benign" syndrome, the clinical course of CAE is variable, and remission rates are far lower than in other classic benign idiopathic epilepsies such as benign rolandic epilepsy (Teoh and Chan 1975). Although a complete discussion of the adverse effects is beyond the scope of this review – the reader is referred to a recent review by Gören and Onat (2007) – the usage of Etx has been associated with commonly observed dose-dependent side effects related to the gastrointestinal tract, central nervous system, and hematopoiesis.

Most of the recent gained insights in the neurobiology of absence epilepsy were obtained in the genetic rodent models, in which GAERS and WAG/Rij rats were most commonly used. Both models mimic each other, and both are well described and have construct, face, and predictive validity (van Luijtelaar and Coenen 1986; Coenen and van Luijtelaar 2003; van Luijtelaar and Sitnikova 2006; Depaulis and van Luijtelaar 2006). In the last 10 years, a new dimension to the pharmacological treatment of absence epilepsy has been added by the proposal that Etx has, in WAG/Rij and in GAERS besides anti-absence, also antiepileptogenic effects when treatment started early and lasted 4 months (Blumenfeld et al. 2008; Sarkisova 2010; Russo et al. 2010; van Luijtelaar et al. 2013; Dezsi et al. 2013) and that the comorbidity of absence epilepsy with a mild form of depression-like behavior, typically for the WAG/Rij strain (Sarkisova and van Luijtelaar 2011), was prevented as a consequence of antiepileptogenesis by Etx (Sarkisova et al. 2010). Most relevant from a theoretical but also from a translational point of view is that remission has been found to be more than likely in high-compliance absence patients that were treated with Etx (Berg et al. 2014). It can be concluded that Etx is the most precious tool in the treatment of absence epilepsy and that it is certainly not without problems, including the relative high failure rate, and its putative adverse effects suggest that there must be room for new and better anti-absence drugs. Since Etx is the drug of choice, new and better anti-absence drugs must be compared with Etx.

11.2 The Assumed Working Mechanisms of ETX

It has long been thought that the action of Etx is to block I_t currents in thalamic-cortical (TC) relay cells and, consequently, the burst firing mode of these cells. The burst firing of TC cells was assumed to be the neurophysiological substrate of SWD activity in cortico-thalamo-cortical pathways. These ideas were developed in the beginning of 1990s based on in vitro neurophysiologic studies using voltage-clamp techniques in acutely isolated neurons of the ventrobasal complex of the thalamus from rats and guinea pigs. Therapeutically relevant concentrations of Etx induced a reduction of the low-threshold Ca^{2+} current, which was most pronounced at more hyperpolarized potentials first in the thalamic relay nuclei with no change in its kinetics or steady-state properties (Coulter et al. 1989; Macdonald and Kelly 1993; White 1999); the later was also found for neurons in the reticular thalamic nucleus (RTN) (Huguenard and Prince 1994). It was generally felt that virtually all thalamic neurons are endowed with a prominent low-threshold Ca^{2+} conductance that is of sufficient magnitude to generate a low-threshold Ca^{2+} spike (LTS) with its accompanying burst firing mode of thalamic cells (Jahnsen and Llinás 1984a, b; Deschênes et al. 1984). Later, it was established that this conductance is crucial for the generation of normal thalamo-cortical oscillations such as sleep spindles (Steriade and Llinás 1988) and for the occurrence of delta waves, typical of deep slow sleep (Domich et al. 1986; Crunelli et al. 2014). The possibility was raised that this conductance may also play a key role in the development of pathological SWDs, and this idea was thought to be a sufficient explanation for the working mechanism of Etx in reducing SWDs in absence epilepsy patients.

However, Etx also decreases persistent Na^+- and Ca^{2+}-activated K^+ currents in thalamic and layer V cortical pyramidal neurons (Leresche et al. 1998; Crunelli and Leresche 2002a), which may also explain the decrease in burst and increase in tonic firing. This newly discovered mechanism of Etx casts doubts on the hypothesis that only a reduction of I_T in thalamic neurons underlies the therapeutic action of this anti-absence medicine (Crunelli and Leresche 2002b). In addition, there is evidence in a genetic absence epilepsy rat model that Etx reduces cortical γ-aminobutyric acid (GABA) levels (Greenhill et al. 2012). Also, elevated glutamate levels in the primary motor cortex of rats with absence epilepsy (but not in normal animals) are reduced by Etx. In addition, whole-cell patch-clamp studies revealed an increase in spontaneous GABA release by Etx concurrent with no change in glutamate release at synapses in the rat's entorhinal cortex in vitro (Greenhill et al. 2012). This was reflected in a substantial rise in the ratio of network inhibition to excitation and a concurrent decrease in excitability of neurons embedded in this network. These authors concluded that Etx directly elevates synaptic inhibition in the cortex next to its well-established effects on ion channels and that both factors contribute to the well-known anti-absence effects. Interestingly, an increase in synaptic inhibition of cortical cells was also found after tiagabine administration by the same authors.

The notion that Etx facilitates cortical inhibition is relevant considering that there is now good evidence that generalized absence seizures in the genetic models, but also in a model in which systemic administration of low-dose PTZ induces absence-like seizures, are initiated at a cortical focus in the deep layers of the peri-oral region of the somatosensory cortex (Meeren et al. 2002; Polack et al. 2007). Subsequent studies revealed that Etx exerted an immediate and strong anti-absence action when injected in the somatosensory cortex but not the cortex in general (Manning et al. 2004; Polack and Charpier 2009). Moreover, in vivo studies showed that the majority (60%) of cat TC neurons are completely silent during SWDs (Steriade and Contreras 1995), and in GAERS, no LTSs were recorded in TC neurons during the majority (90%) of SWDs (Pinault et al. 1998; Slaght et al. 2002). However, hyperpolarized mediated LTSs and the burst firing mode were seen in RTN during SWDs in GAERS (Slaght et al. 2002), and indeed microinfusion of Etx into the ventral basal complex of the thalamus produced only a weak and delayed anti-absence effect, while the effects in the RTN were of a larger magnitude. In all, the outcomes of these studies question the notion that Etx exerts its therapeutic effect to a large extent on the relay nuclei of the thalamus (Manning et al. 2004) and only by blocking I_T channels.

Hypotheses on the working mechanisms of Etx can also be inferred from drug interaction studies. Moreover, drug interaction studies are clinically relevant considering that monotherapy often fails, and seizure control can be only achieved with polytherapy. In order to evaluate the combination of Etx with other anti-absence drugs, the single and combined effects of VPA acid and Etx have been studied. The incidence of SWDs was the readout variable. The median effective doses (ED_{50} values) in this model were 121 and 21.5 mg/kg for VPA and Etx, respectively (van Rijn et al. 2004). When the agents were administered together, the interaction was shown to be infra-additive: the combination was less effective in diminishing the incidence of SWDs in WAG/Rij rats than could be anticipated and based on the net effect of the drugs administered alone. This is considered as indirect evidence that the two anti-absence drugs share some common cellular or molecular mechanism, perhaps blocking Na^+ channels or increasing GABA levels in the cortex.

11.3 New Anti-absence Drugs

The search for new anti-absence drugs has been concentrated on drugs that target the circuitry in which the absence epilepsy characteristic SWDs are elicited, maintained, and aborted. For typical absence epilepsy, it is undoubtedly the reciprocally interconnected cortico-thalamo-cortical network including the RTN (van Luijtelaar and Sitnikova 2006; Blumenfeld 2005; Lüttjohann and van Luijtelaar 2015; Stefan and Lopes da Silva 2013); for atypical absence epilepsy, it is more likely a hippocampal, thalamo-cortical circuit (Onat et al. 2013). GABA and glutamate are the major neurotransmitter systems within the interconnected cortex and thalamus. Therefore, it seems logical to investigate the role of these neurotransmitter systems on the incidence of SWDs. In the mid-1990s, drugs affecting ionotropic glutamate receptors were in the focus of interest, and as was expected, all subclasses of antagonists reduced SWDs, while agonists enhanced SWDs in the WAG/Rij absence model (Peeters et al. 1990; Ramakers et al. 1991; Jakus et al. 2004; van Luijtelaar and Zobeiri 2014). However, many of the ionotropic receptor antagonist compounds had toxic effects, and despite their anti-absence and anticonvulsant action, most of the ionotropic glutamate antagonists were not further developed as antiepileptic drugs.

Since targeting mGluRs might be less toxic than targeting ionotropic glutamate receptors, we decided to focus on the role of mGluRs, also considering that drugs targeting mGluR have been shown to exert pro- or antiepileptic effects in various seizure and epilepsy models (Alexander and Godwin 2006). More specifically, orthosteric antagonists of group I mGluRs reduced motor seizures induced by sound stimulation in DBA mice as well as nonconvulsive (absence) seizures in the lethargic mice (Chapman et al. 1999, 2000) – the latter is a genetic absence model with ataxia and hypoactivity – while the nonselective orthosteric agonist DHPG targeting both type 1 and 5 receptors enhanced convulsions elicited by audiogenic stimulation in DBA mice (Moldrich et al. 2003). The role of orthosteric agonists on nonconvulsive epilepsy has not been investigated. Interestingly, besides orthosteric agonists and antagonists, also selective positive and negative allosteric modulators (PAMs and NAMs) are currently developed and available. These PAMS and NAMS might be even more subtle in inducing changes. A second reason was that mGluRs are widely distributed in the cortico-thalamo-cortical networks (Ngomba et al. 2011a), the brain circuitry in which SWDs originate, are maintained, and end (Cheong et al. 2009; Lüttjohann and van Luijtelaar 2015). These networks consist of a large number of subtypes of mGluRs, and they are selectively present in different parts of the circuitry, offering possibilities for selective targeting of disease-modified (sub)systems. Here we focus specifically on the role of group I mGluRs, with their two subtypes (1 and 5), in the pathogenesis of absence epilepsy and whether drugs affecting this system might be used for anti-absence therapy. It is known that group I mGluRs are involved in absence epilepsy because knockdown as well as whole-animal knockout of PLCβ4, a protein involved in the group I signaling pathway, induced spontaneous SWDs in mice with simultaneous behavioral arrest. In addition, the susceptibility to drug-induced SWDs was increased in these knockout mice, indicating that the deletion of thalamic PLCβ4 leads to the genesis of absence seizures (Cheong et al. 2009). Interestingly, PLCβ4 colocalizes with mGlu1 receptors and mediates mGlu1 receptor signaling in thalamic nuclei (Miyata et al. 2003; Watanabe et al. 1998). Also, the pharmacological studies of Chapman et al. (1999, 2000; Moldrich et al. 2003) on the lethargic mice absence model suggest a role of group I receptors in nonconvulsive seizures mediated in thalamo-cortical networks.

11.4 Neurochemical Studies

As mentioned above, mGluRs are widely distributed in the cortico-thalamo-cortical networks (Ngomba et al. 2011a). In the thalamus, group I mGluRs are located on relay neurons in the ventrobasal thalamus postsynaptic to the cortical inputs (Ferraguti et al. 2008; Romano et al. 1995; Liu et al. 1998). Moderate-to-low mGlu5 receptor mRNA and protein levels are expressed in RTN neurons (Romano et al. 1995; Lourenço Neto et al. 2000), whereas these neurons do not express mGlu1R mRNA (Shigemoto et al. 1992). MGlu1Rs in the cortex are located postsynaptically on GABAergic interneurons (Stinehelfer et al. 2000), and mGlu5Rs are expressed by pyramidal neurons postsynaptic to thalamo-cortical projections (Wijetunge et al. 2008), as well as by interneurons (Romano et al. 1995; Sun et al. 2009). Group I mGluRs are coupled to G_q/G_{11} proteins, and their activation stimulates polyphosphoinositide hydrolysis with formation of inositol-1,4,5-trisphosphate and diacylglycerol and also regulates the activity of different types of Ca^{2+} and K^+ channels (Nicoletti et al. 2011).

Several studies on expression and function were performed in the WAG/Rij model in order to assess the role of mGluRs in the pathophysiology of nonconvulsive epilepsy. In the WAG/Rij rat absence model, the SWDs start to become present in the cortical EEG from 2 to 3 months of age and onward. The analysis of expression of mGluRs was carried out by using symptomatic WAG/Rij rats (minimally 6 months of age) and presymptomatic WAG/Rij rats (not more than 2 months of age) which were compared with age-matched control non-epileptic ACI rats. Various parts of the cortico-thalamo-cortical network, such as RTN, ventrobasal thalamic nuclei, somatosensory cortex, and motor cortex, were inspected (Ngomba et al. 2011b; D'Amore et al. 2013). The functional activity of group I mGluRs was examined by using an in vivo method that allowed measurements of agonist-stimulated PI hydrolysis after incorporation of [3H]inositol into the phospholipids of living rats (Molinaro et al. 2009). mGlu1 or mGlu5 receptor function was evaluated in the thalamus and in the somatosensory cortex and compared to age-matched controls.

Both the expression and function of mGlu1 and mGlu5 receptors were reduced in the thalamus of symptomatic WAG/Rij rats compared to age-matched non-epileptic ACI rats. Lower levels of group I subtype protein receptors in the thalamus of the symptomatic WAG/Rij rats were confirmed by immunohistochemical analysis (Ngomba et al. 2011b). Moreover, these data showed that the reduced expression of mGlu1α receptors found in symptomatic WAG/Rij rats was observed in the dorsal and medial nuclei, but not in the ventroposterolateral thalamus. Furthermore, no mGlu1α receptor mRNA was observed in the RTN, and no change in mGlu1α receptor signaling was detected in the somatosensory cortex of symptomatic WAG/Rij rats compared to age-matched control ACI rats. The level of mGlu1α receptors in the thalamus and the developmental changes in expression in different thalamic nuclei were also investigated by Karimzadeh et al. (2016) in presymptomatic and symptomatic WAG/Rij rats and compared with age-matched non-epileptic wistar controls. These authors found that the protein level of mGlu1α receptors in the thalamus of the symptomatic WAG/Rij rats was lower than in non-epileptic animals. In addition, the number of mGlu1α receptors in different thalamic nuclei was lower in the 6-month-old WAG/Rij compared to age-matched wistar control rats. The gene expression of mGlu1α receptor was also significantly lower in 6-month-old WAG/Rij rats in the lateral dorsal part of the thalamus compared to all other groups. In contrast, the expression of mGlu5Rs was increased either in the somatosensory or motor cortex of symptomatic WAG/Rij rats, as assessed by immunohistochemical and Western blot analysis (D'Amore et al. 2013). In all, these neurochemical data demonstrate that absence epilepsy in this genetic model is accompanied by significant changes in expression and function of group I mGluR in relevant parts of the SWD-generating circuit.

11.5 Acute and Subchronic Pharmacological Studies

Considering our aim to evaluate putative new treatments for absence epilepsy, the effects of selected group I and subgroup 1 PAMs on the occurrence of spontaneous absence seizures were first investigated (Ngomba et al. 2011b). Adult symptomatic rats of the WAG/Rij strain were treated systemically with RO0711401 (3, 10, or 30 mg/kg, s.c.), a selective PAM at mGlu1Rs (Vieira et al. 2009). The outcomes indicated that 3 mg/kg of RO0711401 could only abolish the early stress-related increase in the incidence of SWDs; 10 mg/kg was required for a substantial reduction in the incidence of SWDs; and 30 mg/kg could reduce the incidence as well as the mean duration of trains of SWDs. To further demonstrate a protective role for mGlu1Rs against SWDs, the same rats were systemically injected with JNJ16259685 (2.5 or 5 mg/kg, i.p.), a selective NAM of mGlu1 receptors. Treatment with JNJ16259685 increased the incidence of SWDs in a dose-dependent manner (Ngomba et al. 2011b). In fentanyl-anesthetized WAG/Rij rats, a microinjection of the selective mGlu1α receptor agonist DHPG in the lateral dorsal thalamus reduced the mean duration of SWDs, and the selective antagonist LY367385 increased the amplitude of the spikes of the SWDs and the mean duration of SWDs in symptomatic WAG/Rij rats. These results were in line with the results obtained in the free-moving WAG/Rij rats (Karimzadeh et al. 2016).

Group I (mGluR1 and mGluR5) receptors show a different pattern of distribution in the C-T-C network, which suggests distinct rather than complementary functions of these two receptor subtypes. Therefore, our pharmacological investigation was extended to mGlu5 receptors. The pharmacological enhancement of mGluR5 activity with the PAM VU0360172 at doses of 3 and 10 mg/kg, s.c., known to be centrally active (Rodriguez et al. 2010), caused a robust and dose-dependent reduction in the incidence and mean duration of SWDs. The acute treatment with MTEP (10 or 30 mg/kg, i.p.), a selective NAM of mGlu5 receptors (Anderson et al. 2002; Cosford et al. 2003), did not change the incidence and mean duration of the SWDs (D'Amore et al. 2013).

From a therapeutic standpoint, it is important to establish whether tolerance is developed to the action of these putative anti-absence drugs. Effective doses of VU0360172 (3 mg/kg, s.c.) and RO0711401 (10 mg/kg, s.c.) were administered to rats twice daily for 10 days, and incidence of SWDs was determined across this period (D'Amore et al. 2014, 2013). As expected (Ngomba et al. 2011b), both mGlu1 and mGlu5 receptor PAMs suppressed the incidence of SWDs in the first 2 days of treatment. The anti-absence effect of the mGlu5 receptor PAM (VU0360172) persisted largely over the 10-day administration period with only a small sign of tolerance. In contrast, rats quickly developed complete tolerance to RO0711401 from the third day of treatment. Next, it was wondered whether chronic treatment with RO0711401 or VU0360172 would cause desensitization of mGlu1 and mGlu5Rs in symptomatic WAG/Rij rats and in non-epileptic age-matched wistar control rats (D'Amore et al. 2014). Chronic administration of VU0360172 increased the expression of mGlu5 receptors in the thalamus and in the cortex of WAG/Rij rats without changing the expression of mGlu1α receptors. Treatment with RO0711401 enhanced the expression of both mGlu1α and mGlu5 receptors in the thalamus and cortex of WAG/Rij rats. Opposite data were obtained in non-epileptic wistar rats, in which repeated injections of the two PAMs downregulated the expression of mGlu1α and mGlu5Rs. RO0711401 changed the expression of both the mGlu1α and the mGlu5Rs in WAG/Rij and wistar rats, whereas VU0360172 selectively changed the expression of mGlu5Rs only (D'Amore et al. 2014). These pharmacodynamic and accompanying pharmacokinetic studies could not explain the large differences between VU0360172 and RO0711401 in terms of tolerance to the SWD-suppressing effects.

All these outcomes contribute to highlight the key role of group I mGluRs in the pathophysiology of absence seizures, at least in the spontaneous absence seizure model rats of the WAG/Rij strain. Moreover, they suggest that chronic administration of group I PAMS may change the expression of mGlu1a and mGlu5Rs in the cortex and thalamus, that these changes might be different for the two different PAMS, and that the genetic context (WAG/Rij vs lethargic mice) may also be crucial for the changes in receptor function after repeated administration.

11.6 Local Injection Studies

The next step was a C-T-C circuit demarcation strategy adopted by independent intracortex or intra-thalamus (ventral basal complex) microinfusions to investigate site-specific effects on the regulation of SWDs. WAG/Rij rats received bilateral microinfusion of RO0711401 or VU030172. The two drugs were equally effective in reducing the incidence of SWDs when injected into the cortex. Both drugs were also effective in reducing SWD incidence when they were infused in the thalamus, although the mGlu5 PAM VU036012 displayed somewhat greater thalamic efficacy than the mGlu1 PAM RO0711401 (D'Amore et al. 2015).

Dysfunction of glutamatergic or GABAergic neurotransmission is supposed to be a cause of the initiation and spread of seizures. Moreover, glutamate and GABA interact at many locations within the C-T-C network. Therefore, the effects of tiagabine – a GABA reuptake inhibitor at the neuronal or glial GAT-1 transporter – locally administered to the cortex and thalamus was investigated. Cortical administration of tiagabine resulted in a dose-dependent decrease in the incidence of SWDs, an effect similar to the effects of the group I PAMs. In contrast, intra-thalamic injections of tiagabine showed a dose-dependent increase in the incidence of SWDs, an effect opposite to that obtained with the PAMs.

Next, the interaction between GABA and glutamate was investigated. More precisely, it was established whether and in which direction an increased availability of (extra)synaptic GABA influences responses to VU0360172 in the thalamus and cortex. Combined microinjections of VU0360172 and tiagabine were carried out. Co-administration of VU0360172 and tiagabine reduced the incidence of SWDs when injected in the cortex, an effect similar to the effects of both compounds alone. Moreover, an indication was obtained for subadditive effects of the two drugs. In contrast, completely opposite effects were observed after co-administration to the thalamus: a decrease followed by an increase in the incidence of SWDs was seen, suggesting that the effects of VU0360172 were completely blocked in the presence of increased GABA (D'Amore et al. 2015).

In all the above mentioned pharmacological experiments with PAMs and NAMs, the behavior of the rats was quantified with the aid of a calibrated infrared movement detector; it allowed the quantification of the amount of activity during the EEG recording session. In none of the experiments described above did VU0360172 affect the behavior of the rats.

It can be concluded that these pharmacological studies demonstrate dose-dependent SWD decreasing effects of both PAMs, minimal tolerance in a twice daily 10-day administration protocol of VU0360172, and the quick development of tolerance to RO0711401. Local injections of both PAMs demonstrated that the compounds are effective in reducing SWDs in the cortex and thalamus. Effects on behavior were not noticed. In all, VU0360172 has a good preclinical profile as an anti-absence drug in the WAG/Rij absence model.

11.7 Comparison Between Ethosuximide and VU0360172

The preclinical profile of VU0360172 as an anti-absence drug, as described here, seems promising, but this specific PAM has not been tested yet in other assays and models. Etx has a 60-year-long tradition, and much is known regarding its clinical efficacy and adverse effects. Etx has the best clinical profile among available anti-absence drugs, and new anti-absence drugs have to be compared with it. Recent reviews also point toward the limitations of Etx. These limitations clearly demonstrate the need for other treatment options. Many earlier in vivo and in vitro neurophysiologic studies on healthy animals, such as rats, cats, and ferrets, with pharmacologically induced seizures (e.g. Coulter et al. 1989, Huguenard and Prince 1994; Deschênes et al. 1984; von Krosigk et al 1993; Porcello et al. 2003) have found that Etx acts in the thalamus and by blocking Ca^{2+} channels, both in the relay nuclei and in the RTN. The comparative cortex-thalamus local injection studies of Manning et al. (2004) in the GAERS model and the outcomes of the neurophysiologic studies of the Pinault and Crunelli groups (Leresche et al. 1998; Crunelli and Leresche 2002b; Pinault et al. 1998) have yielded a somewhat different focus, regarding the location where Etx exerts its anti-absence effect, the assumed role of thalamic bursting neurons during SWDs, and the mechanisms of action. Relevant in this context is that these authors did most of their experiments in vivo in the well-characterized GAERS model and not on slices from healthy animals. Etx exerts its action predominantly in the cortical focal area in the somatosensory cortex, not in other cortical regions in the GAERS model. The action of Etx was also demonstrated in the thalamic ventral basal complex and RTN (Manning et al. 2004; Richards et al. 2003), although its efficacy in the thalamus was clearly less (Fig. 11.1).

The comparison of the local (cortical and thalamic) effects of both drugs is presented in Fig. 11.1. Both Etx and VU0360172 show a good SWD-suppressing effect in the cortex; however, the SWD-suppressing effects of VU0360172 in the thalamus seem to be more pronounced than those of Etx, although it is risky to compare results of different studies.

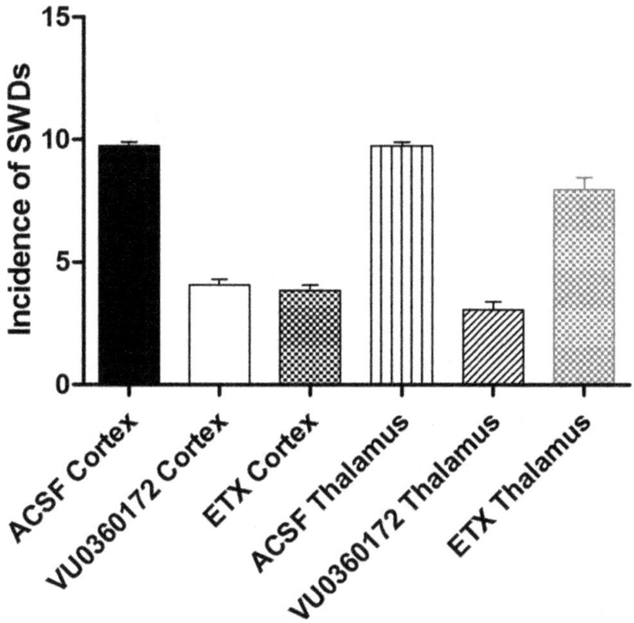

Fig. 11.1 Representative figure: Incidence of SWDs 30 min after bilateral microinfusion of artificial cerebrospinal fluid (ACSF), Ethosuximide (ETX), or VU3601726 in the thalamus and cortex (After Manning et al. 2004; D'Amore et al. 2015)

Our results show that VU0360172 acts well in two target areas, while this seems less to be the case for Etx. VU0360172 and RO0711401 were equally effective in the somatosensory cortex and ventral basal part of the thalamus. This strongly suggests that both PAMs may target both locations equally and effectively and that this may contribute to successful seizure suppression of both group I PAMs.

11.8 Discussion

There is plenty of evidence that group I mGluR PAMs aggravate glutamatergic neurotransmission: binding of these PAMs increases the affinity of glutamate-site agonists at its extracellular N-terminal binding site, group I PAMs potentiate synaptically evoked mGlu1 receptor responses in rat brain slices (Knoflach et al. 2001), and mGlu1R stimulation enhances excitability in layer V cortical neurons without any effect on GABAergic neurotransmission (Bandrowski et al. 2003). Therefore, at first glance, the SWD-reducing effects of the PAMs VU0360172 and RO0711401 seem puzzling considering that the glutamatergic ionotropic receptor agonists aggravate SWDs in the genetic rodent models (Peeters et al. 1990; Ramakers et al.

1991; Jakus et al. 2004; van Luijtelaar and Zobeiri 2014). However, the outcomes of our local injection study might give two clues why this PAM reduces SWDs. The first regard the outcomes of cortical injections of the PAM alone, tiagabine alone, and the combination of both drugs. The PAM VU0360172 had, when injected in the focal area, the same action as tiagabine, a drug known to block GABAergic reuptake. The combined administration showed subadditive effects, suggesting that both drugs targeted the same mechanism. We proposed that VU0360172 and RO0711401, its receptors being present postsynaptically on GABAergic interneurons, stimulate the activity of GABAergic interneurons, facilitating the release of GABA and diminishing SWDs. This would fit nicely with the view that SWDs in the rodent models are due to enhanced local cortical excitation and reduced local inhibition (van Luijtelaar and Sitnikova 2006; Lüttjohann et al. 2011). Whether this mechanism of VU0360172 is only valid for the cortex of genetic epileptic rats, with their epilepsy-related changes in cortical mGlu5 receptor expression and function, as mentioned in the section on biochemistry, needs to be established. The second clue comes from the thalamic injection studies. Group I PAMs infused in the thalamus suppress SWDs. Here it is thought that they reduce the assumed increased tonic inhibition (Cope et al. 2009), a second reason for SWDs to occur in the genetic rat models. Also, our data on local infusion of tiagabine, showing a dose-dependent increase in SWDs, are in full agreement with the tonic inhibition hypothesis of Cope and colaborators.

From a drug development perspective, it is important to mention that none of the local injection studies were effects on the motor system of the rats observed, suggesting that the motor system does not seem to be largely affected by group I PAMs. This is in clear contrast to the large effects of some of the ionotropic glutamatergic agonists, such as MK801. The lack of behavioral effects in our EEG study suggests that the anti-absence effects of the PAMs are not secondary to changes in behavior or sleep–wake behavior, as far as this could be inferred from our quantitative behavioral measurements. Interestingly, VU0360172 did not affect spontaneous locomotor activity in an open field, but it did dose-dependently reduce hyperlocomotion induced by amphetamine (Rodriguez et al. 2010). Tolerance toward the anti-absence action of a twice daily 10-day drug administration regimen was minimal with VU0360172 in contrast to RO0711401. With the latter drug, tolerance was absent in the first 2 days, but on the third day (after five injections), it was complete. Neurotoxic studies with higher doses of VU0360172, such as the measurement of a motor coordination in a rotor-rod task, measures on sedation, and analgesia of higher doses of the drug, are indicated. To establish whether the compound affects cognition is challenging. Fortunately, group I PAMs have already been proposed for the treatment of psychosis and cognitive disturbances in schizophrenic patients, since they enhance certain forms of neural plasticity via LTD and LTD, learning, and memory processes (Rush et al. 2002). The latter properties were ascribed to the ability of PAMs to indirectly potentiate the function of NMDA receptors; our cortical injection studies suggest that the action of group I PAMs might, in addition to this well-known effect, facilitate cortical GABAergic transmission. The drug deserves and needs to be investigated for its putative antiepileptic action in other models of seizure and epilepsy. This is also indicated considering the interaction between VU0360172 and tiagabine, as found in the thalamus. The balance between glutamate and GABA is disturbed in many types of epilepsy, including mesial temporal lobe epilepsy. Also in these models, group I PAMs may increase the diminished GABAergic inhibition. In vivo imaging of mGluR5 via microPET/CT revealed regional changes of mGluR5-binding potential of the rat brain in a pilocarpine-induced epilepsy model. The temporal and spatial changes in mGluR5 availability suggest an abnormal glutamatergic network during epileptogenesis (Choi et al. 2014).

Antiepileptic drugs are often given in combination, considering that monotherapy might often be insufficient for full seizure control. However, many antiepileptic drugs are enzyme inducing and lower the efficacy of other medications. Etx showed an infra-additive effect with VPA (van Rijn et al. 2004); drug combination studies with group I PAMs are necessary. A first drug interaction study on VU0360172 with chronically administered Etx showed that it kept its anti-absence action but without time and dose dependency (D'Amore et al. 2016). The data also suggested that the PAM can be used as an adjunct therapy in patients with absence epilepsy, as well as being used in monotherapy. Whether our group I PAM has disease-modifying properties that stop epileptogenesis is also important. Beneficial effects on cell proliferation by a mGluR1α agonist after episodes of early-life status epilepticus were recently described (Friedman et al. 2016). It might be relevant to investigate the disease-modifying properties of group I PAMs in other models as well.

It needs to be added that the results obtained on the direction of the effects of group I PAMs and NAMs on absence epilepsy as described here (PAMs reduce, NAMs enhance, or have no effect) are opposite to what has been described in another genetic model: the lethargic mice. A discussion on the validity of the GAERS and WAG/Rij models vs the lethargic mouse model is out of the scope of this paper; it is well known that the latter has a complex phenotype (besides absences seizures, it presents with chronic ataxia, hypoactivity, and transient attacks of severe dyskinetic motor behavior) next to a large GAD(67) expression in thalamic cells and disturbances in the $GABA_B$ receptor system accompanied by an involvement of T-type voltage-gated Ca^{2+} channels. It is obvious, though, that differences in the direction of the effects of the mGlu modulators in the two genetic absence models could be due to the genotype and to different causes of the absence seizures. Another possibility is that the PAMs and NAMs act differently in the two species because of changes in group I receptor expression and function in absence epilepsy-relevant parts of the brain. In order to solve the issue regarding the opposite direction of the effects in the two models, the same compounds should be investigated in the two models.

Although the used PAM is considered selective for mGlu5R, VU0360172 does not interact in a fully competitive manner with the prototypic allosteric binding site, in contrast to some other recently synthetized mGlu5 PAMs (Rook et al. 2015). Whether this property contributes to its anti-absence action in vivo remains to be established. If so, this would offer unique chances for discovering new mechanisms of absence seizures.

References

Alexander GM, Godwin DW (2006) Metabotropic glutamate receptors as a strategic target for the treatment of epilepsy. Epilepsy Res 71:1–22

Anderson JJ, Rao SP, Rowe B, Giracello DR, Holtz G, Chapman DF, Tehrani L, Bradbury MJ, Cosford ND, Varney MA (2002) [3H]Methoxymethyl-3-[(2-methyl-1,3-thiazol-4-yl)ethynyl] pyridine binding to metabotropic glutamate receptor subtype 5 in rodent brain: in vitro and in vivo characterization. J Pharmacol Exp Ther 303:1044–1051

Bandrowski AE, Huguenard JR, Prince DA (2003) Baseline glutamate levels affect group I and II mGluRs in layer V pyramidal neurons of rat sensorimotor cortex. J Neurophysiol 89:1308–1316

Berg AT, Levy SR, Testa FM, Blumenfeld H (2014) Long-term seizure remission in childhood absence epilepsy: might initial treatment matter? Epilepsia 55:551–593

Blumenfeld H (2005) Cellular and network mechanisms of spike-wave seizures. Epilepsia 46(Suppl 9):21–33

Blumenfeld H, Klein JP, Schridde U, Vestal M, Rice T, Khera DS, Bashyal C, Giblin K, Paul-Laughinghouse C, Wang F, Phadke A, Mission J, Agarwal RK, Englot DJ, Motelow J, Nersesyan H, Waxman SG, Levin AR (2008) Early treatment suppresses the development of spike-wave epilepsy in a rat model. Epilepsia 49:400–409

Chapman AG, Yip PK, Yap JS, Quinn LP, Tang E, Harris JR, Meldrum BS (1999) Anticonvulsant actions of LY 367385 ((+)-2-methyl-4-carboxyphenylglycine) and AIDA ((RS)-1-aminoindan-1,5-dicarboxylic acid). Eur J Pharmacol 368:17–24

Chapman AG, Nanan K, Williams M, Meldrum BS (2000) Anticonvulsant activity of two metabotropic glutamate group I antagonists selective for the mGlu5. receptor: 2-methyl-6-(phenylethynyl)-pyridine (MPEP), and (E)-6-methyl-2-styryl-pyridine (SIB 1893). Neuropharmacol 39:1567–1574

Cheong E, Zheng Y, Lee K, Lee J, Kim S, Sanati M, Lee S, Kim YS, Shin HS (2009) Deletion of phospholipase C beta4 in thalamocortical relay nucleus leads to absence seizures. Proc Natl Acad Sci U S A 106:21912–21917

Choi H, Kim YK, Oh SW, Im HJ, Hwang DW, Kang H, Lee B, Lee YS, Jeong JM, Kim EE, Chung JK, Lee DS (2014) In vivo imaging of mGluR5 changes during epileptogenesis using [11C] ABP688 PET in pilocarpine-induced epilepsy rat model. PLoS One 9(3):e92765

Coenen AL, van Luijtelaar EL (2003) Genetic animal models of absence epilepsy: a review of the WAG/Rij strain of rats. Behav Genet 33:635–655

Cope DW, Di Giovanni G, Fyson SJ, Orbán G, Errington AC, Lorincz ML, Gould TM, Carter DA, Crunelli V (2009) Enhanced tonic GABAA inhibition in typical absence epilepsy. Nat Med 15:1392–1398

Cosford ND, Roppe J, Tehrani L, Schweiger EJ, Seiders TJ, Chaudary A, Rao S, Varney MA (2003) [3H]-methoxymethyl-MTEP and [3H]-methoxy-PEPy: potent and selective radioligands for the metabotropic glutamate subtype 5 (mGlu5) receptor. Bioorg. Med. Chem Lett 13:351–354

Coulter DA, Huguenard JR, Prince DA (1989) Characterization of ethosuximide reduction of low-threshold calcium current in thalamic neurons. Ann Neurol 25:582–593

Crunelli V, Leresche N (2002a) Childhood absence epilepsy: genes, channels, neurons and networks. Nat Rev Neurosci 3:371–382

Crunelli V, Leresche N (2002b) Block of thalamic T-Type $Ca^{(2+)}$ channels by Ethosuximide is not the whole story. Epilepsy Curr 2:53–56

Crunelli V, David F, Leresche N, Lambert RC (2014) Role for T-type Ca^{2+} channels in sleep waves. Pflugers Arch 466:735–745

D'Amore V, Santolini I, van Rijn CM, Biagioni F, Molinaro G, Prete A, Conn PJ, Lindsley CW, Zhou Y, Vinson PN, Rodriguez AL, Jones CK, Stauffer SR, Nicoletti F, van Luijtelaar G, Ngomba RT (2013) Potentiation of mGlu5 receptors with the novel enhancer, VU0360172, reduces spontaneous absence seizures in WAG/Rij rats. Neuropharmacol 66:330–338

D'Amore V, Santolini I, Celli R, Lionetto L, De Fusco A, Simmaco M, van Rijn CM, Vieira E, Stauffer SR, Conn PJ, Bosco P, Nicoletti F, van Luijtelaar G, Ngomba RT (2014) Head-to head comparison of mGlu1 and mGlu5 receptor activation in chronic treatment of absence epilepsy in WAG/Rij rats. Neuropharmacol 85:91–103

D'Amore V, von Randow C, Nicoletti F, Ngomba RT, van Luijtelaar G (2015) Anti-absence activity of mGlu1 and mGlu5 receptor enhancers and their interaction with a GABA reuptake inhibitor: effect of local infusions in the somatosensory cortex and thalamus. Epilepsia 56:1141–1151

D'Amore V, Raaijmakers RH, Santolini I, van Rijn CM, Ngomba RT, Nicoletti F, van Luijtelaar G (2016) The anti-absence effect of mGlu5 receptor amplification with VU0360172 is maintained during and after antiepileptogenesis. Pharmacol Biochem Behav 146-147:50–59

Depaulis A, van Luijtelaar G (2006) Genetic animal models of absence epilepsy n the rat. In: Pitkanen A, Schwartzkroin PA, Moshe SL (eds) Models of seizures and epilepsy. Elsevier, Amsterdam, pp 233–248

Deschênes M, Paradis M, Roy JP, Steriade M (1984) Electrophysiology of neurons of lateral thalamic nuclei in cat: resting properties and burst discharges. J Neurophysiol 51:1196–1219

Dezsi G, Ozturk E, Stanic D, Powell KL, Blumenfeld H, O'Brien TJ, Jones NC (2013) Ethosuximide reduces epileptogenesis and behavioral comorbidity in the GAERS model of genetic generalized epilepsy. Epilepsia 54:635–643

Domich L, Oakson G, Steriade M (1986) Thalamic burst patterns in the naturally sleeping cat: a comparison between cortically projecting and reticularis neurones. J Physiol Lond 379:429–449

Ferraguti F, Crepaldi L, Nicoletti F (2008) Metabotropic glutamate 1 receptor: current concepts and perspectives. Pharmacol Rev 60:536–581

Frank LM, Enlow T, Holmes GL, Manasco P, Concannon S, Chen C, Womble G, Casale EJ (1999) Lamictal (lamotrigine) monotherapy for typical absence seizures in children. Epilepsia 40:973–979

Friedman LK, Sharma A, Corcia M, Webster T, Qazi L, Simsovits D, Khalil Y, Hu S, Kantrowitz M, Hong H (2016) Selective inhibition of metabotropic glutamate type 1 alpha receptor (mGluR1α) reduces cell proliferation and migration following status epilepticus in early development. Int J Dev Neurosci 54:6–21

Glauser TA, Cnaan A, Shinnar S, Hirtz DG, Dlugos D, Masur D, Clark PO, Capparelli EV, Adamson PC, Childhood Absence Epilepsy Study Group (2010) Ethosuximide, valproic acid, and lamotrigine in childhood absence epilepsy. N Engl J Med 362:790–799

Gören MZ, Onat F (2007) Ethosuximide: from bench to bedside. CNS Drug Rev 13:224–239

Greenhill SD, Morgan NH, Massey PV, Woodhall GL, Jones RS (2012) Ethosuximide modifies network excitability in the rat entorhinal cortex via an increase in GABA release. Neuropharmacol 62:807–814

Huguenard JR, Prince DA (1994) Intrathalamic rhythmicity studied in vitro: nominal T-current modulation causes robust antioscillatory effects. J Neurosci 14:5485–5502

Jahnsen H, Llinás R (1984a) Electrophysiological properties of guinea-pig thalamic neurones: an in vitro study. J Physiol 349:205–226

Jahnsen H, Llinás R (1984b) Voltage-dependent burst-to-tonic switching of thalamic cell activity: an in vitro study. Arch Ital Biol 122:73–82

Jakus R, Graf M, Ando RD, Balogh B, Gacsalyi I, Levay G, Kantor S, Bagdy G (2004) Effect of two noncompetitive AMPA receptor antagonists GYKI 52466 and GYKI 53405 on vigilance, behavior and spike-wave discharges in a genetic rat model of absence epilepsy. Brain Res 1008:236–244

Karimzadeh F, Modarres Mousavi SM, Ghadiri T, Jafarian M, Soleimani M, Sadeghi SM, Mesgari M, Joghataei MT, Gorji A (2016) The modulatory effect of metabotropic glutamate receptor type-1α on spike-wave discharges in WAG/Rij Rats. Mol Neurobiol 16. [Epub ahead of print]

Knoflach F, Mutel V, Jolidon S, Kew JN, Malherbe P, Vieira E, Wichmann J, Kemp JA (2001) Positive allosteric modulators of metabotropic glutamate 1 receptor: characterization, mechanism of action, and binding site. Proc Natl Acad Sci U S A 98:13402–13407

Leresche N, Parri HR, Erdemli G, Guyon A, Turner JP, Williams SR, Aprodini E, Crunelli V (1998) On the action of the anti-absence drug ethosuximide in the rat and cat thalamus. J Neurosci 18:4842–4853

Liu XB, Muñoz A, Jones EG (1998) Changes in subcellular localization of metabotropic glutamate receptor subtypes during postnatal development of mouse thalamus. J Comp Neurol 395:450–465

Lourenço Neto F, Schadrack J, Berthele A, Zieglgänsberger W, Tölle TR, Castro-Lopes JM (2000) Differential distribution of metabotropic glutamate receptor subtype mRNAs in the thalamus of the rat. Brain Res 854:93–105

Lüttjohann A, van Luijtelaar G (2015) Dynamics of networks during absence seizure's on- and offset in rodents and man. Front Physiol 6:16

Lüttjohann A, Zhang S, de Peijper R, van Luijtelaar G (2011) Electrical stimulation of the epileptic focus in absence epileptic WAG/Rij rats: assessment of local and network excitability. Neurosci 188:125–134

Macdonald RL, Kelly KM (1993) Antiepileptic mechanisms of action. Epilepsia 34(Suppl 5):S1–S8

Manning JP, Richards DA, Leresche N, Crunelli V, Bowery NG (2004) Cortical-area specific block of genetically determined absence seizures by ethosuximide. Neuroscience 123:5–9

Meeren HK, Pijn JP, Van Luijtelaar EL, Coenen AM, Lopes da Silva FH (2002) Cortical focus drives widespread corticothalamic networks during spontaneous absence seizures in rats. J Neurosci 22:1480–1495

Miyata M, Kashiwadani H, Fukaya M, Hayashi T, Wu D, Suzuki T, Watanabe M, Kawakami Y (2003) Role of thalamic phospholipase C[beta]4 mediated by metabotropic glutamate receptor type 1 in inflammatory pain. J Neurosci 23:8098–8108

Moldrich RX, Chapman AG, De Sarro G, Meldrum BS (2003) Glutamate metabotropic receptors as targets for drug therapy in epilepsy. Eur J Pharmacol 476(1-2):3–16

Molinaro G, Traficante A, Riozzi B, Di Menna L, Curto M, Pallottino S, Nicoletti F, Bruno V, Battaglia G (2009) Activation of mGlu2/3 metabotropic glutamate receptors negatively regulates the stimulation of inositol phospholipid hydrolysis mediated by 5-hydroxytryptamine2A serotonin receptors in the frontal cortex of living mice. Mol Pharmacol 76:379–387

Ngomba RT, Santolini I, Salt TE, Ferraguti F, Battaglia G, Nicoletti F, van Luijtelaar G (2011a) Metabotropic glutamate receptors in the thalamocortical network: strategic targets for the treatment of absence epilepsy. Epilepsia 52:1211–1222

Ngomba RT, Santolini I, Biagioni F, Molinaro G, Simonyi A, van Rijn CM, D'Amore V, Mastroiacovo F, Olivieri G, Gradini R, Ferraguti F, Battaglia G, Bruno V, Puliti A, van Luijtelaar G, Nicoletti F (2011b) Protective role for type-1 metabotropic glutamate receptors against spike and wave discharges in the WAG/Rij rat model of absence epilepsy. Neuropharmacol 60:1281–1291

Nicoletti F, Bockaert J, Collingridge GL, Conn PJ, Ferraguti F, Schoepp DD, Wroblewski JT, Pin JP (2011) Metabotropic glutamate receptors: from the workbench to the bedside. Neuropharmacology 60:1017–1041

Onat FY, van Luijtelaar G, Nehlig A, Snead OC 3rd (2013) The involvement of limbic structures in typical and atypical absence epilepsy. Epilepsy Res 103:111–123

Peeters BW, van Rijn CM, Vossen JM, Coenen AM (1990) Involvement of NMDA receptors in non-convulsive epilepsy in WAG/Rij rats. Life Sci 47:523–529

Pinault D, Leresche N, Charpier S, Deniau JM, Marescaux C, Vergnes M, Crunelli V (1998) Intracellular recordings in thalamic neurones during spontaneous spike and wave discharges in rats with absence epilepsy. J Physiol 509:449–456

Polack PO, Charpier S (2009) Ethosuximide converts ictogenic neurons initiating absence seizures into normal neurons in a genetic model. Epilepsia 50:1816–1820

Polack PO, Guillemain I, Hu E, Deransart C, Depaulis A, Charpier S (2007) Deep layer somatosensory cortical neurons initiate spike-and-wave discharges in a genetic model of absence seizures. J Neurosci 27:6590–6599

Porcello DM, Smith SD, Huguenard JR (2003) Actions of U-92032, a T-type Ca2+ channel antagonist, support a functional linkage between I(T) and slow intrathalamic rhythms. J Neurophysiol 89:177–185

Posner EB, Mohamed K, Marson AG (2005) Systematic review of treatment of typical absence seizures in children and adolescents with ethosuximide, sodium valproate or lamotrigine. Seizure 14:117–122

Ramakers GM, Peeters BW, Vossen JM, Coenen AM (1991) CNQX, a new non-NMDA receptor antagonist, reduces spike wave discharges in the WAG/Rij rat model of absence epilepsy. Epilepsy Res 9:127–131

Richards DA, Manning JP, Barnes D, Rombola L, Bowery NG, Caccia S, Leresche N, Crunelli V (2003) Targeting thalamic nuclei is not sufficient for the full anti-absence action of ethosuximide in a rat model of absence epilepsy. Epilepsy Res 54:97–107

Rodriguez AL, Grier MD, Jones CK, Herman EJ, Kane AS, Smith RL, Williams R, Zhou Y, Marlo JE, Days EL, Blatt TN, Jadhav S, Menon UN, Vinson PN, Rook JM, Stauffer SR, Niswender CM, Lindsley CW, Weaver CD, Conn PJ (2010) Discovery of novel allosteric modulators of metabotropic glutamate receptor subtype 5 reveals chemical and functional diversity and in vivo activity in rat behavioral models of anxiolytic and antipsychotic activity. Mol Pharmacol 78:1105–1123

Romano C, Sesma MA, McDonald CT, O'Malley K, Van den Pol AN, Olney JW (1995) Distribution of metabotropic glutamate receptor mGluR5 immunoreactivity in rat brain. J Comp Neurol 355:455–469

Rook JM, Tantawy MN, Ansari MS, Felts AS, Stauffer SR, Emmitte KA, Kessler RM, Niswender CM, Daniels JS, Jones CK, Lindsley CW, Conn PJ (2015) Relationship between in vivo receptor occupancy and efficacy of metabotropic glutamate receptor subtype 5 allosteric modulators with different in vitro binding profiles. Neuropsychopharmacology 40:755–765

Rush AM, Wu J, Rowan MJ, Anwyl R (2002) Group I metabotropic glutamate receptor (mGluR)-dependent long-term depression mediated via p38 mitogen-activated protein kinase is inhibited by previous high-frequency stimulation and activation of mGluRs and protein kinase C in the rat dentate gyrus in vitro. J Neurosci 22:6121–6128

Russo E, Citraro R, Scicchitano F, De Fazio S, Di Paola ED, Constanti A, De Sarro G (2010) Comparison of the antiepileptogenic effects of an early long-term treatment with ethosuximide or levetiracetam in a genetic animal model of absence epilepsy. Epilepsia 51:1560–1569

Sarkisova K, van Luijtelaar G (2011) The WAG/Rij strain: a genetic animal model of absence epilepsy with comorbidity of depression [corrected]. Prog Neuropsychopharmacol Biol Psychiatry 35:854–876

Sarkisova KY, Kuznetsova GD, Kulikov MA, van Luijtelaar G (2010) Spike-wave discharges are necessary for the expression of behavioral depression-like symptoms. Epilepsia 51:146–160

Shigemoto R, Nakanishi S, Mizuno N (1992) Distribution of the mRNA for a metabotropic glutamate receptor (mGluR1) in the central nervous system: an in-situ hybridization study in adult and developing rat. J Comp Neurol 322:121–135

Slaght SJ, Leresche N, Deniau JM, Crunelli V, Charpier S (2002) Activity of thalamic reticular neurons during spontaneous genetically determined spike and wave discharges. J Neurosci 22:2323–2334

Stefan H, Lopes da Silva FH (2013) Epileptic neuronal networks: methods of identification and clinical relevance. Front Neurol 4:8

Steriade M, Contreras D (1995) Relations between cortical and thalamic cellular events during transition from sleep patterns to paroxysmal activity. J Neurosci 15(1 Pt 2):623–624

Steriade M, Llinás RR (1988) The functional states of the thalamus and the associated neuronal interplay. Physiol Rev 68:649–742

Stinehelfer S, Vruwink M, Burette A (2000) Immunolocalization of mGluR1alpha in specific populations of local circuit neurons in the cerebral cortex. Brain Res 861:37–44

Sun QQ, Zhang Z, Jiao Y, Zhang C, Szab G, Erdelyi F (2009) Differential metabotropic glutamate receptor expression and modulation in two neocortical inhibitory networks. J Neurophysiol 101:2679–2692

Tenney JR, Glauser TA (2013) The current state of absence epilepsy: can we have your attention? Epilepsy Curr 13:135–140

Teoh PC, Chan HL (1975) Lupus-scleroderma syndrome induced by ethosuximide. Arch Dis Child 50:58–61

van Luijtelaar EL, Coenen AM (1986) Two types of electrocortical paroxysms in an inbred strain of rats. Neurosci Lett 70:393–397

van Luijtelaar G, Sitnikova E (2006) Global and focal aspects of absence epilepsy: the contribution of genetic models. Neurosci Biobehav Rev 30:983–1003

van Luijtelaar G, Zobeiri M (2014) Progress and outlooks in a genetic absence epilepsy model (WAG/Rij). Curr Med Chem 21:704–721

van Luijtelaar G, Mishra AM, Edelbroek P, Coman D, Frankenmolen N, Schaapsmeerders P et al (2013) Anti-epileptogenesis: electrophysiology, diffusion tensor imaging and behaviour in a genetic absence model. Neurobiol Dis 60:126–138

van Rijn CM, Sun MS, Deckers CL, Edelbroek PM, Keyser A, Renier W, Meinardi H (2004) Effects of the combination of valproate and ethosuximide on spike wave discharges in WAG/Rij rats. Epilepsy Res 59:181–189

Vieira E, Huwyler J, Jolidon S, Knoflach F, Mutel V, Wichmann J (2009) Fluorinated 9H-xanthene-9-carboxylic acid oxazol-2-yl-amides as potent, orally available mGlu1 receptor enhancers. Bioorg Med Chem Lett 19:1666–1669

von Krosigk M, Bal T, McCormick DA (1993) Cellular mechanisms of a synchronized oscillation in the thalamus. Science 261:361–364

Wallace SJ (1986) Use of ethosuximide and valproate in the treatment of epilepsy. Neurol Clin 4:601–616

Watanabe M, Nakamura M, Sato K, Kano M, Simon MI, Inoue Y (1998) Patterns of expression for the mRNA corresponding to the four isoforms of phospholipase C beta in mouse brain. Eur J Neurosci 10:2016–2025

White HS (1999) Comparative anticonvulsant and mechanistic profile of the established and newer antiepileptic drugs. Epilepsia 40(Suppl 5):S2–10

Wijetunge LS, Till SM, Gillingwater TH, Ingham CA, Kind PC (2008) mGluR5 regulates glutamate-dependent development of the mouse somatosensory cortex. J Neurosci 28:13028–13137

Chapter 12
Regulation of Hippocampal mGluR-Dependent Long-Term Depression by GluA2-Dependent Cofilin-Mediated Actin Remodeling

Zhengping Jia and Graham Collingridge

Abstract The AMPA glutamate receptors (AMPARs) are the principal mediators of the fast excitatory synaptic transmission, and they are important for various forms of synaptic plasticity. The GluA2 subunit of AMPARs is particularly interesting because it controls a number of key properties of AMPARs, including Ca^{2+} permeability and receptor trafficking. In this chapter, we will discuss the critical involvement of GluA2 in mGluR-dependent long-term depression (mGluR-LTD) at the hippocampal CA1 synapse. We will summarize the data showing that GluA2 is essential for the expression of mGluR-LTD, but interestingly, the contribution of GluA2 in this form of LTD requires GluA2 N-terminal domain and its interaction with N-cadherin. In addition, we will evaluate the role of cofilin and its upstream regulator LIMK1 in mGluR-LTD and associated spine plasticity. Finally we will discuss interactions between GluA2 and LIMK1/cofilin and propose a molecular model to coordinate structural and functional plasticity through GluA2-dependent, cofilin-mediated actin remodeling.

Keywords mGluR-LTD • Spine • GluA2 • Actin • Cofilin • N-cadherin • Structural and functional plasticity

---

Z. Jia, Ph.D. (✉)
Neurosciences & Mental Health Program, The Hospital for Sick Children, Department of Physiology, University of Toronto, Toronto, Canada
e-mail: zhengping.jia@sickkids.ca

G. Collingridge, Ph.D.
Centre for Synaptic Plasticity, School of Physiology & Pharmacology, University of Bristol, Bristol, UK
e-mail: glcollingridge@gmail.com

R.T. Ngomba et al. (eds.), *mGLU Receptors*, The Receptors 31,
DOI 10.1007/978-3-319-56170-7_12

12.1 Introduction

Synaptic transmission and plasticity at glutamatergic synapses are fundamentally important for both brain development and function, and deficits at these synapses are responsible for a wide range of neurodevelopmental and neuropsychiatric disorders such as autism, schizophrenia, intellectual disability, and Alzheimer's diseases. Long-lasting synaptic plasticity, a key mechanism for neural network remodeling and memory formation, involves changes in both synaptic morphology and electrical transmission (often referred to as morphological and functional plasticity, respectively). Although enormous progress has been made in understanding the molecular mechanisms underlying these two forms of plasticity individually, whether and how these forms of plastic changes are coupled to achieve persistent and input-specific modifications of the synapse, which are thought to be essential for precise information processing and various aspects of brain function, remains a major unresolved question. In this respect, we have been identifying and characterizing synaptic proteins that are required for both morphological and functional regulation at the synapse. These include AMPA glutamate receptors (AMPARs) and signaling molecules involving the Rho family GTPases. AMPARs are the primary mediators of fast excitatory synaptic transmission, whereas the Rho GTPases are the central mediators of the actin cytoskeleton, the predominant morphological determinant of the excitatory synapse. Specifically, we have studied a number of protein-protein interactions mediated by the GluA2 subunit of the AMPARs and shown that they potently regulate cofilin and the Arp2/Arp3 complex, both of which are common targets of the Rho GTPases and key players in actin regulation. It is our hypothesis that these interactions, by virtue of their ability to directly mediate both actin-based remodeling and synaptic transmission, may serve as a mechanism by which the morphological and functional plasticity is coupled during synaptic plasticity. In this chapter, we will discuss the role of GluA2-dependent cofilin in metabotropic glutamate receptor (mGluR)-dependent long-term depression (LTD) in the hippocampus.

12.2 AMPA Glutamate Receptors Mediate Synaptic Plasticity

In the mammalian CNS, glutamate is the major excitatory neurotransmitter. It is released from the presynaptic terminal and triggers transmission by activating various glutamate receptors at the postsynaptic neurons. Two types of glutamate receptors have been identified: ionotropic receptors that include NMDA receptors (NMDARs) and AMPARs and metabotropic glutamate receptors (mGluRs) that are directly coupled to G-protein signaling (Hollmann and Heinemann 1994; Collingridge et al. 2009). Under most physiological conditions, the vast majority of fast excitatory synaptic transmission is mediated by AMPARs. NMDARs are also important for synaptic transmission, but their involvement is subject to regulation by a strong, voltage-dependent extracellular Mg^{2+} blockade. mGluRs may also contribute to synaptic transmission, by either mediating a slow excitatory response or modulating the release of neurotransmitters. However, appropriate patterns of neuronal activities, such as those occurring during learning experience, can readily activate both mGluRs and NMDARs, which can lead to various forms of synaptic plasticity, including long-term potentiation (LTP) and depression (LTD), the most extensively studied forms of long-lasting synaptic plasticity widely regarded as cellular models for learning and memory (Malenka and Bear 2004). The mGluR-LTD, the main plasticity model to be discussed in this chapter, is of particular importance not only because of its critical role in the normal physiological and cognitive processes but also because of its causative association with a number of neurodevelopmental and mental disorders, including intellectual disability, autism, and schizophrenia (Lüscher and Huber 2010; Collingridge et al. 2010). For example, in mouse models of the Fragile X syndrome, mGluR-LTD is selectively altered, and importantly, manipulations of mGluR signaling rescue the cognitive deficits associated with the disorder (Dölen and Bear 2008; Waung and Huber 2009). The mechanisms underlying synaptic plasticity are complex, but it has been accepted that both mGluRs and NMDARs induce plasticity by elevating intracellular Ca^{2+} and/or interacting with other proteins to trigger various signal transduction pathways, including protein phosphorylation and dephosphorylation, which ultimately modify the number and/or biophysical properties of AMPARs, resulting in changes in synaptic strength (i.e., AMPAR-mediated synaptic transmission referred to as functional plasticity here) (Malinow and Malenka 2003; Bredt and Nicoll 2003; Collingridge et al. 2004, 2010; Malenka and Bear 2004; Shepherd and Huganir 2007; Lüscher and Huber 2010). Therefore, it becomes clear that AMPARs are a major locus of modifications underlying functional plasticity.

12.3 Rho Signaling Pathways Mediate Spine Plasticity

On the morphological side, most research has been focused on small neuronal protrusions called dendritic spines, highly specialized postsynaptic structures where most excitatory synapses are formed (Bourne and Harris 2008). Spine properties (e.g., spine size, morphology, and their plastic changes) are closely associated with synaptic properties and cognition (Lamprecht and LeDoux 2004; Carlisle and Kennedy 2005; Alvarez and Sabatini 2007). For instance, spine head size is positively correlated with the abundance of AMPARs, synaptic strength, and learning and memory, and spine deficits are associated with numerous neurological and mental disorders (Engert and Bonhoeffer 1999; Zhou et al. 2004; Lamprecht and LeDoux 2004; Alvarez and Sabatini 2007; Bourne and Harris 2008). Therefore, regulated changes in spine properties such as spine morphology and density are considered to be an integral component of overall synaptic and behavioral plasticity and thought to be molecularly coupled with the functional plasticity in order to achieve persistent and synapse-/spine-specific modifications. The mechanisms underlying this coupling remain largely unexplored. However, it is known that spine properties and their plasticity are mainly controlled by the actin cytoskeleton, the predominant structural component of the spine, and indeed, perturbations of this cytoskeleton profoundly affect not only spine properties (including spine formation, morphogenesis, and plasticity) but also overall synaptic transmission and plasticity (Luo 2002; Cingolani and Goda 2008). Therefore, understanding the molecular processes that regulate spine actin is of critical importance, not only for spine regulation but also for exploring the molecular mechanisms coordinating spine and functional plasticity. Although many molecules (e.g., various actin-binding proteins) are involved in the regulation of spine actin, the Rho family small GTPases are considered to be the master regulators. The Rho GTPases (including RhoA, Rac1, and Cdc42) are small GTP-binding proteins that act as molecular switches that are activated, either directly or indirectly, via GTP exchange factors (GEF) and GTP-activating proteins (GAP), by various synaptic proteins including glutamate receptors, to relay synaptic signals to actin. Indeed, perturbations of these GTPases have been shown to affect both spines and synaptic function (Govek et al. 2005). The importance of Rho proteins in human cognition is clearly demonstrated by genetic studies showing that mutations in many genes of Rho signaling proteins are linked to mental disorders, including intellectual disability and autism (Van. Galen and Ramakers 2005; Gilman et al. 2011). Exactly how the Rho GTPases achieve their spine regulation is not completely clear, but recent studies, including our own, indicate that at least a number of downstream interacting pathways are involved. These include protein kinases such as p21-activated kinase 1 and 3 (PAK1 and 3) and Rho-kinase 2 (ROCK2) that are directly activated by the Rho GTPases (Meng et al. 2005; Zhou et al. 2009; Asrar et al. 2009a; Huang et al. 2011). Here, we will focus our discussion on the role of LIMK1/cofilin, a downstream, converging point of several Rho-dependent signaling processes (Bamburg 1999; Bernard 2007; DesMarais et al. 2005; Jia et al. 2009). Although both cofilin and its upstream regulators are also present in the presynaptic terminal and accumulating evidence supports their role in regulating neurotransmitter release (Bamburg 1999; Meng et al. 2002; Huang et al. 2011; Wolf et al. 2014; Zimmermann et al. 2014; Rust 2015), we will specifically discuss their postsynaptic roles.

12.4 Cofilin Regulates Spine Actin and Basal Spine Morphology

Cofilin is a family of actin-binding proteins whose function is well conserved in various cell types and species (Bamburg 1999). An important aspect of cofilin function is its ability to regulate both actin polymerization and depolymerization; therefore it alone may be sufficient to induce actin remodeling (Bamburg 1999; DesMarais et al. 2005; Andrianantoandro and Pollard 2006; Van Troys et al. 2008). At low concentrations, cofilin binds to and causes disassembly of actin filaments (F-actin) by promoting (1) the dissociation of monomeric actin (G-actin) from the pointed end of the filament and (2) the severing of the filaments. At high concentrations however, cofilin is able to initiate nucleation and promote actin assembly. In addition to being differentially regulated by the protein level, the activity of cofilin is subject to regulation by various post-translational mechanisms, among which protein phosphorylation is the most extensively studied (Bamburg 1999). The key phosphorylation site on cofilin is serine 3, and importantly its phosphorylation renders cofilin completely inactive to bind to, thus unable to regulate actin. Thus, phosphorylation/dephosphorylation at serine 3 is a key determinant and reliable indicator of cofilin activity when the amount of the protein is constant or limiting. This confers cofilin with the ability of being rapidly deactivated and reactivated within specific subcellular compartments (e.g., synapse) without protein synthesis. In vitro studies indicate that the main kinases and phosphatases that directly phosphorylate and dephosphorylate cofilin at serine 3 are, respectively, LIM-domain kinases (LIMK, Yang et al. 1998; Arber et al. 1998) and slingshot (SSH, Niwa et al. 2002), although other kinases (e.g., TEST1/2) and phosphatases (e.g., PP1 and PP2) are also effective (Van Troys et al. 2008).

One of the earliest indications that cofilin is involved in the regulation of spine actin, spine morphology, synaptic function, and behavioral responses came from studies using knockout (KO) mice lacking LIMK1 (Meng et al. 2002, 2004). It was shown that in these mice, the level of phosphorylated forms, but not total protein level of cofilin, is significantly reduced, confirming that LIMK1 is indeed a key player responsible for cofilin phosphorylation in vivo. In accordance with enhanced cofilin activity, the amount of F-actin in the spine head is significantly reduced in LIMK1 KO mice indicating a critical role of cofilin phosphorylation in the assembly and/or stabilization of spine F-actin. Interestingly LIMK1 KO neurons also exhibit abnormal accumulation of both cofilin and F-actin in the dendrites and growth cones, as observed in certain physiological and pathological conditions (Bernstein and Bamburg 2010), suggesting that cofilin may play differential roles in different subcellular compartments within the neurons. Consistent with the changes in F-actin, LIMK1 KO neurons have altered spine morphology with a smaller spine head and thicker neck, characteristics of immature or pathological spines (Bourne and Harris 2008). The size of the postsynaptic density is also reduced in the KO neurons. The spine density however remains unaltered. Taken together, these results suggest that, under basal conditions, regulation of cofilin phosphorylation by LIMK1, although not important for initial spine formation, is critical for the maturation and/maintenance of the spines.

12.5 Cofilin Regulates Both Spine Plasticity and mGluR-LTD

Our early work on LIMK1 KO mice also suggests that cofilin is involved in spine plasticity and this process is regulated by LIMK1-mediated phosphorylation. In wild-type neurons, treatment of hippocampal slices by glutamate or NMDA induces rapid and robust changes (within min) in the level of phosphorylated cofilin, but not total amount of cofilin (Meng et al. 2002). Interestingly, these changes are bidirectional; glutamate induces a decrease, whereas NMDA induces an increase in the amount of phosphorylated cofilin, and importantly both changes are abolished in LIMK1 KO mice (Meng et al. 2002). These results suggest that LIMK-mediated cofilin activation/deactivation is involved in spine plasticity. Subsequent studies by our and other groups demonstrate that cofilin phosphorylation/dephosphorylation is regulated by synaptic activity and is indeed required for spine plasticity (Zhou et al. 2004; Chen et al. 2007; Fedulov et al. 2007; Asrar et al. 2009a; b; Zhou et al. 2009, 2011; Gu et al. 2010; Huang et al. 2011; Bosch et al. 2014).

The direct evidence that supports the role of cofilin in spine plasticity during mGluR-LTD comes from our recent studies on hippocampal slices and cultured neurons (Zhou et al. 2011). In these preparations, treatment by DHPG, a group I mGluR agonist commonly used to induce mGluR-LTD (Palmer et al. 1997), causes a rapid redistribution of cofilin to the dendritic spine that is accompanied by a decrease in phosphorylated cofilin (i.e., activation). Again total cofilin is not altered by this treatment. Parallel to the activation and spine accumulation of cofilin, DHPG treatment induces elongation followed by elimination of the spines as observed in previous studies (Vanderklish and Edelman 2002), and these spine changes require cofilin because a cofilin peptide, called pS3, which prevents cofilin from being dephosphorylated by SSH (Aizawa et al. 2001; Eibert et al. 2004; Huang et al. 2011), abolishes these changes. It is interesting to note that DHPG-induced cofilin activation and redistribution to the spine is transient and returns to the base level within 5–10 min after the treatment (Zhou et al. 2011). Importantly, these dynamic changes of cofilin are also observed during spine enlargement and LTP (Gu et al. 2010; Bosch et al. 2014), suggesting that cofilin may play a general role in spine changes during synaptic remodeling. It is possible that during the initial period (within seconds to minutes) of synaptic remodeling, following both LTP and LTD inductions, increased cofilin activity is necessary to depolymerize and/or severe F-actin to generate a sufficient pool of monomeric actin before any subsequent actin reorganization is possible (given that most actin in the spine is in its filamentous form) (Okamoto et al. 2004; Matsuzaki et al. 2004). It would be interesting to investigate the precise temporal and spatial properties of cofilin activation/deactivation in association with actin remodeling in the spine by using coffin and actin FRET.

Early studies suggest that actin reorganization is involved in mGluR-LTD. Thus, application of F-actin perturbation agents has been shown to block mGluRLTD, but not NMDAR-LTD (Xiao et al. 2001; Zhou et al. 2004; Morishita et al. 2005; Moult et al. 2006). Our recent studies indicate that cofilin-mediated actin reorganization is particularly important for this process (Zhou et al. 2011; Huang et al. 2011). In both hippocampal slices and cultured neurons, while neither actin depolymerization nor polymerization inhibitor (i.e., phalloidin or latrunculin A) affects basal synaptic responses or NMDAR-LTD, they each individually blocks mGluR-LTD, indicating that both actin polymerization and depolymerization are necessary for mGluR-LTD. These results are consistent with those obtained by using two separate cofilin inhibitory peptides that respectively prevent cofilin from binding to F-actin (Q peptide) and block its activation through dephosphorylation (pS3 peptide).

Both peptides block mGluR-LTD without affecting basal responses (Zhou et al. 2011). Given the role of cofilin in both actin polymerization and depolymerization, it is reasonable to conclude that cofilin is actually mediating the actin reorganization required for mGluR-LTD.

In summary, cofilin is required for both mGluR-LTD and associated spine plasticity (Fig. 12.1a, b).

12.6 GluA2 Is Necessary for mGluR-LTD and Associated Spine Plasticity

AMPARs are hetero-tetrameric complexes assembled from GluA1-4 (formerly called GluR1-4) subunits (Hollmann and Heinemann 1994; Collingridge et al. 2009; Lu et al. 2009). The GluA2 subunit is particularly important because it dictates some of the key properties of AMPARs, including Ca^{2+} permeability, channel conductance, and receptor trafficking (Jia et al. 1996; Meng et al. 2003; Isaac et al. 2007; Asrar et al. 2009b). Although GluA2 has been extensively investigated and shown to be critical for NMDAR-LTD (Isaac et al. 2007; Collingridge et al. 2010), its role in mGluR-LTD remains unresolved until recently where we show that it is also important for this form of LTD (Zhou et al. 2011). In these studies we show that in wild-type mice, either bath application of DHPG or paired-pulse low-frequency stimulation (PP-LFS, in the presence of an NMDA receptor antagonist) induces persistent synaptic depression. This form of LTD is blocked by the group I mGluR antagonist MPEP and protein synthesis inhibitors, consistent with previous studies (Palmer et al. 1997; Kemp and Bashir 1999; Huber et al. 2000, 2001; Fitzjohn et al. 2001). In GluA2 KO slices however, this form of plasticity, induced by either DHPG or PP-LFS, is completely absent. The lack of mGluR-LTD in GluA2 KO mice is not likely due to a developmental or chronic effect of the genetic deletion because re-expression of the full-length GluA2, by local viral injections specifically into the adult hippocampus, fully rescues the LTD. In cultured hippocampal neurons, similar results are obtained; DHPG treatment causes a persistent decrease in the frequency, but not amplitude, of spontaneous and miniature excitatory currents (sEPSCs and mEPSCs), but these changes are absent in GluA2 KO-cultured neurons. Importantly, our data show that the DHPG-induced LTD in cultured neurons share the same mechanisms as in hippocampal slices (Kemp and Bashir 1999; Huber et al. 2000, 2001; Gallagher et al. 2004; Mao et al. 2005) being dependent on protein synthesis, ERK signaling, actin dynamics, and cofilin. These results demonstrate that both in vitro and in vivo, GluA2 is indispensable for mGluR-LTD.

Aside from being required for mGluR-LTD, GluA2 is also involved in spine changes associated with this form of plasticity. In GluA2 KO neurons, DHPG-induced spine elongation and elimination that occur in wild-type neurons are not observed. Therefore GluA2 is required for both morphological and functional plasticity induced by activation of mGluRs (Fig. 12.1c, d).

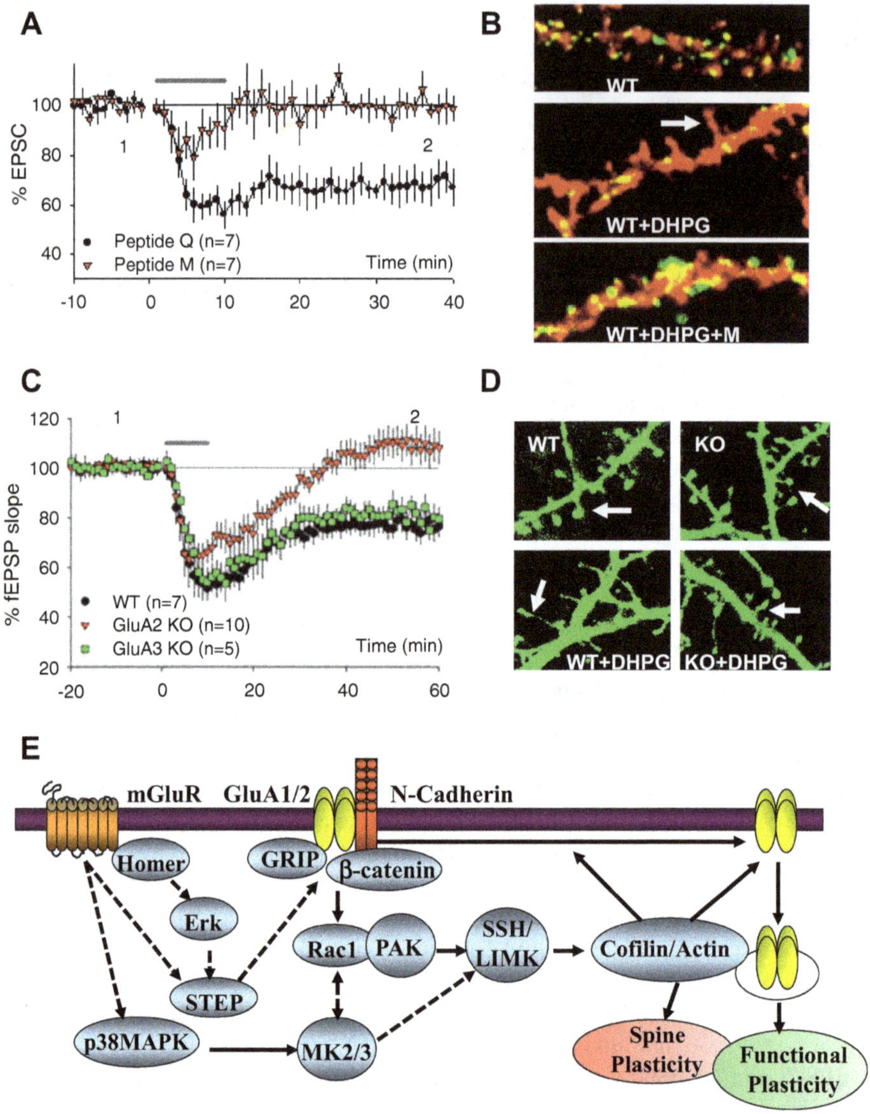

Fig. 12.1 GluA2 regulates mGluR-LTD through cofilin-dependent actin reorganization. (**a**) DHPG-induced LTD is blocked by cofilin inhibitory peptide M, but not by control peptide Q. (**b**) DHPG-induced spine plasticity is blocked by the peptide M. (**c**) DHPG-induced LTD is absent in GluA2 KO mice, but normal in GluA3 KO and wild-type mice. (**d**) DHPG-induced spine plasticity is absent in GluA2 KO mice. (**e**) A hypothetic model. Under basal conditions, GluA2-containing AMPARs form a stable complex with N-cadherin/β-catenin and GRIP/PICK1 linked to less dynamic actin cytoskeleton to maintain spine morphology and synaptic transmission. During mGluR-LTD, stimulation of mGluRs leads to at least two parallel pathways: protein tyrosine phosphatase STEP and p38MAPK. Dephosphorylation of β-catenin and GluA2 by STEP results in disruption of the GluA2/N-cadherin/β-catenin/GRIP/PICK1 complex and release of both AMPARs and β-catenin.

12.7 GluA2 Regulation of mGluR-LTD and Spines Is Mediated by Cofilin

Although above discussion indicates that both GluA2 and cofilin-mediated actin reorganization are critical for mGluR-LTD and associated spine morphological changes, an important question remains as to whether these two processes are related or independent. The following results however suggest that the role of GluA2 in both mGluR-LTD and spine plasticity is mediated by cofilin. First, mGluR-induced activation of cofilin and its upstream regulator Rac1 requires GluA2. Thus, while DHPG treatment induces rapid activation of both cofilin and Rac1 (i.e., decrease in phosphorylated cofilin and increase in GTP-bound Rac1), these changes are not observed in GluA2KO slices or cultured neurons. In fact, DHPG treatment results in significant inhibition of both cofilin and Rac1 (i.e., increase in phosphorylated cofilin and decrease in GTP-bound Rac1) in the KO samples (Zhou et al. 2011). It is important to note that certain other signaling processes activated by mGluR including ERK and RhoA are not affected by GluA2 deletion, suggesting that Rac1 signaling targeting cofilin is specifically regulated by GluA2. Second, mGluR-induced spine accumulation of cofilin and actin requires GluA2. Thus, the increase in the amount of cofilin and decrease in F-actin in the spine induced by DHPG treatment in wild-type cultured neurons are absent in GluA2 KO samples (Zhou et al. 2011). Third, spine plasticity associated with mGluR-LTD is also dependent on the presence of GluA2; as mentioned earlier, DHPG-induced spine elongation and elimination that are found in wild-type neurons are not observed in GluA2 KO neurons (Zhou et al. 2011). Finally, the functional rescue experiments confirm that the requirement of GluA2 in mGluR-LTD is mediated by cofilin (Zhou et al. 2011). This conclusion is based on: (1) inclusion of LIMK antibodies or cofilin peptide S3 in the recording electrode aimed to promote cofilin activation, while having no effect on basal synaptic responses, fully restores DHPG-induced LTD in GluA2 KO mice; (2) double KO mice lacking both LIMK1 and GluA2, which have elevated cofilin activity, show normal DHPG-induced LTD. Taken together, it can be concluded that the role of GluA2 in mGluR-LTD and associated spine plasticity is mediated by cofilin.

Fig. 12.1 (continued) The released AMPARs are free for lateral diffusion and/or internalization, resulting in synaptic depression. The released β-catenin causes activation of Rac1 that ultimately targets LIMK1/cofilin to promote actin reorganization, leading to spine structural changes. In addition to promoting spine structural changes, cofilin-mediated actin can also regulate AMPAR diffusion and internalization. The STEP pathway may be subject to regulation by ERK-dependent protein synthesis. The p38MAKP pathway stimulates MK2/MK3, which in turn activates Rac1 and LIMK1 to target cofilin. Therefore, the GluA2/N-cadherin/β-catenin/GRIP/PICK1 interaction may act as a brake mechanism to keep both spine structure and synaptic physiology stable, and disruption of this interaction would allow both types of plasticity to occur

12.8 GluA2 Regulates Cofilin Through Its N-terminal Interaction with N-cadherin and Subsequent Activation of Rac1

The next question is then how GluA2 regulates cofilin. Although GluA2 can interact with a number of proteins via its C-terminal domain, including PICK1, which is shown to be involved in spine morphology, actin regulation, and NMDAR-LTD (Isaac et al. 2007; Rocca et al. 2008, 2013; Nakamura et al. 2011), our data suggest that GluA2 N-terminal interaction with N-cadherin is particularly important in activating cofilin during mGluR-LTD. First, the N-terminal domain, specifically the binding motif for N-cadherin (Passafaro et al. 2003; Saglietti et al. 2007) of GluA2, is required for mGluR-LTD; thus viral expression of GluA2 lacking this motif fails to rescue the mGluR-LTD deficit in GluA2 KO mice (Zhou et al. 2011). Second, activation of mGluR induces functional changes of N-cadherin, as judged by N-cadherin interaction with its intracellular partner β-catenin and the level of tyrosine phosphorylation of β-catenin, and this process requires GluA2 (Zhou et al. 2011). Third, the disruption of the N-cadherin function by a short peptide (INPISGQ), which is mimetic of the extracellular ectodomain 1 of N-cadherin shown to have a highly specific and potent disruptive effect on N-cadherin function (Williams et al. 2000), impairs mGluR-LTD, and this inhibitory effect of the peptide is rescued by enhancing cofilin activity with S3 peptides or LIMK1 antibodies (Zhou et al. 2011). These results together suggest that GluA2 regulates mGluR-LTD through interacting with the N-cadherin/β-catenin complex. This conclusion is consistent with other studies demonstrating that the N-cadherin/β-catenin complex is a potent regulator of Rac1 signaling, actin reorganization, and synaptic plasticity (Fukata and Kaibuchi 2001; Togashi et al. 2002; Bamji 2005; Elia et al. 2006; Takeichi 2007: Tai et al. 2008).

12.9 Concluding Remarks and Future Studies

In summary, we have summarized evidence to support that GluA2 regulates mGluR-LTD and associated spine plasticity via N-cadherin-dependent, cofilin-mediated actin reorganization. A hypothetical model is shown in Fig. 12.1e. Given the trans-synaptic nature of N-cadherin molecules, the GluA2/N-cadherin interaction may also have important implications in coordinating pre- and postsynaptic communications and remodeling. By placing N-cadherin and subsequent cofilin-mediated actin reorganization under the direct control of GluA2, postsynaptic AMPARs are coupled not only to spine morphological changes but also to the presynaptic element, which ensure coordinated structural and functional changes during long-lasting synaptic plasticity. Perhaps, relevant to this is the evidence that mGluR-LTD is also expressed in part via presynaptic mechanisms (Fitzjohn et al. 2001). Some of the future investigations would include the following key questions. (1) How is the

N-cadherin/β-catenin complex regulated by mGluR activation? In this respect, activation of ERK and p38 MAPK signaling pathways as well as tyrosine phosphorylation/dephosphorylation may play an important role as both mGluR-LTD and the N-cadherin/β-catenin complex are regulated by these processes (Pulido et al. 1998; Mao et al. 2005; Gallagher et al. 2004; Lilien and Balsamo 2005; Moult et al. 2006; Huang and Hsu 2006; Zhang et al. 2008; Gladding et al. 2009; Collingridge et al. 2010; Corre^a and Eales 2012). Consistent with this possibility, a recent study demonstrates that p38 MAPK-activated kinases 2 and 3 are required for DHPG-induced cofilin dephosphorylation, spine plasticity, and LTD (Eales et al. 2014). (2) How is cofilin activated and accumulated in the spine? Although LIMK1 is important, other kinases and phosphatases including SSH may also be involved not only in activation/deactivation but also trafficking and stabilization of cofilin in the spine. (3) How does cofilin regulate spine actin and AMPAR trafficking? In this respect, it is important to resolve the precise temporal and spatial properties of cofilin in relation to actin rearrangement and AMPAR trafficking (Gu et al. 2010; Rust et al. 2010). These analyses may provide a valuable venue for understanding molecular mechanisms underlying long-lasting synaptic plasticity at the central synapse. Because cofilin is involved in many physiological and pathological processes, including neuronal development and brain diseases (Gurniak et al. 2005; Bellenchi et al. 2007; Bernstein and Bamburg 2010; Flynn et al. 2012; Rust 2015), investigations of cofilin regulation during synaptic plasticity may also provide insight into the understanding and treatment of these processes.

Acknowledgments This work was supported by grants from the Canadian Institutes of Health Research (CIHR, MOP119421, ZPJ), Canadian Natural Science and Engineering Research Council (NSERC, RGPIN341498, ZPJ), CIHR, and China Joint Health Research Initiative Program (81161120543, ZPJ) and the MRC (GLC).

References

Aizawa H, Aizawa H, Wakatsuki S, Ishii A, Moriyama K, Sasaki Y, Ohashi K, Sekine-Aizawa Y, Sehara-Fujisawa A, Mizuno K, Goshima Y, Yahara I (2001) Phosphorylation of cofilin by LIM-kinase is necessary for semaphorin 3A-induced growth cone collapse. Nat Neurosci 4:367–373

Alvarez VA, Sabatini BL (2007) Anatomical and physiological plasticity of dendritic spines. Annu Rev Neurosci 30:79–97

Andrianantoandro E, Pollard TD (2006) Mechanism of actin filament turnover by severing and nucleation at different concentrations of ADF/cofilin. Mol Cell 24:187–195

Arber S, Barbayannis FA, Hanser H, Schneider C, Stanyon CA, Bernard O, Caroni P (1998) Regulation of actin dynamics through phosphorylation of cofilin by LIM-kinase. Nature 393:805–809

Asrar S, Meng Y, Zhou Z, Todorovski Z, Huang WW, Jia Z (2009a) Regulation of hippocampal long-term potentiation by p21-activated protein kinase 1 (PAK1). Neuropharmacology 56:73–80

Asrar S, Zhou Z, Ren W, Jia Z (2009b) Ca-permeable AMPA receptor induced long-term potentiation requires PI3/MAP kinases but not Ca/CaM-dependent kinase II. PLoS One 4:e4339

Bamburg JR (1999) Proteins of the ADF/cofilin family: essential regulators of actin dynamics. Annu Rev Cell Dev Biol 15:185–230

Bamji SX (2005) Cadherins: actin with the cytoskeleton to form synapses. Neuron 47:175–178
Bellenchi GC, Gurniak CB, Perlas E, Middei S, Ammassari-Teule M, Witke W (2007) N-cofilin is associated with neuronal migration disorders and cell cycle control in the cerebral cortex. Genes Dev 21:2347–2357
Bernard O (2007) Lim kinases, regulators of actin dynamics. Int J Biochem Cell Biol 39:1071–1076
Bernstein BW, Bamburg JR (2010) ADF/cofilin: a functional node in cell biology. Trends Cell Biol 20:187–195
Bosch M, Castro J, Saneyoshi T, Matsuno H, Sur M, Hayashi Y (2014) Structural and molecular remodeling of dendritic spine substructures during long-term potentiation. Neuron 82:444–459
Bourne JN, Harris KM (2008) Balancing structure and function at hippocampal dendritic spines. Annu Rev Neurosci 31:47–67
Bredt DS, Nicoll RA (2003) AMPA receptor trafficking at excitatory synapses. Neuron 40:361–379
Carlisle HJ, Kennedy MB (2005) Spine architecture and synaptic plasticity. Trends Neurosci 28:182–187
Chen LY, Rex CS, Casale MS, Gall CM, Lynch G (2007) Changes in synaptic morphology accompany actin signaling during LTP. J Neurosci 27:5363–5372
Cingolani LA, Goda Y (2008) Actin in action: the interplay between the actin cytoskeleton and synaptic efficacy. Nat Rev Neurosci 9:344–356
Collingridge GL, Isaac JT, Wang YT (2004) Receptor trafficking and synaptic plasticity. Nat Rev Neurosci 5:952–962
Collingridge GL, Olsen RW, Peters J, Spedding M (2009) A nomenclature for ligand-gated ion channels. Neuropharmacology 56:2–5
Collingridge GL, Peineau S, Howland JG, Wang YT (2010) Long-term depression in the CNS. Nat Rev Neurosci 11:459–473
Corre^a SA, Eales KL (2012) The role of p38 MAPK and its substrates in neuronal plasticity and neurodegenerative disease. J Signal Transduct 2012(6):649079
DesMarais V, Ghosh M, Eddy R, Condeelis J (2005) Cofilin takes the lead. J Cell Sci 118:19–26
Dölen G, Bear MF (2008) Role for metabotropic glutamate receptor 5 (mGluR5) in the pathogenesis of fragile X syndrome. J Physiol 586:1503–1508
Eales KL, Palygin O, O'Loughlin T, Rasooli-Nejad S, Gaestel M, Müller J, Collins DR, Pankratov Y, Corrêa SA (2014) The MK2/3 cascade regulates AMPAR trafficking and cognitive flexibility. Nat Commun 19(5):4701
Eibert SM, Lee KH, Pipkorn R, Sester U, Wabnitz GH, Giese T, Meuer SC, Samstag Y (2004) Cofilin peptide homologs interfere with immunological synapse formation and T cell activation. Proc Natl Acad Sci U S A 101:1957–1962
Elia LP, Yamamoto M, Zang K, Reichardt LF (2006) p120 catenin regulates dendritic spine and synapse development through Rho-family GTPases and cadherins. Neuron 51:43–56
Engert F, Bonhoeffer T (1999) Dendritic spine changes associated with hippocampal long-term synaptic plasticity. Nature 399:66–70
Fedulov V, Rex CS, Simmons DA, Palmer L, Gall CM, Lynch G (2007) Evidence that long-term potentiation occurs within individual hippocampal synapses during learning. J Neurosci 27:8031–8039
Fitzjohn SM, Palmer MJ, May JE, Neeson A, Morris SA, Collingridge GL (2001) A characterisation of long-term depression induced by metabotropic glutamate receptor activation in the rat hippocampus in vitro. J Physiol 537(Pt 2):421–430
Flynn KC, Hellal F, Neukirchen D, Jacob S, Tahirovic S, Dupraz S, Stern S, Garvalov BK, Gurniak C, Shaw AE, Meyn L, Wedlich-Söldner R, Bamburg JR, Small JV, Witke W, Bradke F (2012) ADF/cofilin-mediated actin retrograde flow directs neurite formation in the developing brain. Neuron 76:1091–1107
Fukata M, Kaibuchi K (2001) Rho-family GTPases in cadherin-mediated cell-cell adhesion. Nat Rev Mol Cell Biol 2:887–897
Gallagher SM, Daly CA, Bear MF, Huber KM (2004) Extracellular signal-regulated protein kinase activation is required for metabotropic glutamate receptor-dependent long-term depression in hippocampal area CA1. J Neurosci 24:4859–4864

Gilman SR, Iossifov I, Levy D, Ronemus M, Wigler M, Vitkup D (2011) Rare de novo variants associated with autism implicate a large functional network of genes involved in formation and function of synapses. Neuron 70:898–907

Gladding CM, Collett VJ, Jia Z, Bashir ZI, Collingridge GL, Molnár E (2009) Tyrosine dephosphorylation regulates AMPAR internalisation in mGluR-LTD. Mol Cell Neurosci 40:267–279

Govek EE, Newey SE, Van Aelst L (2005) The role of the Rho GTPases in neuronal development. Genes Dev 19(1):1–49

Gu J, Lee CW, Fan Y, Komlos D, Tang X, Sun C, Yu K, Hartzell HC, Chen G, Bamburg JR, Zheng JQ (2010) ADF/cofilin-mediated actin dynamics regulate AMPA receptor trafficking during synaptic plasticity. Nat Neurosci 13:1208–1215

Gurniak CB, Perlas E, Witke W (2005) The actin depolymerizing factor n-cofilin is essential for neural tube morphogenesis and neural crest cell migration. Dev Biol 278:231–241

Hollmann M, Heinemann S (1994) Cloned glutamate receptors. Annu Rev Neurosci 17:31–108

Huang CC, Hsu KS (2006) Sustained activation of metabotropic glutamate receptor 5 and protein tyrosine phosphatases mediate the expression of (S)-3,5-dihydroxyphenylglycine-induced long-term depression in the hippocampal CA1 region. J Neurochem 96:179–194

Huang W, Zhou Z, Asrar S, Henkelman M, Xie W, Jia Z (2011) p21-activated kinases 1 and 3 control brain size through coordinating neuronal complexity and synaptic Properties. Mol Cell Biol 31:388–403

Huber KM, Kayser MS, Bear MF (2000) Role for rapid dendritic protein synthesis in hippocampal mGluR-dependent long-term depression. Science 288:1254–1257

Huber KM, Roder JC, Bear MF (2001) Chemical induction of mGluR5- and protein synthesis--dependent long-term depression in hippocampal area CA1. J Neurophysiol 86:321–325

Isaac JTR, Ashby MC, McBain CJ (2007) The role of the GluR2 subunit in AMPA receptor function and synaptic plasticity. Neuron 54:859–871

Jia ZP, Agopyan N, Miu P, Xiong Z, Henderson J, Gerlai R, Taverna F, Velumian A, MacDonald J, Carlen P, Abramow-Newerly W, Roder J (1996) Enhanced LTP in mice deficient in the AMPA receptor GluR2. Neuron 17:945–956

Jia ZP, Todorovski Z, Meng YH, Asrar S, Wang LY (2009) LIMK-1 and actin regulation of spine and synaptic function. In: Squire LR (ed) New encyclopedia of neuroscience, vol 5. Academic Press, Oxford, pp 467–472

Kemp N, Bashir ZI (1999) Induction of LTD in the adult hippocampus by the synaptic activation of AMPA/kainate and metabotropic glutamate receptors. Neuropharmacology 38:495–504

Lamprecht R, LeDoux J (2004) Structural plasticity and memory. Nat Rev Neurosci 5:45–54

Lilien J, Balsamo J (2005) The regulation of cadherin-mediated adhesion by tyrosine phosphorylation/dephosphorylation of beta-catenin. Curr Opin Cell Biol 17:459–465

Lu W, Shi Y, Jackson AC, Bjorgan K, During MJ, Sprengel R, Seeburg PH, Nicoll RA (2009) Subunit composition of synaptic AMPA receptors revealed by a single-cell genetic approach. Neuron 62:254–268

Luo L (2002) Actin cytoskeleton regulation in neuronal morphogenesis and structural plasticity. Annu Rev Cell Dev Biol 18:601–635

Lüscher C, Huber KM (2010) Group 1 mGluR-dependent synaptic long-term depression: mechanisms and implications for circuitry and disease. Neuron 65:445–459

Malenka RC, Bear MF (2004) LTP and LTD: an embarrassment of riches. Neuron 44:5–21

Malinow R, Malenka RC (2003) AMPA receptor trafficking and synaptic plasticity. Annu Rev Neurosci 25:103–126

Mao L, Yang L, Tang Q, Samdani S, Zhang G, Wang JQ (2005) The scaffold protein Homer1b/c links metabotropic glutamate receptor 5 to extracellular signal-regulated protein kinase cascades in neurons. J Neurosci 25:2741–2752

Matsuzaki M, Honkura N, Ellis-Davies GC, Kasai H (2004) Structural basis of long-term potentiation in single dendritic spines. Nature 429:761–766

Meng Y, Zhang Y, Tregoubov V, Janus C, Cruz L, Jackson M, Lu WY, MacDonald JF, Wang JY, Falls DL, Jia Z (2002) Abnormal spine morphology and enhanced LTP in LIMK-1 knockout mice. Neuron 35:121–133

Meng Y, Zhang Y, Jia Z (2003) Synaptic transmission and plasticity in the absence of AMPA glutamate receptor GluR2 and GluR3. Neuron 39:163–176

Meng Y, Takahashi H, Meng J, Zhang Y, Lu G, Asrar S, Nakamura T, Jia Z (2004) Regulation of ADF/cofilin phosphorylation and synaptic function by LIM-kinase. Neuropharmacology 47:746–754

Meng J, Meng Y, Hanna A, Janus C, Jia Z (2005) Abnormal long-lasting synaptic plasticity and cognition in mice lacking the mental retardation gene Pak3. J Neurosci 25:6641–6650

Morishita W, Marie H, Malenka RC (2005) Distinct triggering and expression mechanisms underlie LTD of AMPA and NMDA synaptic responses. Nat Neurosci 8:1043–1050

Moult PR, Gladding CM, Sanderson TM, Fitzjohn SM, Bashir ZI, Molnar E, Collingridge GL (2006) Tyrosine phosphatases regulate AMPA receptor trafficking during metabotropic glutamate receptor-mediated long-term depression. J Neurosci 26:2544–2554

Nakamura Y, Wood CL, Patton AP, Jaafari N, Henley JM, Mellor JR, Hanley JG (2011) PICK1 inhibition of the Arp2/3 complex controls dendritic spine size and synaptic plasticity. EMBO J 30:719–730

Niwa R, Nagata-Ohashi K, Takeichi M, Mizuno K, Uemura T (2002) Control of actin reorganization by Slingshot, a family of phosphatases that dephosphorylate ADF/cofilin. Cell 108:233–246

Okamoto K, Nagai T, Miyawaki A, Hayashi Y (2004) Rapid and persistent modulation of actin dynamics regulates postsynaptic reorganization underlying bidirectional plasticity. Nat Neurosci 7:1104–1112

Palmer MJ, Irving AJ, Seabrook GR, Jane DE, Collingridge GL (1997) The group I mGlu receptor agonist DHPG induces a novel form of LTD in the CA1 region of the hippocampus. Neuropharmacology 36:1517–1532

Passafaro M, Nakagawa T, Sala C, Sheng M (2003) Induction of dendritic spines by an extracellular domain of AMPA receptor subunit GluR2. Nature 424:677–681

Pulido R, Zúñiga A, Ullrich A (1998) PTP-SL and STEP protein tyrosine phosphatases regulate the activation of the extracellular signal-regulated kinases ERK1 and ERK2 by association through a kinase interaction motif. EMBO J 17:7337–7350

Rocca DL, Martin S, Jenkins EL, Hanley JG (2008) Inhibition of Arp2/3-mediated actin polymerization by PICK1 regulates neuronal morphology and AMPA receptor endocytosis. Nat Cell Biol 10:259–271

Rocca DL, Amici M, Antoniou A, Suarez EB, Halemani N, Murk K, McGarvey J, Jaafari N, Mellor JR, Collingridge GL, Hanley JG (2013) The small GTPase Arf1 modulates Arp2/3-mediated actin polymerization via PICK1 to regulate synaptic plasticity. Neuron 79:293–307

Rust MB (2015) ADF/cofilin: a crucial regulator of synapse physiology and behavior. Cell Mol Life Sci 72(18):3521–3529

Rust MB, Gurniak CB, Renner M, Vara H, Morando L, Görlich A, Sassoè-Pognetto M, Banchaabouchi MA, Giustetto M, Triller A, Choquet D, Witke W (2010) Learning, AMPA receptor mobility and synaptic plasticity depend on n-cofilin-mediated actin dynamics. EMBO J 29:1889–1902

Saglietti L, Dequidt C, Kamieniarz K, Rousset MC, Valnegri P, Thoumine O, Beretta F, Fagni L, Choquet D, Sala C, Sheng M, Passafaro M (2007) Extracellular interactions between GluR2 and N-cadherin in spine regulation. Neuron 54:461–477

Shepherd JD, Huganir RL (2007) The cell biology of synaptic plasticity: AMPA receptor trafficking. Annu Rev Cell Dev Biol 23:613–643

Tai CY, Kim SA, Schuman EM (2008) Cadherins and synaptic plasticity. Curr Opin Cell Biol 20:567–575

Takeichi M (2007) The cadherin superfamily in neuronal connections and interactions. Nat Rev Neurosci 8:11–20

Togashi H, Abe K, Mizoguchi A, Takaoka K, Chisaka O, Takeichi M (2002) Cadherin regulates dendritic spine morphogenesis. Neuron 35:77–89

Vanderklish PW, Edelman GM (2002) Dendritic spines elongate after stimulation of group 1 metabotropic glutamate receptors in cultured hippocampal neurons. Proc Natl Acad Sci U S A 99:1639–1644

Van Galen EJ, Ramakers GJ (2005) Rho proteins, mental retardation and the neurobiological basis of intelligence. Prog Brain Res 147:295–317

Van Troys M, Huyck L, Leyman S, Dhaese S, Vandekerkhove J, Ampe C (2008) Ins and outs of ADF/cofilin activity and Regulation. Eur J Cell Biol 87:649–667

Waung MW, Huber KM (2009) Protein translation in synaptic plasticity: mGluR-LTD, Fragile X. Curr Opin Neurobiol 19:319–326

Williams EJ, Williams G, Gour B, Blaschuk O, Doherty P (2000) INP, a novel N-cadherin antagonist targeted to the amino acids that flank the HAV motif. Mol Cell Neurosci 15:456–464

Wolf M, Zimmermann AM, Görlich A, Gurniak CB, Sassoè-Pognetto M, Friauf E, Witke W, Rust MB (2014) ADF/Cofilin controls synaptic actin dynamics and regulates synaptic vesicle mobilization and exocytosis. Cereb Cortex 25(9):2863–2875. [Epub ahead of print]

Xiao MY, Zhou Q, Nicoll RA (2001) Metabotropic glutamate receptor activation causes a rapid redistribution of AMPA receptors. Neuropharmacology 41:664–671

Yang N, Higuchi O, Ohashi K, Nagata K, Wada A, Kangawa K, Nishida E, Mizuno K (1998) Cofilin phosphorylation by LIM-kinase 1 and its role in Rac-mediated actin reorganization. Nature 393:809–812

Zhang Y, Venkitaramani DV, Gladding CM, Kurup P, Molnar E, Collingridge GL, Lombroso PJ (2008) The tyrosine phosphatase STEP mediates AMPA receptor endocytosis after metabotropic glutamate receptor stimulation. J Neurosci 28:10561–10566

Zhou Q, Homma KJ, Poo MM (2004) Shrinkage of dendritic spines associated with long-term depression of hippocampal synapses. Neuron 44:749–757

Zhou Z, Meng Y, Asrar S, Todorovski Z, Jia Z (2009) A critical role of Rho-kinase ROCK2 in the regulation of spine and synaptic function. Neuropharmacology 56:81–89

Zhou Z, Hu J, Passafaro M, Xie W, Jia Z (2011) GluA2 (GluR2) regulates metabotropic glutamate receptor-dependent long-term depression through N-cadherin-dependent and cofilin-mediated actin reorganization. J Neurosci 31:819–833

Zimmermann AM, Jene T, Wolf M, Görlich A, Gurniak CB, Sassoè-Pognetto M, Witke W, Friauf E, Rust MB (2014) Attention-deficit/hyperactivity disorder-like phenotype in a mouse model with impaired actin dynamics. Biol Psychiatry 78:95–106

Chapter 13
Metabotropic Glutamate Receptors in Amygdala Functions

Francesco Ferraguti

Abstract Exactly three decades ago, metabotropic glutamate receptors (mGluRs) were discovered as glutamate receptors coupled to heterotrimeric G-proteins and able to induce polyphosphoinositide hydrolysis. Since then, eight different mGluRs have been identified, which activate a variety of second messenger systems and trigger complex physiological responses in nearly all brain areas. A participation of mGluRs in amygdala networks was shown to regulate emotional-affective behaviours. In this chapter, I will examine the localization of mGluRs in the amygdaloid complex and how genetic and/or pharmacological manipulation of these receptors affects different amygdala-dependent emotional behaviours including fear learning and extinction. Furthermore, I will discuss the growing literature addressing the influence of mGluRs on affective aspects of pain processing within the nociceptive amygdala. Finally, the potential of drugs acting at mGluRs as novel targets for neuropsychiatric disorders characterized by impaired amygdala functions will be examined.

Keywords Amygdala • Fear • Extinction • Learning • Anxiety

Abbreviations

AAA	Anterior amygdala area
ACo	Anterior cortical nuclei
AHA	Amygdalohippocampal area
ASDs	Autism spectrum disorders
BA	Basal nucleus
BLA	Basolateral complex
BM	Basomedial nucleus
BNST	Bed nucleus of the stria terminalis
CeA	Central nucleus

F. Ferraguti, M.D. (✉)
Department of Pharmacology, Medical University of Innsbruck,
Peter Mayr Strasse 1A, Innsbruck 6020, Austria
e-mail: francesco.ferraguti@i-med.ac.at

R.T. Ngomba et al. (eds.), *mGLU Receptors*, The Receptors 31,
DOI 10.1007/978-3-319-56170-7_13

CeC	Capsular subdivision of the central nucleus
CeL	Lateral subdivision of the central nucleus
CeM	Medial subdivision of the central nucleus
CS	Conditioned stimulus
CxA	Periamygdaloid cortex
DCPG	(S)-3,4-Dicarboxyphenylglycine
DHPG	3,5-Dihydroxy-phenylglycine
FMRP	Fragile X mental retardation protein
FXS	Fragile X syndrome
GAD	Generalized anxiety disorders
ITC	Intercalated cell masses
LA	Lateral nucleus
L-AP4	L-2-Amino-4-phosphonobutyric acid
LTD	Long-term depression
LTP	Long-term potentiation
MeA	Medial nucleus
Met-1	N-Benzhydrylethane-1,2-diamine
MGluRs	Metabotropic glutamate receptors
MPEP	2-Methyl-6-(phenylethynyl)-pyridine
NLG3	Neuroligin 3
PCo	Posterior cortical nuclei
PSD-95	Postsynaptic density-95
PTSD	Post-traumatic stress disorder
t-ACPD	trans-1-Aminocyclopentane-1,3-dicarboxylic acid
US	Unconditioned stimulus

13.1 Introduction

The amygdaloid complex, located in the medial temporal lobe, encompasses a heterogeneous group of distinct nuclear and cortical structures, which are involved in a bewildering variety of behavioural functions. These include regulation of the autonomic and neuroendocrine systems, reproduction and aggression, social behaviours, decision-making, attention and, perhaps most important, the formation and storage of memories associated with emotional events (Phelps and LeDoux 2005). Fear learning has been the function most associated with the amygdala. Malfunctioning of the amygdala can bring to a variety of neuropsychiatric disorders such as anxiety disorders, post-traumatic stress disorder (PTSD), autism and depression (Anand and Shekhar 2003; Mahan and Ressler 2012; Schumann et al. 2011).

Metabotropic glutamate receptors (mGluRs) are widely expressed in the amygdala and have been reported to affect many of the physiological functions of this area. In this chapter, the current knowledge on how mGluRs participate in amygdala networks to regulate emotional-affective behaviours will be reviewed. In particular, I will examine mGluRs in fear pathways and their possible roles in fear learning and extinction. Furthermore, I will discuss the growing literature addressing the influence of mGluRs on affective aspects of pain processing within the nociceptive amygdala and addiction. At first, however, I will provide a brief overview on the basic aspects of the anatomical organization of the amygdala. Particular emphasis will be given to the specialized synaptic localization of different mGluR subtypes in distinct amygdala circuits and their contribution to neural plasticity. Finally, I will discuss the potential of drugs acting at mGluRs as novel targets for neuropsychiatric disorders characterized by impaired amygdala functions, including anxiety, sleep disorders and autism.

13.2 Basic Anatomical Organization of the Amygdala

The distinct nuclei and cortical areas of which the amygdala is constituted have been grouped based on ontogenetic, cytoarchitectonic and connectivity criteria into three main divisions: (1) the basolateral complex (BLA), evolutionarily related to the neocortex; (2) the centromedial nuclear group, which shares many features with the basal ganglia; and (3) the superficial corticomedial areas, which have many cortical characteristics, such as a layered organization. Several structures, however, remain somehow refractory to be grouped with any of these main divisions. The major nuclei and divisions of the amygdala are shown in schematic sections from the mouse brain in Fig. 13.1. It should be kept in mind, however, that in recent years several other alternative organizational schemes of the amygdala have been proposed (Heimer 2003; Swanson and Petrovich 1998).

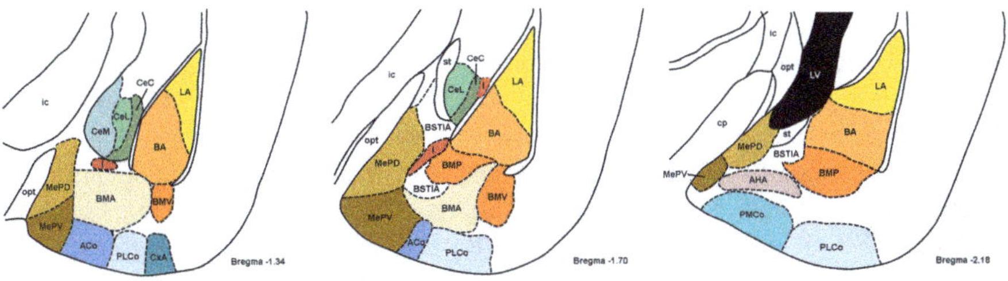

Fig. 13.1 Nuclear divisions and subdivisions of the mouse amygdaloid complex. Most of the different amygdaloid nuclei are classified into three groups as described in the text. The basolateral complex is shown in different shades of *orange*; the centromedial nuclear group is shown in different shades of *green* or *brown*; the superficial corticomedial areas are shown in different shades of *blue*. Abbreviations: *ACo*, anterior cortical amygdaloid area, *AHA* amygdalohippocampal area, *BA* basal nucleus, *BMA* basomedial nucleus anterior part, BMP basomedial nucleus posterior part, *BMV* basomedial nucleus ventral part, *BSTIA* intra-amygdala part of the bed nucleus of the stria terminalis, *CeC* capsular subdivision of the central nucleus, *CeL* lateral subdivision of the central nucleus, *CeM* medial subdivision of the central nucleus, *cp* cerebral peduncle, *CxA* periamygdaloid cortex, *I* intercalated cell clusters, *ic* internal capsule, *LA* lateral nucleus, *LV* lateral ventricle, *MePD* medial nucleus posterodorsal subdivision, *MePV* medial amygdala posteroventral subdivision, *opt* optic tract, *PLCo* posterolateral cortical area, *PMCo* posteromedial cortical area, *st* stria terminalis

Most of the distinct amygdaloid nuclei are further divided into subdivisions that have extensive inter- and intranuclear connections (Pitkänen et al. 1997). An extensive discussion of the structure and connectivity of amygdala nuclei is beyond the scope of this chapter. The reader, therefore, is referred to other specialized reviews (Pitkänen 2000; Sah et al. 2003). Briefly, the basolateral complex, which includes the lateral (LA), the basal (BA) and the basomedial or accessory basal (BM) nuclei, receives a large part of the incoming inputs from auditory, visual and somatosensory (including pain) systems. The centromedial nuclear group is located in the dorsomedial portion of the amygdaloid complex and is the main output structure of the amygdala. It comprises the central (CeA) and medial (MeA) nuclei as well as the amygdaloid part of the bed nucleus of the stria terminalis (BNST). The CeA can be further divided into three subdivisions: capsular (CeC), lateral (CeL) and medial (CeM). The CeM, in particular, connects with several hypothalamic and brainstem nuclei to control the expression of emotional responses. The MeA is found medially to the optic tract and is strongly connected with the olfactory system. The superficial corticomedial areas comprise the anterior and posterior cortical areas (ACo and PCo, respectively) and the periamygdaloid cortex (CxA). Finally, the set of nuclei that do not easily fall into any of these three divisions include the anterior amygdala area (AAA), the amygdalohippocampal area (AHA) and the intercalated cell masses (ITC). The ITC are formed by different clusters of GABAergic medium spiny neurons surrounding the basolateral complex (Busti et al. 2011).

13.3 Historical Overview

Despite the first report about a metabotropic effect of glutamate analogues in the amygdala dates back to nearly 30 years ago, when ibotenic acid was shown to increment inositol phospholipid hydrolysis in slices of the rat amygdala (Akiyama et al. 1987), for years the presence and effects of mGluRs in this brain area have been largely neglected. In 1992, Shigemoto and coworkers reported on the mRNA expression of the first cloned mGluR in various nuclei of the amygdaloid complex (Shigemoto et al. 1992). In the same year, Rainnie and Shinnick-Gallagher (1992) also provided the first physiological evidence of the effects of mGluR activation in the BLA. By applying the conformationally restricted analogue of glutamate trans-1-aminocyclopentane-1,3-dicarboxylic acid (t-ACPD) to acute slices, they observed in BLA neurons a dose-dependent reduction in the EPSP amplitude evoked by stimulation of the stria terminalis (Rainnie and Shinnick-Gallagher 1992). They also reported a similar effect with L-2-amino-4-phosphonobutyric acid (L-AP4), an effect that was additive to that of t-ACPD, hence suggesting the presence of distinct mGluRs in presynaptic terminals. Later, these authors showed, using the same preparation, that t-ACPD also activates postsynaptic mGluRs resulting in a unique hyperpolarizing response mediated through calcium-dependent large conductance potassium channels (Rainnie et al. 1994).

The last decade of the twentieth century saw the molecular cloning of eight distinct mGluRs and of several alternatively spliced variants of these receptors in several species, including humans (Nicoletti et al. 2011). Transcripts and proteins for all mGluRs, with the exception of mGluR6, were detected in the amygdala. However, a systematic analysis of their distribution, particularly at the cellular level, is still missing. Only recently, we have reported the identification and detailed characterization of GABAergic projection neurons in the rodent amygdala characterized by the expression of mGluR1α (Bienvenu et al. 2015).

Functionally, the first observation that mGluRs are implicated in amygdala-dependent emotional behaviours was made by Nakanishi, van der Putten and colleagues, who showed that mice with gene-targeted deletion of mGluR7 have an impaired fear response (Masugi et al. 1999). Later, also group I mGluRs were shown to be required for fear memory formation and long-term potentiation (LTP) in the LA (Rodrigues and LeDoux 2002). Besides the participation of mGluRs in fear learning, accumulating evidence indicates that mGluRs in amygdala microcircuits are significantly involved in anxiety (Palucha and Pilc 2007). In recent years, amygdala mGluRs were also described as having important modulatory roles on nociception and pain behaviour (Palazzo et al. 2014; Ren et al. 2011) as well as on addiction, including alcohol reinforcement and relapse-like behaviour (Bäckström and Hyytiä 2005; Cannady et al. 2011).

13.4 Distribution of mGluRs in the Amygdala

Despite many studies have addressed the cellular and subcellular localization of mGluRs in a large number of areas in the central nervous system, only very few have focused on the amygdaloid complex. Notwithstanding, each cloned mGluR, with the only exception of mGluR6 that is restricted to the retina, was found expressed at some degrees in the amygdala. Tables 13.1, 13.2 and 13.3 summarize the distribution patterns of mRNA and immunoreactivity for the distinct mGluRs in the different nuclei and cortical areas of the adult rodent (mouse or rat) amygdala. Only a few studies explored the expression of mGluRs in the human amygdala (Emile et al. 1996; Makoff et al. 1996; Malherbe et al. 1999), providing only a scant description of their distribution. Therefore, these data will not be reviewed.

The subcellular localization of mGluRs in the amygdala appears to follow the general pattern observed elsewhere. Group I mGluRs are located postsynaptically and are enriched in the perisynaptic area of type I asymmetric synapses (Baude et al. 1993), whereas they are probably intrasynaptic at type II symmetric synapses as observed in the cerebellar cortex (Mansouri et al. 2015) and basal ganglia (Hanson and Smith 1999). Group II mGluRs are present both pre- and postsynaptically (Petralia et al. 1996), and group III mGluRs are exclusively placed at the active zone of axon terminals of both asymmetric and symmetric synapses (Ferraguti and Shigemoto 2006). In the case of mGluR3, clear evidence has been provided for its expression on the plasma membrane of astrocytes in most amygdala nuclei (Fig. 13.2) (Hetzenauer et al. 2008; Muly et al. 2007).

Table 13.1 Distribution of group I mGluR mRNA and immunoreactivity in the adult rodent brain

	mGluR1		mGluR5	
	mRNA	IHC	mRNA	IHC
Basolateral complex				
Lateral nucleus (LA)	+	+	++	++
Basal nucleus (BA)	+	+/++	++	++
Basomedial nucleus (BM)	−/+	−/+	+	++
Centromedial nuclear group				
Central nucleus (CeA)	+	+	+/++	++
Capsular division (Cec)	−/+	−/+		++
Lateral division (CeL)	−/+	−/+		++
Medial division (CeM)	+	+		+
Bed nucleus of the stria terminalis intra-amygdaloid division (BSTIA)	++	+++		+
Medial nucleus (MeA)	+	−/+	+	−/+
Superficial corticomedial areas				
Anterior cortical nucleus (ACo)	+	−/+	++	+
Posterior cortical nucleus (PCo)	+	−/+		+
Periamygdaloid cortex (CxA)	+	+		++
Other areas				
Anterior amygdaloid area (AAA)	+	++		+
Amygdalohippocampal area (AHA)	++	+		+
Intercalated clusters (ITC)	++	+++		+++
Amygdalo-striatal transition area (AStr)	−	−		+++

Intensities: +++ intense, ++ moderate, + weak, −/+ very weak or contains only scattered labelled cells, − absent

Data were taken from the following references and from personal unpublished observations: Baude et al. (1993), Dobi et al. (2013), Romano et al. (1995) and Shigemoto et al. (1992)

13.5 Expression of Group I mGluRs in Amygdala

High expression of the main splice variant mGluR1α (Ferraguti et al. 2008) was recently shown to characterize a class of large neurons in the amygdala surrounding ITC clusters (Bienvenu et al. 2015; Busti et al. 2011; Dobi et al. 2013). Moreover, an unknown population of BLA GABAergic interneurons, as well as a few neurons within the CeA, also display moderate to weak immunoreactivity for this mGluR (personal communication). Neurons in the BNST possess considerable levels of mGluR1α as well as of the shorter variant mGluR1β (Ferraguti et al. 1998).

BLA pyramidal neurons, which appear devoid of mGluR1, express high levels of mGluR5 at their dendritic spines (Rodrigues et al. 2002; Romano et al. 1995), including those targeted by afferents from the medial geniculate nucleus (Rodrigues et al. 2002). This indicates that this receptor mediates auditory information processing in the amygdala. Medium spiny neurons of the ITC and capsular subdivision of the CEA are also highly enriched in mGluR5 (personal communication).

Table 13.2 Distribution of group II mGluR mRNA and immunoreactivity in the adult rodent brain

	mGluR2		mGluR3	
	mRNA	IHC	mRNA	IHC
Basolateral complex				
Lateral nucleus (LA)	+++	++	++	+++
Basal nucleus (BA)	+++	+++	+/++	+++
Basomedial nucleus (BM)	++	+	+	−/+
Centromedial nuclear group				
Central nucleus (CeA)	−	++	+	++
Capsular division (Cec)		++		++
Lateral division (CeL)		++		++
Medial division (CeM)		−		++
Bed nucleus of the stria terminalis intra-amygdaloid division (BSTIA)	++	+	+	−
Medial nucleus (MeA)	+	+	+	+
Superficial corticomedial areas				
Anterior cortical nucleus (ACo)	++	++	+	+
Posterior cortical nucleus (PCo)	++	++	+	+
Periamygdaloid cortex (CxA)		+++		+
Other areas				
Anterior amygdaloid area (AAA)	−/+	+	+	+
Amygdalohippocampal area (AHA)		+		+
Intercalated clusters (ITC)	−/+	+	+	+
Amygdalo-striatal transition area (AStr)	−	++	+	++

Intensities: +++ intense, ++ moderate, + weak, −/+ very weak or contains only scattered labelled cells, − absent

Data were taken from the following references and from personal unpublished observations: Gu et al. (2008), Hetzenauer et al. (2008), Muly et al. (2007), Ohishi (1993a, b), Petralia et al. (1996) and Tamaru et al. (2001)

13.6 Expression of Group II mGluRs in Amygdala

Previous in situ hybridization studies have reported a widespread expression of mGluR2 and mGluR3 in BLA neurons and a lack in neuronal profiles of the CeA (Gu et al. 2008; Ohishi et al. 1993a, b). The BLA showed a remarkable high density of mGluR2 mRNA expression whereas moderate levels were observed in the superficial corticomedial areas. On the other hand, transcripts for mGluR3 were more broadly distributed and showed lower intensity levels. Immunohistochemical (Hetzenauer et al. 2008; Petralia et al. 1996) and radioligand binding studies (Wright et al. 2001), besides robust expression in the neuropil of the BLA, revealed low to moderate expression of both subtypes in the CeA, with a prevailing presence of mGluR3 in the CeM (Fig. 13.2). In the BLA and BNST, ultrastructural analysis showed a predominant presynaptic localization of mGluR2 in asymmetric synaptic contacts on dendritic spines (Muly et al. 2007). Conversely, immunoreactivity for mGluR3 was detected both pre- and postsynaptically as well as in astroglial cells

Table 13.3 Distribution of group III mGluR mRNA and immunoreactivity in the adult rodent brain

	mGluR4		mGluR7		mGluR8	
	mRNA	IHC	mRNA	IHC	mRNA	IHC
Basolateral complex						
Lateral nucleus (LA)	+	+	+++	+	+/++	
Basal nucleus (BA)	+	+	+++	+++	+/++	+
Basomedial nucleus (BM)	–	–/+	+++	+		
Centromedial nuclear group						
Central nucleus (CeA)	+	–/+	+++	+		
Capsular division (Cec)				++		
Lateral division (CeL)				+		
Medial division (CeM)				+		
Bed nucleus of the stria terminalis intra-amygdaloid division (BSTIA)	+	+	++	+		++
Medial nucleus (MeA)	–	–/+	+++	+	+	
Superficial corticomedial areas						
Anterior cortical nucleus (ACo)	–	–/+	++	+		
Posterior cortical nucleus (PCo)	–	–/+	++	+		
Periamygdaloid cortex (CxA)		–/+	+	+++		
Other areas						
Anterior amygdaloid area (AAA)						
Amygdalohippocampal area (AHA)	–		+++	+++		
Intercalated clusters (ITC)	++	++	+++	+++		+++
Amygdalo-striatal transition area (AStr)	++	+	++	++		++

Intensities: +++ intense, ++ moderate, + weak, –/+ very weak or contains only scattered labelled cells, – absent

Data were taken from the following references and from personal unpublished observations: Corti et al. (1998, 2002), Dobi et al. (2013), Kinoshita et al. (1998), Masugi et al. (1999), Messenger et al. (2002), Ohishi et al. (1995) and Saugstad et al. (1997)

(Hetzenauer et al. 2008; Muly et al. 2007). Although the source of axons containing group II mGluRs has not been clarified so far, it is likely that the majority arise intrinsically to the amygdala and in particular from the BLA.

13.7 Expression of Group III mGluRs in Amygdala

Several studies have shown that for all group III mGluRs medium to high mRNA levels were detectable in the BLA (Corti et al. 1998; Messenger et al. 2002; Ohishi et al. 1995). High resolution studies revealed that in the amygdala group III mGluRs are located presynaptically (Dobi et al. 2013; Kinoshita et al. 1998). Axon terminals containing these receptors were found particularly enriched along the intermediate and external capsules in association with dendrites of neurons surrounding the ITCs (Fig. 13.3) (Dobi et al. 2013; Masugi et al. 1999). In the BLA, mGluR7 has the highest density of all group III mGluRs (Corti et al. 1998; Kinoshita et al. 1998; Masugi et al. 1999; Shigemoto et al. 1997).

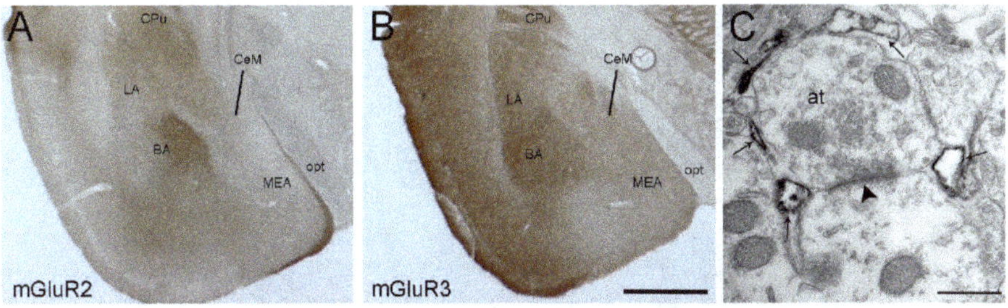

Fig. 13.2 Immunohistochemical localization of mGluR2 and mGluR3 receptors in the CeA. A-B) Low magnification microphotographs of mouse hemisections from (**a**) mGluR2-KO and (**b**) mGluR3-KO immunolabelled for mGluR2/3. Immunoreactivity for mGluR2 and mGluR3 is observed as punctate staining in the neuropil of the CeA and BLA. Moderate staining in the CeM is observed only in mGluR2-KO, thus corresponding to mGluR3. (**c**) Ultrastructural localization of mGluR3 immunoreactivity in the CeM of mGluR2-KO mice. Immunoreactivity for mGluR3 can be observed in small glial processes (indicated by *small arrows*) surrounding pre- and postsynaptic elements forming an asymmetric synapse (*arrowhead*). Abbreviations: *at* axon terminal, *BLA* basolateral complex of the amygdala, *CeM* central nucleus of the amygdala, medial subdivision, *CPu* caudate-putamen, *LA* lateral nucleus of the amygdala, *MeA* medial nucleus of the amygdala, *opt* optic tract. Scale bars: (**a–b**) 1 mm; (**c**) 500 nm (From Hetzenauer et al. 2008 with permission)

At the cellular and subcellular level, little is known about the localization of mGluR4. On the other hand, ultrastructural studies have shown that despite the similar pattern of immunoreactivity between mGluR7 and mGluR8 within the BLA and around the ITCs (Dobi et al. Dobi et al. 2013; Kinoshita et al. 1998; Masugi et al. 1999), these two receptors are largely segregated in different glutamatergic and GABAergic inputs (Dobi et al. 2013). Among glutamatergic afferents, mGluR7 was found on axon terminals of BLA pyramidal neurons and on extrinsic inputs arising from the intralaminar nuclei of the thalamus (Dobi et al. 2013). Although the origin of mGluR7-containing inputs forming symmetric synapses has not been identified so far, it is plausible that they arise in large part from BLA interneurons. However, noradrenergic axons in the BLA are known to form both asymmetric and symmetric synapses primarily with dendrites of pyramidal cells (Li et al. 2001). The noradrenergic innervation of the BLA arises mainly from the locus coeruleus in the brain stem (Fallon et al. 1978; Pickel et al. 1974; Roder and Ciriello 1993), a brain structure expressing high levels of mGluR7 transcripts (Corti et al. 1998; Ohishi et al. 1995). Therefore, besides glutamatergic and GABAergic inputs, mGluR7 may be located also presynaptically on noradrenergic afferents to the BLA and ITCs. So far, mGluR8 have not been detected on a definite pathway, thus leaving open the source of inputs enriched in these receptors. However, mGluR8-containing inputs forming asymmetric synapses most likely arise from areas extrinsic to the amygdala. These may include the somatosensory, perirhinal and/or piriform cortex, as they express significant amounts of mGluR8 mRNA (Corti et al. 1998; Saugstad et al. 1997) and are a significant source of axons to the BLA (Pitkänen 2000). The mGluR8-containing inputs forming symmetric synapses may originate from local interneurons, consistent with the moderate expression of mGluR8 mRNA in the BLA (Corti et al. 1998). However, further tracing studies combined

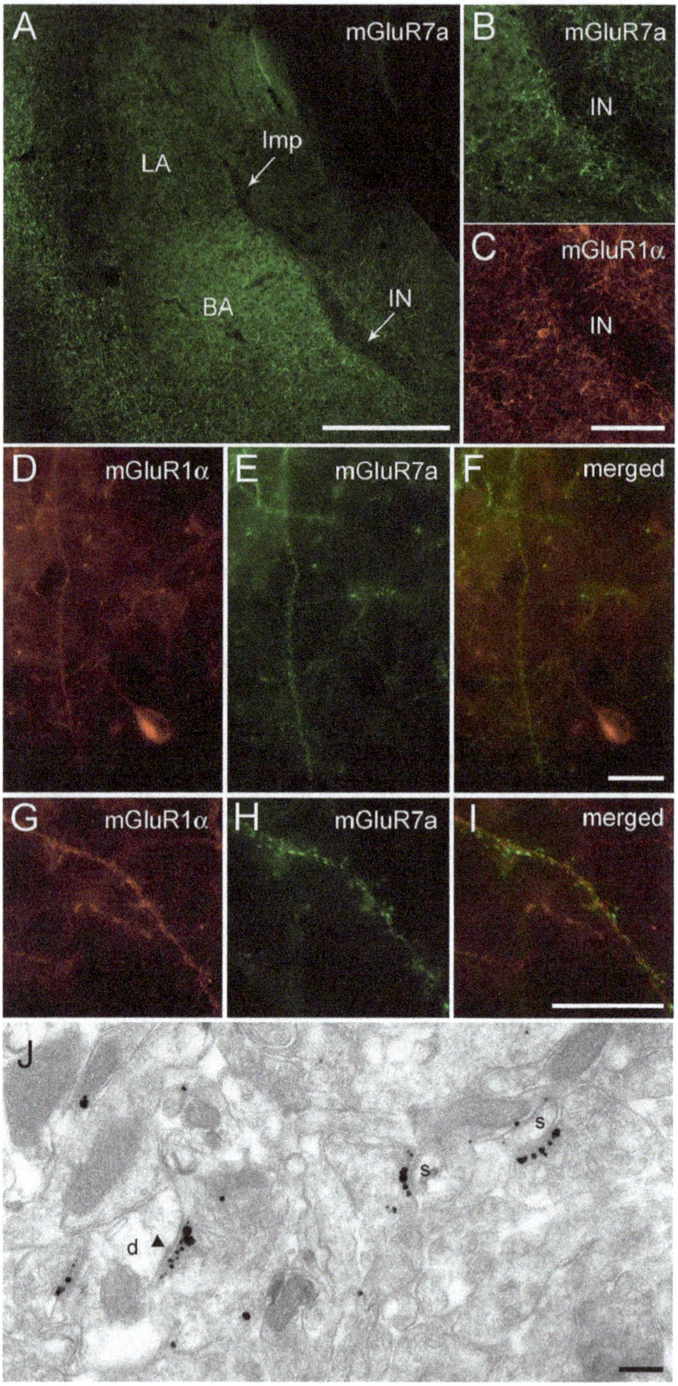

Fig. 13.3 Distribution of mGluR7a in the amygdala and preferential targeting of large mGluR1α+ITC neurons. (**a**) Micrograph showing intense punctate mGluR7a receptor immunoreactivity in the BLA and in particular around the Imp and IN clusters. (**b**) This micrograph shows an enlarged view

with immunoelectron microscopy are needed to finally reveal the exact source of inputs containing group III mGluRs.

13.8 Physiology

Group I mGluRs have been shown to be particularly important for synaptic plasticity in various experimental paradigms (Bashir et al. 1993; Conquet et al. 1994; Huber et al. 1998; Ireland and Abraham 2009), including long-term potentiation (LTP) at thalamic input synapses onto LA principal neurons (Fendt and Schmid 2002; Rodrigues et al. 2002). Thalamic-LA synapses have been extensively studied as a site of learning-induced plasticity and tightly correlated with fear memory (McKernan and Shinnick-Gallagher 1997; Rogan et al. 1997). Conversely, long-term depression (LTD) at these synapses has been proposed as a cellular mechanism underlying fear memory extinction (Lin et al. 2003; Kim et al. 2007a). In accordance, the group I mGluR agonist 3,5-dihydroxy-phenylglycine (DHPG) was reported to induce synaptic depotentiation of thalamic-LA synapses and to induce fear extinction (Kim et al. 2007b). DHPG-induced depotentiation was shown to occur only at synapses whose synaptic strength was consolidated after fear conditioning (Heinbockel and Pape 2000; Kim et al. 2010). This might depend on changes in AMPA receptor composition at postsynaptic sites, similar to what was previously reported for dopaminergic neurons of the ventral tegmental area in which insertion of GluR1 homomeric subunits is a prerequisite for group I mGluR-dependent LTD (Bellone and Lüscher 2006). While the exact contribution of individual group I mGluRs to synaptic plasticity and associative learning in the BLA remains to be established, the influence of mGluR5 might be related to a mutual modulatory activity with NMDA receptors at postsynaptic sites (Alagarsamy et al. 1999; Rodrigues et al. 2002).

Activation of presynaptic group II mGluRs in the BLA inhibited glutamatergic neurotransmission (Rainnie and Shinnick-Gallagher 1992) and induced LTD (Lin et al. 2000, 2005), whereas activation of these receptors postsynaptically hyperpolarised pyramidal cells (Rainnie et al. 1994). However, Womble and Moises (1994) reported that in the BLA, t-ACPD, which preferentially activates group II mGluRs, inhibited two potassium conductances, namely, I_{AHP} and M-currents, in pyramidal neurons producing a sustained cellular depolarization. The reason for this discrepancy remains unclear.

Fig. 13.3 (continued) of the IN where the intense labelling around the cluster, but not inside, can be better appreciated. (**c**) Immunofluorescence micrograph taken at the same focal plane as in (**b**) and showing mGluR1α immunoreactivity. (**d-f**) Low and (**g-i**) high magnification images depicting mGluR1α + dendrites and spines of large ITC neurons decorated by dense mGluR7a positive terminals. (**j**) Preembedding immunogold/silver labelling for mGluR7a in the BA. Immunometal particles identifying mGluR7a are selectively concentrated in the presynaptic active zone of axon terminals, which establish primarily excitatory synapses with postsynaptic dendrites (*solid arrowhead*) and spines. Abbreviations: *BA* basal nucleus, *d* dendrite, *Imp* medial paracapsular ITC cluster, *IN* main ITC nucleus, *LA* lateral nucleus, *s* spine. Scale bars: (**a**) 500 μm; (**b–c**) 125 μm. (**d–i**) 20 μm, (**j**) 500 nm (From Dobi et al. 2013 with permission)

Two independent forms of LTD have been observed in BLA pyramidal neurons, one at thalamic-LA synapses dependent on both group I mGluRs and NMDA receptors (Wang and Gean 1999) and the other dependent on presynaptic group II mGluRs and occurring at cortical-LA synapses (Hong et al. 2009; Kaschel et al. 2004; Li and Rainnie 2014; Lucas et al. 2013).

Unselective agonists at group III mGluRs, such as L-AP4, were shown to regulate baseline synaptic transmission in the amygdala (Neugebauer et al. 1997) and in particular in the laterocapsular part of the CeA (Han et al. 2004), which is the main target of the spinoparabrachial-amygdaloid pain pathway. So far, nothing is known about the physiological effects of mGluR4 at amygdala synapses. The mGluR7 positive allosteric modulator (PAM) AMN082 (Mitsukawa et al. 2005) was shown to elicit no effects on thalamo-LA field potentials or cortico-LA EPSCs, but to reduce thalamo-LA LTP in acute slices (Fendt et al. 2013). Conversely, activation of mGluR8 by the orthosteric agonist (S)-3,4-dicarboxyphenylglycine (DCPG) strongly attenuated synaptic transmission from both thalamic and cortical afferents to BLA pyramidal neurons and inhibited LTP induced by tetanic stimulation. However, DCPG showed no effects on thalamo-LA LTP (Schmid and Fendt 2006; Fendt et al. 2013). Therefore, group III mGluRs appear to play different roles in modulating sensory inputs in the amygdala consistent with their distinct, although in part overlapping, anatomical localization (Dobi et al. 2013).

Interestingly, we have recently shown that the extensive dendritic arbours of large GABAergic neurons surrounding ITC clusters, located between the BLA and CeA and expressing high levels of mGluR1α, are decorated by terminals enriched in mGluR7 and/or mGluR8 (Bienvenu et al. 2015). A prominent functional feature of these large neurons was their robust, prompt but transient response to noxious stimuli (Bienvenu et al. 2015). Although several factors can explain the transient nature of this response, including short activation of excitatory synaptic inputs or the recruitment of inhibitory ones, the enrichment in mGluR7 and/or mGluR8 at axon terminals innervating these large neurons strongly suggests an autoreceptor-mediated short-term inhibition of glutamate release. Therefore, presynaptic group III mGluRs may control in time the activation of these large neurons upon noxious stimulation.

13.9 Relevance of mGluRs for Amygdala-Dependent Emotional Behaviours

Several amygdala-dependent emotional behaviours appear to be significantly influenced by modulation of mGluR activity. In particular, mGluRs have been strongly implicated in fear learning (Table 13.4).

Table 13.4 mGluR modulation of fear and extinction learning

Receptor	Treatment	Route	Species	Method	Fear		Extinction		References
					Acquis.	Memory	Acquis.	Memory	
Group I									
mGluR1	**Antagonist**								
	EMQMCM	i.p.	Rat	FPS	↓				Pietraszeck et al. (2005)
	EMQMCM	i.p.	Rat	FC-cue	↓				Pietraszeck et al. (2005)
	EMQMCM	i.p.	Rat	FC-cue	↓	↓			Gravius et al. (2006)
	CPCCOEt	BLA	Rat	FC-cue	=	=	↓		Kim et al. (2007b)
	AIDA	i.p.	Mouse	FC-cue		↑			Clem and Huganir (2010)
mGluR5	**Agonist**								
	DHPG	BLA	Rat	FC-cue	↑				Rudy and Matus-Amat (2009)
	DHPG	BLA	Rat	FC-context	↑				Rudy and Matus-Amat (2009)
	DHPG	BLA	Rat	FC-cue		↓			Kim et al. (2010)
	Antagonist								
	MPEP	i.p.	Rat	FC-cue				=	Toth et al. (2012)
	MPEP	BLA	Rat	FC-cue	↓	=			Rodrigues et al. (2002)
	MPEP	BLA	Rat	FC-context	↓	=			Rodrigues et al. (2002)
	MPEP	p.o.	Rat	FPS	↓				Schulz et al. (2001)
	MPEP	i.p.	Rat	FPS	↓				Brodkin et al. (2002)
	MTEP	i.p.	Rat	FPS	↓				Pietraszeck et al. (2005)
	MTEP	i.p.	Rat	FC-cue	↓				Gravius et al. (2006)
	KO		Mouse	FC-cue	↓	=	↓		Xu et al. (2009)
			Mouse	FC-context	↓	=/↓	↓		Xu et al. (2009)
Group II									
mGluR2/3	**Antagonist**								
	MCPG	BLA	Rat	CTA	↓				Yasoshima et al. (2000)
	LY3414495	BLA	Rat	FC-cue			=	↓	Kim et al. (2015)
	LY3414495	CeA	rat	FC-cue			=	=	Kim et al. (2015)

(continued)

Table 13.4

Receptor	Treatment	Route	Species	Method	Fear		Extinction		References
					Acquis.	Memory	Acquis.	Memory	
Group III									
mGluR4	**Agonist**								
	LSPI-2111	i.p.	Mouse	FC-cue	↓		=		Davis et al. (2013)
	ADX88178	i.p.	Mouse	FC-cue	↓	=	=	=	Kalinichev et al. (2013)
	KO		Mouse	FC-cue	=	=	=	↑	Davis et al. (2013)
mGluR7	**Agonist**								
	AMN082	BLA	Rat	FC-cue	=	↓			Siegl et al. (2008)
	AMN082	i.p.	Rat	FPS	↓		↑		Fendt et al. (2008)
	AMN082	i.p.	Mouse	CTA			↑		Fendt et al. (2008)
	AMN082	i.p.	Rat	FC-cue			=	↑	Toth et al. (2012)
	AMN082	BLA	Mouse	FC-context			↑	=	Dobi et al. (2013)
	AMN082	BLA	Mouse	FC-cue	=	=	=	↓	Fendt et al. (2013)
	AMN082	BLA	Mouse	FC-context	=	=	↓	=	Fendt et al. (2013)
	AMN082	p.o.	Mouse	FC-cue			↑		Whittle et al. (2013)
	KO		Mouse	FC-cue	↓	↓			Fendt et al. (2013)
			Mouse	FC-context	↓	↓			Masugi et al. (1999)
			Mouse	FC-context	↓				Goddyn et al. (2008)
			Mouse	CER	=	=	↑		Callaerts-Vegh et al. (2006)
			Mouse	FC-context	↓				Fendt et al. (2013)
mGluR8	**Agonist**								
	DCPG	BLA	Rat	FPS	↓	↓			Schmidt and Fendt (2006)
	DCPG	BLA	Mouse	FC-context			↓	=	Dobi et al. (2013)
	DCPG	BLA	Mouse	FC-cue	↓	=	=	=	Fendt et al. 2013)
	DCPG	BLA	Mouse	FC-context	↓	=	↓	=	Fendt et al. (2013)
	KO		Mouse	FC-cue	=	=			Fendt et al. (2013)
				FC-context	↓	↓			Fendt et al. (2010, 2013)
				CDP-facil.			↓		Fendt et al. (2010)

CDP facil chlordiazepoxide-induced facilitation of extinction, *CER* instrumental learning superimposed to fear conditioning, *CTA* conditioned taste avoidance, *FC* fear conditioning, *FPS* fear-potentiated startle

13.10 Role of mGluRs in Fear Learning

The amygdala is central for the formation of fear memories. The acquisition and consolidation of a fear memory is often studied using classical Pavlovian auditory fear conditioning (Maren 2001; Pape and Pare 2010). During fear conditioning, a neutral conditioned stimulus (CS, e.g. a tone) is paired with an aversive unconditioned stimulus (US, e.g. a mild footshock). Training leads to a CS-US association and a conditioned fear response that is elicited by the presentation of the CS alone. Conditioned fear responses can be modified by a parallel and independent form of learning known as extinction (Herry et al. 2010). During extinction training, the CS is repetitively presented in the absence of the US. This leads to the disconfirmation of the CS-associated aversive expectation and to the inhibition of the conditioned fear response. The BLA appears as a key region for the acquisition and storage of fear memories although other nuclei, such as the CeA and ITC, also play important roles (Pape and Pare 2010).

13.11 Influence of Group I mGluRs on Fear Learning

Activation of group I mGluRs with the non-selective orthosteric agonist DHPG, directly injected into the BLA prior to a conditioning experience, was found to facilitate the formation of a conditioned response (Rudy and Matus-Amat 2009).

Injection into the BLA of a mGluR1 antagonist impaired both short- and long-term extinction, but not fear acquisition (Kim et al. 2007b). Moreover, systemic pharmacologic blockade of mGluR1 before extinction retrieval facilitated relapse phenomena, such as renewal and reinstatement (Clem and Huganir 2010). Taken together these data suggest that the activity of mGluR1 is linked specifically to mechanisms underlying fear extinction. However, the exact cellular and molecular mechanisms underlying the role of mGluR1 on extinction of fear remain to be established.

Gene-targeted deletion (Xu et al. 2009) and administration of mGluR5 antagonists, both systemically (Schulz et al. 2001) and intra-amygdala (Rodrigues et al. 2002), impaired fear conditioning. These data indicate that mGluR5 is important for the acquisition of fear memories. However, given that post-training infusions of mGluR5 antagonists did not influence either short or long term memories, it can be surmised that mGluR5 is not involved in the consolidation nor in the retrieval of fear memory (Rodrigues et al. 2002). Conditional allele ablation of mGluR5 showed also impaired fear extinction (Xu et al. 2009). Activation of mGluR5 induced a depolarization-induced suppression of inhibition in the BLA (Zhu and Lovinger 2005) through the release of endocannabinoids and activation of presynaptic cannabinoid 1 receptors (CB1). Interestingly, CB1 null mice showed significant fear extinction deficits (Marsicano et al. 2002), suggesting that a potential cellular basis for extinction learning might depend on the depression of inhibitory synapses mediated through the activation of mGluR5 and the release of endocannabinoids. Moreover, prolonged extinction training resulted in a mGluR5-dependent, but not mGluR1-dependent, long-term inhibition of fear recovery, also known as spontaneous recovery (Mao et al. 2013).

Associative learning appears to regulate mGluR5 expression, as fear conditioning was shown to increase it at least in the hippocampus (Riedel et al. 2000). Whether this also occurs in amygdala remains to be verified.

13.12 Influence of Group II mGluRs on Fear Learning

Activation of group II mGluRs in the amygdala was shown to impair fear memory formation (Lin et al. 2005; Walker et al. 2002). On the other hand, activation of mGluR2/3 in the LA is required for the consolidation, but not the acquisition, of extinction by selectively weakening synaptic transmission at cortical-LA synapses (Kim et al. 2015). Conversely, group II mGluRs in the CeA appears to have no effects on extinction (Kim et al. 2015). Therefore, agonists at mGluR2/3 may be good candidates for developing novel drugs for treating aberrant fear-related disorders such as PTSD or phobias.

13.13 Influence of Group III mGluRs on Fear Learning

Activation or blockade (including gene-targeted deletion) of group III mGluRs was consistently shown to alter fear responses. Although only a limited number of studies have addressed the role of mGluR4 in fear learning, mice carrying a gene-targeted deletion of mGluR4 showed an enhanced amygdala-dependent acquisition of cued fear conditioning (Davis et al. 2012). Mice deficient in mGluR4 were also reported to have enhanced extinction of cued fear (Davis et al. 2013). In accordance with these genetic ablation studies, acute pharmacological stimulation with the orthosteric mGluR4 agonist LSP1–2111 inhibited the acquisition of cued fear conditioning in wild-type male mice. However, LSP1–2111 showed no effects on extinction (Davis et al. 2013).

Mice deficient in mGluR7 exhibited a marked reduction in the acquisition of appetitive and aversive conditioning (Callaerts-Vegh et al. 2006; Masugi et al. 1999), as well as delayed extinction learning in both aversive and appetitive experimental paradigms (Callaerts-Vegh et al. 2006; Goddyn et al. 2008) and a facilitated extinction of cue-induced fear-potentiated startle (Fendt et al. 2008). Systemic as well as intra-amygdala administration of AMN082 in rats impaired the acquisition of conditioned fear (Fendt et al. 2008; Siegl et al. 2008). It is puzzling that both genetic ablation and pharmacological activation of mGluR7 showed similar effects on conditioned fear. Extinction of fear was also affected by systemic administration of AMN082, which facilitated between-session but attenuated within-session extinction, contrary to what was observed in mGluR7 null mice (Fendt et al. 2008, 2013; Toth et al. 2012). We observed a facilitatory effect of AMN082 on the extinction of contextual fear when microinjected into the BLA (Dobi et al. 2013). Interestingly, oral administration of AMN082 to 129/SvImJ mice, an animal model of impaired extinction (Hefner et al. 2008), produced a significant improvement of both acquisition and consolidation of fear extinction (Whittle et al. 2013). The altered acquisition of conditioned fear mediated by AMN082 could be ascribed to a functional blockade of mGluR7 due to their internalization (Pelkey et al. 2007), although this appears unlikely given that AMN082 elicited opposite effects on extinction compared to genetic ablation or selective downregulation of mGluR7 mRNA expression by means of short interfering RNAs (Fendt et al. 2008). On the other hand, the extinction promoting effect of mGluR7 activation is difficult to explain. Activation of this receptor induces an activity-dependent depression of both glutamatergic and GABAergic synaptic transmission (Perroy et al. 2002; Ugolini et al. 2008). Given the widespread distribution of mGluR7 within amygdala microcircuits, its cumulative excitatory and inhibitory effects are hard to predict. The limited knowledge on the neural pathway underlying fear extinction further prevents to formulate a mechanistic hypothesis on how mGluR7 activation facilitates extinction retention.

Earlier studies on mGluR8 deficient mice gave rise to highly inconsistent results regarding their influence on fear learning and measures of anxiety, possibly due to the different genetic background of the different mutant mouse lines investigated (Duvoisin et al. 2005; Gerlai et al. 2002; Linden et al. 2002; Robbins et al. 2007). Using mGluR8 null mice backcrossed into the C57BL/6 J line, Fendt and colleagues could reveal a robust deficit in the expression of contextual conditioned fear and in chlordiazepoxide-induced facilitation of extinction (Fendt et al. 2010). Acquisition and expression of conditioned fear were dose-dependently inhibited by DCPG microinjections into the amygdala measured by fear-potentiated startle in rats (Schmid and Fendt 2006). However, immediately after its intra-amygdala application, this drug was shown to induce in mice a reduced spontaneous locomotor activity and a short-lasting immobility (Dobi et al. 2013; Fendt et al. 2013). These effects may be interpreted as an anxiogenic-like activity of this drug independent from motor impairment or sedation or an activity related to unconditioned fear (Duvoisin et al. 2010). When injected before extinction training, DCPG precluded the attenuation of the freezing response during the extinction training, but did not prevent the formation and consolidation of an extinction memory (Dobi et al. 2013; Fendt et al. 2013), suggesting that activation of mGluR8 does not significantly influence extinction of fear learning.

In conclusion, agonists at mGluR7 may be seen as potentially effective add-on drugs to exposure therapy to facilitate extinction learning in individuals with pathological fear, whereas antagonizing mGluR8 may find therapeutic application in forms of context-dependent anxiety.

13.14 Role of mGluRs in Social Behaviours

The amygdala has long been involved in social behaviours (Brothers 1990; Bunnell et al. 1970; Kling and Steklis 1976), and increased neuronal activity in the BLA has been recorded during social interaction (Katayama et al. 2009). However, little is known about the neural circuits that within the amygdala contribute to social behaviours and whether mGluRs influence synaptic function at these circuits.

Increased social and nonsocial exploration were observed following inhibition of mGluR5 with the non-competitive antagonist 2-methyl-6-(phenylethynyl)-pyridine (MPEP) administered systemically (Navarro et al. 2006). MPEP also reduced offensive behaviour without affecting motor activity (Navarro et al. 2006). Similarly, systemic administration of the mGluR7 PAM AMN082 was shown to produce a marked reduction of offensive behaviours, such as attacks and threats in mice (Navarro et al. 2009).

While the involvement of other mGluRs in social behaviours using pharmacological tools has not been investigated, mGluR8 null mice were shown to have an enhanced social interaction (Duvoisin et al. 2011).

Increased social behaviours and reduced aggression elicited by mGluR5 antagonists and mGluR7 agonists are consistent with their anxiolytic activity (Fendt et al. 2013; Spooren and Gasparini 2004; see below) and could, therefore, impinge on receptors located in amygdala circuits. The suggested contribution of amygdala mGluRs in social behaviours indicates these receptors as potential targets for the treatment of antisocial personality disorders and/or autism spectrum disorders (ASDs; see below for further details).

13.15 Relevance of Amygdala mGluRs for Human Neuropsychiatric Disorders

A variety of psychiatric disorders are characterized by abnormal amygdala activation (Etkin and Wager 2007), in particular in response to sensory stimuli, and impaired glutamatergic transmission. These include anxiety disorders, PTSD, ASDs and schizophrenia (Rauch et al. 2010; Schneider et al. 1998; Stein et al. 2007).

13.16 Anxiety Disorders

Anxiety disorders, which include generalized anxiety disorders (GAD), panic disorder and specific and social phobias, are a major public health problem worldwide (Lepine Lépine 2002; Whiteford et al. 2013). Anxiety disorders appear to originate from a disrupted inhibitory/excitatory balance within a distributed network of brain areas that includes the amygdala (Linden et al. 2006; Wierońska and Pilc 2009). Given the limited efficacy and significant side effects of the anxiolytic drugs currently in clinical use, such as monoamine reuptake inhibitors and benzodiazepines, there is a strong medical need for new therapeutic agents acting through different mechanisms of action.

13.17 Modulation of Anxiety-Like Behaviours by Group I mGluRs

Mice lacking group I mGluRs show altered measures of anxiety such as a reduced stress-induced hyperthermia (Brodkin et al. 2002). Pharmacologically, drugs that block group I mGluRs were also invariably reported to exert anxiolytic-like activity in a variety of animal models including the elevated plus maze, conflict tests and stress-induced hyperthermia (Spooren et al. 2010; Steckler et al. 2005). The anxiolytic activity obtained with mGluR1 antagonists appears task-specific affecting more the conflict component than the exploratory activity (Klodzinska et al. 2004; Steckler et al. 2005; Tatarczynska et al. 2001). How mGluR1 antagonists exert anxiolytic effects remains currently unknown, but it can be postulated that they influence the activity of GABAergic neurons in the amygdala and possibly in the hippocampus and the bed nucleus of the stria terminalis. Possible limitations to the development of mGluR1 antagonists as anxiolytic drugs are the reported amnesic effects of such compounds (Spooren et al. 2010) and a weaker anxiolytic activity in comparison with benzodiazepines (Varty et al. 2005).

Blockade of mGluR5 might be considered a better option for the clinical development of novel anxiolytic drugs. Competitive and non-competitive mGluR5 antagonists as well as negative allosteric modulators (NAMs) consistently showed anxiolytic-like effects in animal models of anxiety (Busse et al. 2004; Pietraszek et al. 2005; Spanka et al. 2010; Spooren et al. 2000; Stachowicz et al. 2007; Tatarczynska et al. 2001). Moreover, Fenobam, a clinically validated anxiolytic in humans, was found to be a potent and selective mGluR5 NAM (Porter et al. 2005). A potential advantage in the repeated use of mGluR5 NAMs is the apparent lack of tolerance with respect to benzodiazepines, at least in preclinical models (Nordquist et al. 2007). Fenobam, however, was shown to produce anxiolytic activity at doses that also impaired spatial learning (Jacob et al. 2009) and produced psychostimulant side effects. It remains to be established whether these unwanted effects of Fenobam depend on its activity on mGluR5 or are due to its binding to other receptors (off-target effects).

13.18 Modulation of Anxiety-Like Behaviours by Group II mGluRs

Neuronal hyperexcitability in limbic areas including the amygdala has been implicated in the pathophysiology of anxiety disorders (Swanson et al. 2005). Activation of presynaptic group II and group III mGluRs limits the release of glutamate, hence can prevent excessive neural activation (Ferraguti and Shigemoto 2006). These receptors are also present at heterosynapses, where they regulate the release of other neurotransmitters, including GABA, monoamines and neuropeptides (Cartmell and Schepp 2000).

Agonists at group II mGluRs have consistently shown anxiolytic activity both in rodents and humans (Grillon et al. 2003; Helton et al. 1998; Monn et al. 1997; Shekhar and Keim 2000; Walker and Davis 2002), whereas antagonists at these receptors, such as LY341495, elicited anxiogenic-like responses (Linden et al. 2005). The antagonist LY341495 also induced neuronal activation in the lateral and medial subdivisions of the CeA (Hetzenauer et al. 2008; Linden et al. 2005) that appears predominantly mediated via mGluR3 (Hetzenauer et al. 2008). However, studies involving mGluR2 or mGluR3 deficient mice indicated that LY354740 should activate both subtypes in order to exert anxiolytic effects (Linden et al. 2005). The efficacy of group II mGluR agonists as anxiolytics has been evaluated in a series of clinical studies. In particular, LY354740 (as prodrug) was reported to be efficacious in the treatment of GAD (Dunayevich et al. 2008), but not in reducing panic symptoms (Bergink and Westernberger 2005). On the other hand, a derivative of LY354740 with higher bioavailability (LY544344) was found effective both in the treatment of GAD and in decreasing cholecystokinin tetrapeptide-induced panic symptoms in healthy volunteers (Dunayevich et al. 2008; Kellner et al. 2005). However, clinical development of these compounds has been stopped due to the occurrence of seizures observed in toxicological studies after repeated treatments with high doses of these drugs (Wierowska and Pilc 2013).

13.19 Modulation of Anxiety-Like Behaviours by Group III mGluRs

Drugs acting at group III mGluRs have received in recent years increasing attention as potential novel targets for anxiety disorders (Spooren et al. 2010). Specifically, positive allosteric modulation of mGluR4, both after systemic administration and following microinjection in the BLA, showed modest but significant anxiolytic effects in a wide range of preclinical animal models of anxiety (Duvoisin et al. 2011; Sławińska et al. 2013; Stachowitz et al. 2004; Wierońska et al. 2010). However, mice lacking mGluR4 showed a complex and sex-dependent phenotype concerning measures of anxiety. While male mice presented higher levels of anxiety in the open field and elevated zero maze, in comparison with wild-type mice, female mice showed reduced measures of anxiety in these tests (Davis et al. 2012). These data indicate a complex role of mGluR4 in anxiety-like behaviour that warrant further investigations.

Mice deficient in mGluR7 revealed a dysregulated hypothalamic–pituitary–adrenal axis and increased hippocampal BDNF levels, which were also observed following chronic treatment with anxiolytics (Mitsukawa et al. 2006). Likewise, knockdown of mGluR7 in the amygdala (and hippocampus), using short interfering RNAs, resulted in a robust anxiolytic phenotype (O′Connor et al. 2013). Anxiolytic effects were also elicited by the mGluR7 agonist AMN082 in the four-plate test and stress-induced hyperthermia (Stachowicz et al. 2008). As already discussed above for fear learning, there are several potential explanations, which may explain this paradox. First, AMN082-induced activation of the mGluR7 may lead to rapid receptor internalization, thus resulting in a functional antagonistic effect (Pelkey et al. 2007). Second, a metabolite of AMN082, namely, N-benzhydrylethane-1,2-diamine (Met-1), inhibits the serotonin, noradrenaline and dopamine transporters (Sukoff-Rizzo et al. 2011). Thus data obtained in animals with systemic administration of AMN082 have to be interpreted with caution. Third, genetic manipulations may trigger compensatory changes in circuits relevant to anxiety, which could produce paradoxical effects. However, recent data using the novel mGluR7 NAM ADX71743 showed that blockade of mGluR7 produces an anxiolytic effect in the elevated plus maze and marble burying tests (Kalinichev et al. 2013). Taken together these findings indicate that blockade of mGluR7 might be a suitable strategy for the development of novel therapeutics designed to treat anxiety disorders.

Acute pharmacological activation of mGluR8 with the orthosteric agonist DCPG or the PAM AZ12216052 was found to reduce measures of anxiety in wild-type mice (Duvoisin et al. 2010, 2011). On the other hand, data obtained from mGlu8-deficient mice on anxiety-like behaviour gave rise to largely conflicting results. In fact, some of the studies reported no effects of the genotype on measures of anxiety (Fendt et al. 2010; Gerlai et al. 2002), whereas others found an anxiogenic-like phenotype (Duvoisin et al. 2005; Linden et al. 2002; Robbins et al. 2007), in line with the pharmacological findings. However, neither the transgenic models nor the systemic administration of mGluR8 agonists can provide insights on the role played by this receptor in amygdala microcircuits related to anxiety-like behaviour (Gosnell et al. 2011). Microinjection of DCPG in the rat BLA showed no effect in the elevated plus maze and Vogel test (Palazzo et al. 2008; Stachowicz et al. 2005). Although these data might question a participation of amygdala mGluR8 on anxiety-like behaviour, further studies are warranted before an influence of this receptor on anxiety can be ruled out.

Compared with group I and group II mGluRs, fewer studies have evaluated the efficacy of group III mGluRs in anxiety disorders. The problem remains the limited availability of potent and selective ligands at each of these receptors. However, the preliminary evidence for an anxiolytic-like activity besides their highly discrete distribution and limited overlap, which may limit the risk for side effects, suggests that these receptors, either modulated individually or in combination, are attractive therapeutic targets for anxiety disorders.

So far, clinical research has evaluated the efficacy of mGluR ligands only in a relatively narrow group of anxiety disorders. For instance, no studies have assessed the efficacy of drugs acting at mGluRs in PTSD, obsessive-compulsive disorder or phobias. Therefore, further studies are needed to address the potential of mGluR modulation in different anxiety disorders.

13.20 Influence of Amygdala mGluRs on Addiction

The amygdaloid complex has been implicated by a large number of studies in the process of associative learning for appetitive reinforcers (Everitt et al. 2000). Recently, interest has grown on the contribution of different amygdala nuclei in the learned association that occur during the process of drug addiction and relapse (See et al. 2003). A growing body of evidence further points toward an important contribution of mGluRs in modulating drug taking and seeking behaviour. In particular, several behavioural analyses have indicated that group I mGluRs are engaged in the rewarding and seeking effects of drugs of abuse. However, the contribution of mGluR1 to the additive potential of abused drugs remains controversial because of the difficulties in separating specific effects of mGluR1 antagonists on rewarding effects from impairments of locomotion and anxiety-like behaviour (Pomierny-Chamioło et al. 2014). A larger body of evidence implicates mGluR5 in drug addiction. Mice lacking mGluR5 did not develop cocaine self-administration (Bird et al. 2014; Chiamulera et al. 2001), and administration of mGluR5 antagonists decreased cocaine self-administration, cue-induced relapse and conditioned place preference (Bäckström and Hyytiä 2006; Kumaresan et al. 2009). Tonic stimulation of mGluR5 also appears to attenuate the behavioural effects of cocaine and alcohol (Besheer et al. 2006; Lee et al. 2005). The activity of mGluR5 in the BLA appears primarily important for cue-induced reinstatement of drug-seeking behaviour (Sinclair et al. 2012). Using high-resolution positron emission tomography with the PET tracer [^{11}C]ABP688, which selectively binds to an mGluR5 allosteric site (Hintermann et al. 2007), a significant downregulation of mGluR5 binding sites in the left amygdala was shown in subjects with cocaine dependence, tested during a period of brief abstinence (Milella et al. 2014). The functional interpretation of this downregulation following cocaine withdrawal might be of a compensatory response to an increase in glutamate release (Miyake et al. 2011), which may underlie the incubation of craving.

Studies on the contribution of group II mGluRs to the behavioural responses of drugs of abuse have consistently shown that mGluR2/3 stimulation reduces the reinforcing properties of most additive drugs not only after systemic administration (Adewale et al. 2006; Pomierny-Chamioło et al. 2014) but also after site-specific microinjection into the amygdala (Cannady et al. 2011). Activation of group II mGluRs in the amygdala with the selective agonist LY379268 was also shown to reduce the discriminative stimulus effects of alcohol, one of the determinants of abuse liability (Cannady et al. 2011).

The CeA has been proposed to play an important role in cue-induced cocaine seeking (Lu et al. 2004). Injection of LY379268 in the CeA attenuated the cue-induced operant behaviour characterizing prolonged cocaine withdrawal (Lu et al. 2007). Interestingly, microinjection of LY379268 in the CeA also prevented cue-induced sucrose seeking in extinction tests after a prolonged withdrawal from oral self-administration of sucrose (Uejima et al. 2007). This suggests that group II mGluRs are important players in the regulation of CeA transmission underlying craving for both drug and natural rewards.

The recent development of more selective compounds for mGluR2 and mGluR3 may provide further insights into the contribution that each of these receptors has in modulating the discriminative stimulus and reinforcing effects of alcohol and other drugs of abuse (Jin et al. 2010).

13.21 Role of Amygdala mGluRs on Emotional–Affective Dimensions of Pain

In recent years, the amygdala has emerged as a key brain area for the emotional–affective dimension of pain (Neugebauer et al. 2004). Hyperactivity in the laterocapsular division of the CeA (CeC, also termed the "nociceptive amygdala") was found to account for pain-related emotional responses and to contribute to anxiety-like behaviour. However, the BLA, which is central for the acquisition and storage of aversive associations, can strongly influence the CeA/CeC both directly via glutamatergic projections and indirectly involving GABAergic ITC neurons, which in turn project to the CeA (Busti et al. 2011).

Activation of group I mGluRs within the amygdala has been shown to be of critical importance for pain-related synaptic plasticity and behaviour (Kolber et al. 2010; Li and Neugebauer 2004; Neugebauer et al. 2003). Activation of mGluR5, but not of mGluR1, enhanced excitatory responses on CeC neurons to both innocuous and noxious stimuli in naïve animals (Ji and Neugebauer 2010; Li and Neugebauer 2004), whereas in the arthritis pain model antagonists at both group I mGluRs decreased the enhanced activity of CeC neurons (Neugebauer et al. 2003). Similarly, blockade of group I mGluRs in the CeA in this model inhibited ultrasonic vocalizations after noxious stimulation (Han and Neugebauer 2005), which appears to depend on the amygdala and to reflect the unpleasant emotional-affective state of the animal (Borszcz and Leaton 2003). Conversely, only mGluR1 antagonists were able to inhibit ultrasonic vocalizations during a noxious stimulation (Han and Neugebauer 2005). Therefore, amygdala mGluR1 appears to be involved in the processing of both nociceptive and affective information. Although the cellular mechanisms by which mGluR1 contributes to the facilitation of excitatory transmission at CeC neurons during sensitization remain to be identified, it has been proposed that mGluR1 may reduce feed-forward inhibition by modulating the activity of ITC neurons (Ren and Neugebauer 2010).

Activation of mGluR5 within the CeA was found to be necessary for the development of mechanical hypersensitivity in response to inflammation and for visceral pain through an intracellular signalling cascade involving mitogen-activated protein kinases (Crock et al. 2012; Kolber et al. 2010).

Group II and III mGluRs have both been shown to modulate synaptic plasticity and to inhibit nociceptive processing at CeC neurons in a model of arthritic pain (Han et al. 2004, 2006). The potency of agonists for both group II and III mGluRs increased in the arthritis pain model (Li and Neugebauer 2006). Whereas the increased sensitivity of group II mGluRs was found to be input and activity dependent, group III mGluRs showed a general change (Li and Neugebauer 2006). The contribution of individual mGluRs at amygdala circuits on pain behaviour remains to be determined. Activation of mGluR7 in the CeC with AMN082 was shown to elicit pro-nociceptive effects in naïve animals (Palazzo et al. 2008), but not in a model of arthritis, by indirectly facilitating excitatory transmission (Ren et al. 2011). Conversely, activation of mGluR8 in the CeC with DCPG showed anti-nociceptive effects in arthritic animals (Palazzo et al. 2008), but not under normal conditions, most likely through presynaptic inhibition of the enhanced glutamate release at BLA-CeC synapses observed in this pain model (Ren et al. 2011). Unlike the non-selective group III mGluR agonist L-AP4, DCPG and AMN082 did not inhibit baseline transmission from BLA to CeC neurons but only pain-related plasticity, which suggests another group III mGluR, most likely mGluR4, is responsible for this effect. The role of mGluR4 in the amygdala on pain behaviour has not been investigated so far, but with the recent introduction of new selective pharmacological agents for this receptor subtype, it is expected that soon its contribution to the regulation of amygdala output and pain behaviour will be elucidated.

13.22 Role of Amygdala mGluRs on Sleep Disorders

The amygdala has been increasingly implicated in the regulation of arousal and sleep (Deboer et al. 1999; Jha et al. 2005; Morrison et al. 2000; Sanford et al. 2002; Tang et al. 2005). A number of brain stem nuclei directly involved in the generation and regulation of REM sleep indeed receive strong inputs from the CeA (Krettek and Price 1978; Takeuchi et al. 1982), and stimulation of the BA increased low-voltage, high-frequency activity in the cortical EEG and decreased NREM and total sleep time (Dringenberg and Vanderwolf 1996).

Glutamatergic neurotransmission has also been postulated to play an essential role in regulating arousal (Jones 2005), by means of its activity on ionotropic receptors as well as mGluRs. Systemic administration of group II mGluR agonists was found to induce a dose-dependent reduction of REM sleep, but also an increase in the REM onset latency (Ahnaou et al. 2009; Feinberg et al. 2002). These effects were reproduced by direct microinjection of the group II mGluR agonist LY379268 into the BA, whereas microinjection of the antagonist LY341495 suppressed NREM and total sleep, thereby increasing wakefulness (Dong et al. 2012). On the other hand, injection of the same agonist or antagonist into the CeA produced no significant alterations in sleep (Dong et al. 2012). Therefore, activation of group II mGluRs in the BA suppresses REM without altering NREM and total sleep time, a pattern common to many hypnotic drugs. On the other hand, antagonists significantly suppress NREM sleep enhancing wakefulness.

It remains to be evaluated whether in patients with sleep disorders, group II mGluR agonists would be able to significantly reduce the time to fall asleep and to increase the duration of sleep, without producing daytime sedation, essential requirements for an effective hypnotic. However, the effective anxiolytic activity and the apparent low abuse potential of group II mGluR agonists represent desired additional features if they were to be developed as novel agents for sleep disorders.

A recent imaging study in humans has revealed in the amygdala, among other brain areas, an increased uptake of ^{11}C–ABP688 after prolonged wakefulness (Hefti et al. 2013). This observation suggests that also mGluR5 may be involved in sleep-wake regulation. Sleep-wake disturbances show high comorbidity with mood disorders, and sleep deprivation has mood-enhancing effects in depressed patients (Bunney and Bunney 2012). Interestingly, reduced levels of mGluR5 in corticolimbic areas were reported in patients with major depression (Deschwanden et al. 2011). Moroeover, it has been suggested that agonists at mGluR5 have antidepressant properties (Sanacora et al. 2008). Therefore, the increased functional availability of mGluR5 after prolonged wakefulness within the amygdala, as well as in other brain areas, may underlie the mood-enhancing effects of sleep deprivation (Hefti et al. 2013).

13.23 Autism Spectrum Disorders

The amygdala is an important component of the neural networks underlying social behaviours (Brothers 1990). Impaired social interaction is a hallmark of ASDs, and neuropathological studies in postmortem autistic brains reported a volume reduction and structural abnormalities of the amygdala (Kemper and Bauman 1993). These findings, along with recent preclinical studies (Amaral et al. 2003; Suvrathan and Chattarji 2011) and functional neuroimaging data (Baron-Cohen et al. 1999), have led to propose that dysfunction of the amygdala may be responsible, at least in part, for the impairment of social behaviour in autism (Baron-Cohen et al. 2000).

Altered signalling downstream of group I mGluRs, and in particular of mGluR5, has been proposed by Bear and Huber (Bear et al. 2004) as playing a pivotal role in the pathogenesis of fragile X syndrome (FXS) and ASDs. FXS is one of the main inherited causes of mental retardation and an identified cause of autism. FXS is caused by a mutation that leads to transcriptional silencing of the FMR1 gene, which encodes the fragile X mental retardation protein (FMRP) (Pieretti et al. 1991) and functions as a negative regulator of protein synthesis (Dolen et al. 2007). Genetic and pharmacological modulation of mGluR5 was indeed shown to ameliorate numerous impairments in animal models of FXS (Dolen and Bear 2008). Fmr1 null mice show a selective amplification of mGluR5-mediated LTD at Schaffer collateral-CA1 pyramidal cell synapses in the hippocampus (Waung and Huber 2009). This mGluR5-mediated LTD appears dependent on protein synthesis occurring in the dendrites of CA1 pyramidal neurons as a result of local mRNA translation. In Fmr1 null mice, mGluR5-dependent aberrant synaptic plasticity was also revealed at thalamic inputs to principal neurons in the lateral amygdala (Suvrathan et al. 2010). FMRP is one of the dendritic proteins synthesized in response to mGluR5 activation and has the specific function of acting as a translational repressor of other LTD-associated proteins, such as postsynaptic density-95 (PSD-95), the elongation factor 1a and the cytoskeleton-associated protein, Arc. In the absence of FMRP, these proteins are constitutively and highly expressed, and this would make the mGluR5-mediated LTD insensitive to protein synthesis inhibitors (Waung and Huber 2009). A series of recent elegant studies have characterized at least in part the signalling pathways that in response to mGluR5 activation lead to dendritic protein synthesis. This includes the association of mGluR5 with Homer proteins and the activation of upstream regulators of dendritic mRNA translation, such as the MAPK-interacting kinase/eukaryotic initiation factor 4E pathway, as well as the mTOR/p70S6K pathway. Many steps along these pathways were found abnormal in Fmr1-KO (Giuffrida et al. 2005; Ronesi and Huber 2008; Sharma et al. 2010). However, the exact participation of group I mGluR-related signalling into the molecular pathology of ASDs remains to be fully elucidated. Therefore, increased group I mGluR signalling may provide a link that connects diverse manifestations of ASDs. However, it remains to be established whether the core of the phenotype of ASDs arises directly from a synaptic dysfunction primarily involving group I mGluRs, in which altered protein synthesis might be a primary cause or a secondary consequence of a broader neuronal dysfunction. Some evidence for synaptic dysregulation dependent from group I mGluRs has come also from the study of the synaptic protein neuroligin 3 (NLG3) and its associated knockout mouse model, which relates to a form of autism not associated with any other symptoms, thus a non-syndromic form. In this model, increased levels of mGluR1 at parallel fibre to Purkinje cell synapses in the cerebellum have been identified (Baudouin et al. 2012). In correlation with this, an occlusion of the mGluR-LTD at these synapses was also found, confirming a convergence of the phenotype with syndromic forms of ASDs (Baudouin et al. 2012). These results further support the hypothesis that synaptic dysregulation of type I mGluR function is an important pathophysiological mechanism in ASDs.

The consequences of the dysregulation of group I mGluR signalling might thus be complex affecting a cluster of associated molecular signalling proteins and could influence several forms of synaptic plasticity and neural pathways, in the amygdala and elsewhere, implicated in ASDs.

13.24 Conclusions

Evidence for the participation of mGluRs in a number of amygdala-dependent emotional behaviours has increased dramatically in the last few years. Hence allosteric modulators of these receptors may hold the promise to develop into novel drugs for the treatment of neuropsychiatric disorders characterized by impaired amygdala functions.

References

Adewale AS, Platt DM, Spealman RD (2006) Pharmacological stimulation of group II metabotropic glutamate receptors reduces cocaine self-administration and cocaine-induced reinstatement of drug seeking in squirrel monkeys. J Pharmacol Exp Ther 318:922–931

Ahnaou A, Dautzenberg FM, Geys H, Imogai H, Gibelin A, Moechars D, Steckler T, Drinkenburg WH (2009) Modulation of group II metabotropic glutamate receptor (mGlu2) elicits common changes in rat and mice sleep-wake architecture. Eur J Pharmacol 603:62–72

Akiyama K, Yamada N, Sato M (1987) Increase in ibotenate-stimulated phosphatidylinositol hydrolysis in slices of the amygdala/pyriform cortex and hippocampus of rat by amygdala kindling. Exp Neurol 98:499–508

Alagarsamy S, Marino MJ, Rouse ST, Gereau RW 4th, Heinemann SF, Conn PJ (1999) Activation of NMDA receptors reverses desensitization of mGluR5 in native and recombinant systems. Nat Neurosci 2:234–240

Amaral DG, Bauman MD, Schumann CM (2003) The amygdala and autism: implications from non-human primate studies. Genes Brain Behav 2:295–302

Anand A, Shekhar A (2003) Brain imaging studies in mood and anxiety disorders: special emphasis on the amygdala. Ann N Y Acad Sci 985:370–388

Bäckström P, Hyytiä P (2005) Suppression of alcohol self-administration and cue-induced reinstatement of alcohol seeking by the mGlu2/3 receptor agonist LY379268 and the mGlu8 receptor agonist (S)-3, 4-DCPG. Eur J Pharmacol 528:110–118

Bäckström P, Hyytiä P (2006) Ionotropic and metabotropic glutamate receptor antagonism attenuates cue-induced cocaine seeking. Neuropsychopharmacology 31:778–786

Baron-Cohen S, Ring HA, Wheelwright S, Bullmore ET, Brammer MJ, Simmons A, Williams SC (1999) Social intelligence in the normal and autistic brain: an fMRI study. Eur J Neurosci 11:1891–1898

Baron-Cohen S, Ring HA, Bullmore ET, Wheelwright S, Ashwin C, Williams SC (2000) The amygdala theory of autism. Neurosci Biobehav Rev 24:355–364

Bashir ZI, Bortolotto ZA, Davies CH, Berretta N, Irving AJ, Seal AJ, Henley JM, Jane DE, Watkins JC, Collingridge GL (1993) Induction of LTP in the hippocampus needs synaptic activation of glutamate metabotropic receptors. Nature 363:347–350

Baude A, Nusser Z, Roberts JD, Mulvihill E, McIlhinney RA, Somogyi P (1993) The metabotropic glutamate receptor (mGluR1 alpha) is concentrated at perisynaptic membrane of neuronal subpopulations as detected by immunogold reaction. Neuron 11:771–787

Baudouin SJ, Gaudias J, Gerharz S, Hatstatt L, Zhou K, Punnakkal P, Tanaka KF, Spooren W, Hen R, De Zeeuw CI, Vogt K, Scheiffele P (2012) Shared synaptic pathophysiology in syndromic and nonsyndromic rodent models of autism. Science 338:128–132

Bear MF, Huber KM, Warren ST (2004) The mGluR theory of fragile X mental retardation. Trends Neurosci 27:370–377

Bellone C, Lüscher C (2006) Cocaine triggered AMPA receptor redistribution is reversed in vivo by mGluR-dependent long-term depression. Nat Neurosci 9:636–641

Bergink V, Westenberg HG (2005) Metabotropic glutamate II receptor agonists in panic disorder: a double blind clinical trial with LY354740. Int Clin Psychopharmacol 20:291–293

Besheer J, Stevenson RA, Hodge CW (2006) mGlu5 receptors are involved in the discriminative stimulus effects of self-administered ethanol in rats. Eur J Pharmacol 551:71–75

Bienvenu TC, Busti D, Micklem BR, Mansouri M, Magill PJ, Ferraguti F, Capogna M (2015) Large intercalated neurons of amygdala relay noxious sensory information. J Neurosci 35:2044–2057

Bird MK, Lohmann P, West B, Brown RM, Kirchhoff J, Raymond CR, Lawrence AJ (2014) The mGlu5 receptor regulates extinction of cocaine-driven behaviours. Drug Alcohol Depend 137:83–89

Borszcz GS, Leaton RN (2003) The effect of amygdala lesions on conditional and unconditional vocalizations in rats. Neurobiol Learn Mem 79:212–225

Brodkin J, Bradbury M, Busse C, Warren N, Bristow LJ, Varney MA (2002) Reduced stress-induced hyperthermia in mGluR5 knockout mice. Eur J Neurosci 16:2241–2244

Brothers L (1990) The social brain: a project for integrating primate behaviour and neurophysiology in a new domain. Concepts Neuroscience 1:27–51

Bunnell BN, Sodetz FJ Jr, Shalloway DI (1970) Amygdaloid lesions and social behavior in the golden hamster. Physiol Behav 5:153–161

Bunney BG, Bunney WE (2012) Rapid-acting antidepressant strategies: mechanisms of action. Int J Neuropsychopharmacol 15:695–713

Busse CS, Brodkin J, Tattersall D, Anderson JJ, Warren N, Tehrani L, Bristow LJ, Varney MA, Cosford ND (2004) The behavioral profile of the potent and selective mGlu5 receptor antagonist 3-[(2-methyl-1,3-thiazol-4-yl)ethynyl] pyridine (MTEP) in rodent models of anxiety. Neuropsychopharmacology 29:1971–1979

Busti D, Geracitano R, Whittle N, Dalezios Y, Manko M, Kaufmann W, Satzler K, Singewald N, Capogna M, Ferraguti F (2011) Different fear states engage distinct networks within the intercalated cell clusters of the amygdala. J Neurosci 31:5131–5144

Callaerts-Vegh Z, Beckers T, Ball SM, Baeyens F, Callaerts PF, Cryan JF, Molnar E, D'Hooge R (2006) Concomitant deficits in working memory and fear extinction are functionally dissociated from reduced anxiety in metabotropic glutamate. Receptor 7-deficient mice. J Neurosci 26:6573–6582

Cannady R, Grondin JJ, Fisher KR, Hodge CW, Besheer J (2011) Activation of group II metabotropic glutamate receptors inhibits the discriminative stimulus effects of alcohol via selective activity within the amygdala. Neuropsychopharmacology 36:2328–2338

Cartmell J, Schoepp DD (2000) Regulation of neurotransmitter release by metabotropic glutamate receptors. J Neurochem 75:889–907

Chiamulera C, Epping-Jordan MP, Zocchi A, Marcon C, Cottiny C, Tacconi S, Corsi M, Orzi F, Conquet F (2001) Reinforcing and locomotor stimulant effects of cocaine are absent in mGluR5 null mutant mice. Nat Neurosci 4:873–874

Clem RL, Huganir RL (2010) Calcium-permeable AMPA receptor dynamics mediate fear memory erasure. Science 330:1108–1112

Conquet F, Bashir ZI, Davies CH, Daniel H, Ferraguti F, Bordi F, Franz-Bacon K, Reggiani A, Matarese V, Condé F et al (1994) Motor deficit and impairment of synaptic plasticity in mice lacking mGluR1. Nature 372:237–243

Corti C, Restituito S, Rimland JM, Brabet I, Corsi M, Pin JP, Ferraguti F (1998) Cloning and characterization of alternative mRNA forms for the rat metabotropic glutamate receptors mGluR7 and mGluR8. Eur J Neurosci 10:3629–3641

Corti C, Aldegheri L, Somogyi P, Ferraguti F (2002) Distribution and synaptic localisation of the metabotropic glutamate receptor 4 (mGluR4) in the rodent CNS. Neuroscience 110:403–420

Crock LW, Kolber BJ, Morgan CD, Sadler KE, Vogt SK, Bruchas MR, Gereau RW 4th. (2012) Central amygdala metabotropic glutamate receptor 5 in the modulation of visceral pain. J Neurosci 32:14217–14226

Davis MJ, Haley T, Duvoisin RM, Raber J (2012) Measures of anxiety, sensorimotor function, and memory in male and female mGluR4−/− mice. Behav Brain Res 229:21–28

Davis MJ, Iancu OD, Acher FC, Stewart BM, Eiwaz MA, Duvoisin RM, Raber J (2013) Role of mGluR4 in acquisition of fear learning and memory. Neuropharmacology 66:365–372

Deboer T, Ross RJ, Morrison AR, Sanford LD (1999) Electrical stimulation of the amygdala increases the amplitude of elicited ponto-geniculo-occipital waves. Physiol Behav 66:119–124

Deschwanden A, Karolewicz B, Feyissa AM, Treyer V, Ametamey SM, Johayem A et al (2011) Reduced metabotropic glutamate receptor 5 density in major depression determined by [(11) C]ABP688 PET and postmortem study. Am J Psychiatry 168:727–734

Dobi A, Sartori SB, Busti D, Van der Putten H, Singewald N, Shigemoto R et al (2013) Neural substrates for the distinct effects of presynaptic group III metabotropic glutamate receptors on extinction of contextual fear conditioning in mice. Neuropharmacology 66:274–289

Dolen G, Bear MF (2008) Role for metabotropic glutamate. Receptor 5 (mGluR5) in the pathogenesis of fragile X syndrome. J Physiol 586:1503–1508

Dolen G, Osterweil E, Rao BS, Smith GB, Auerbach BD, Chattarji S, Bear MF (2007) Correction of fragile X syndrome in mice. Neuron 56:955–962

Dong E, Wellman LL, Yang L, Sanford LD (2012) Effects of microinjections of group II metabotropic glutamate agents into the amygdala on sleep. Brain Res 1452:85–95

Dringenberg HC, Vanderwolf CH (1996) Cholinergic activation of the electrocorticogram: an amygdaloid activating system. Exp Brain Res 108:285–296

Dunayevich E, Erickson J, Levine L, Landbloom R, Schoepp DD, Tollefson GD (2008) Efficacy and tolerability of an mGlu2/3 agonist in the treatment of generalized anxiety disorder. Neuropsychopharmacol 33:1603–1610

Duvoisin R, Zhang C, Pfankuch T, O'Connor H, Gayet-Primo J, Quraishi S, Raber J (2005) Increased measures of anxiety and weight gain in mice lacking the group III metabotropic glutamate receptor mGluR8. Eur J Neurosci:1–13

Duvoisin RM, Pfankuch T, Wilson JM, Grabell J, Chhajlani V, Brown DG, Johnson E, Raber J (2010) Acute pharmacological modulation of mGluR8 reduces measures of anxiety. Behav Brain Res 212:168–173

Duvoisin R, Villasana L, Davis M, Winder DG, Raber J (2011) Opposing roles of mGluR8 in measures of anxiety involving non-social and social challenges. Behav Brain Res 221:50–54

Emile L, Mercken L, Apiou F, Pradier L, Bock MD, Menager J, Clot J, Doble A, Blanchard JC (1996) Molecular cloning, functional expression, pharmacological characterization and chromosomal localization of the human metabotropic glutamate receptor type 3. Neuropharmacol 35:523–530

Etkin A, Wager TD (2007) Functional neuroimaging of anxiety: a meta-analysis of emotional processing in PTSD, social anxiety disorder, and specific phobia. Am J Psychiatry 164:1476–1488

Everitt BJ, Cardinal RN, Hall J, Parkinson JA, Robbins TW (2000) Differential involvement of amygdala subsystems in appetitive conditioning and drug addiction. In: Aggleton JP (ed) The amygdala: a functional analysis. Oxford University Press, Oxford, pp 353–390

Fallon JH, Koziell DA, Moore RY (1978) Catecholamine innervation of the basal forebrain. II. Amygdala, suprarhinal cortex and entorhinal cortex. J Comp Neurol 180:509–532

Feinberg I, Campbell IG, Schoepp DD, Anderson K (2002) The selective group mGlu2/3 receptor agonist LY379268 suppresses REM sleep and fast EEG in the rat. Pharmacol Biochem Behav 73:467–474

Fendt M, Schmid S (2002) Metabotropic glutamate receptors are involved in amygdaloid plasticity. Eur J Neurosci 15:1535–1541

Fendt M, Schmid S, Thakker DR, Jacobson JH, Yamamoto R, Mitsukawa K et al (2008) mGluR7 facilitates extinction of aversive memories and controls amygdala plasticity. Mol Psychiatry 13:970–979

Fendt M, Bürki H, Imobersteg S, van der Putten H, McAllister K, Leslie JC, Shaw D, Hölscher C (2010) The effect of mGlu8 deficiency in animal models of psychiatric diseases. Genes Brain Behav 9:33–44

Fendt M, Imobersteg S, Peterlik D, Chaperon F, Mattes C, Wittmann C, Olpe HR, Mosbacher J, Vranesic I, van der Putten H, McAllister KH, Flor PJ, Gee CE (2013) Differential roles of mGlu(7) and mGlu(8) in amygdala-dependent behavior and physiology. Neuropharmacol 72:215–223

Ferraguti F, Shigemoto R (2006) Metabotropic glutamate receptors. Cell Tissue Res 326:483–504

Ferraguti F, Conquet F, Corti C, Grandes P, Kuhn R, Knopfel T (1998) Immunohistochemical localization of the mGluR1beta metabotropic glutamate receptor in the adult rodent forebrain: evidence for a differential distribution of mGluR1 splice variants. J Comp Neurol 400:391–407

Ferraguti F, Crepaldi L, Nicoletti F (2008) Metabotropic glutamate 1 receptor: current concepts and perspectives. Pharmacol Rev 60:536–581

Gerlai R, Adams B, Fitch T, Chaney S, Baez M (2002) Performance deficits of mGluR8 knockout mice in learning tasks: the effects of null mutation and the background genotype. Neuropharmacol 43:235–249

Giuffrida R, Musumeci S, D'Antoni S, Bonaccorso CM, Giuffrida-Stella AM, Oostra BA, Catania MV (2005) A reduced number of metabotropic glutamate subtype 5 receptors are associated with constitutive homer proteins in a mouse model of fragile X syndrome. J Neurosci 25:8908–8916

Goddyn H, Callaerts-Vegh Z, Stroobants S, Dirikx T, Vansteenwegen D, Hermans D, van der Putten H, D'Hooge R (2008) Deficits in acquisition and extinction of conditioned responses in mGluR7 knockout mice. Neurobiol Learn Mem 90:103–111

Gosnell H, Silberman Y, Grueter B, Duvoisin R, Raber J, Winder DG (2011) mGluR8 modulates excitatory transmission in the bed nucleus of the stria terminalis in a stress-dependent manner. Neuropsychoparmacol 36:1599–1607

Gravius A, Barberi C, Schäfer D, Schmidt WJ, Danysz W (2006) The role of group I metabotropic glutamate receptors in acquisition and expression of contextual and auditory fear conditioning in rats – a comparison. Neuropharmacol 51:1146–1155

Grillon C, Cordova J, Levine LR, Morgan CA 3rd. (2003) Anxiolytic effects of a novel group II metabotropic glutamate receptor agonist (LY354740) in the fear-potentiated startle paradigm in humans. Psychopharmacol 168:446–454

Gu G, Lorrain DS, Wei H, Cole RL, Zhang X, Daggett LP, Schaffhauser HJ, Bristow LJ, Lechner SM (2008) Distribution of metabotropic glutamate 2 and 3 receptors in the rat forebrain: implication in emotional responses and central disinhibition. Brain Res 1197:47–62

Han JS, Neugebauer V (2005) mGluR1 and mGluR5 antagonists in the amygdala inhibit different components of audible and ultrasonic vocalizations in a model of arthritic pain. Pain 113:211–222

Han JS, Bird GC, Neugebauer V (2004) Enhanced group III mGluR-mediated inhibition of pain-related synaptic plasticity in the amygdala. Neuropharmacol 46:918–926

Han JS, Fu Y, Bird GC, Neugebauer V (2006) Enhanced group II mGluR-mediated inhibition of pain-related synaptic plasticity in the amygdala. Mol Pain 2:18

Hanson JE, Smith Y (1999) Group I metabotropic glutamate receptors at GABAergic synapses in monkeys. J Neurosci 19:6488–6496

Hefner K, Whittle N, Juhasz J, Norcross M, Karlsson RM, Saksida LM, Bussey TJ, Singewald N, Holmes A (2008) Impaired fear extinction learning and cortico-amygdala circuit abnormalities in a common genetic mouse strain. J Neurosci 28:8074–8085

Hefti K, Holst SC, Sovago J, Bachmann V, Buck A, Ametamey SM et al (2013) Increased metabotropic glutamate receptor subtype 5 availability in human brain after one night without sleep. Biol Psychiatry 73:161–168

Heimer L (2003) A new anatomical framework for neuropsychiatric disorders and drug abuse. Am J Psychiatry 160:1726–1739

Heinbockel T, Pape HC (2000) Input-specific long-term depression in the lateral amygdala evoked by theta frequency stimulation. J Neurosci 20:RC68

Helton DR, Tizzano JP, Monn JA, Schoepp DD, Kallman MJ (1998) Anxiolytic and side-effect profile of LY354740: a potent, highly selective, orally active agonist for group II metabotropic glutamate receptors. J Pharmacol Exp Ther 284:651–660

Herry C, Ferraguti F, Singewald N, Letzkus JJ, Ehrlich I, Luthi A (2010) Neuronal circuits of fear extinction. Eur J Neurosci 31:599–612

Hetzenauer A, Corti C, Herdy S, Corsi M, Ferraguti F, Singewald N (2008) Individual contribution of metabotropic glutamate receptor (mGlu) 2 and 3 to c-Fos expression pattern evoked by mGlu2/3 antagonism. Psychopharmacol 201:1–13

Hintermann S, Vranesic I, Allgeier H, Brülisauer A, Hoyer D, Lemaire M, Moenius T, Urwyler S, Whitebread S, Gasparini F, Auberson YP (2007) ABP688, a novel selective and high affinity ligand for the labeling of mGlu5 receptors: identification, in vitro pharmacology, pharmacokinetic and biodistribution studies. Bioorg Med Chem 15:903–914

Hong I, Song B, Lee S, Kim J, Kim J, Choi S (2009) Extinction of cued fear memory involves a distinct form of depotentiation at cortical input synapses onto the lateral amygdala. Eur J Neurosci 30:2089–2099

Huber KM, Sawtell NB, Bear MF (1998) Effects of the metabotropic glutamate receptor antagonist MCPG on phosphoinositide turnover and synaptic plasticity in visual cortex. J Neurosci 18:1–9

Ireland DR, Abraham WC (2009) Mechanisms of group I mGluR-dependent long-term depression of NMDA receptor-mediated transmission at Schaffer collateral-CA1 synapses. J Neurophysiol 101:1375–1385

Jacob W, Gravius A, Pietraszek M, Nagel J, Belozertseva I, Shekunova E, Malyshkin A, Greco S, Barberi C, Danysz W (2009) The anxiolytic and analgesic properties of fenobam, a potent mGlu5 receptor antagonist, in relation to the impairment of learning. Neuropharmacol 57:97–108

Jha SK, Ross RJ, Morrison AR (2005) Sleep-related neurons in the central nucleus of the amygdala of rats and their modulation by the dorsal raphe nucleus. Physiol Behav 86:415–426

Ji G, Neugebauer V (2010) Reactive oxygen species are involved in group I mGluR-mediated facilitation of nociceptive processing in amygdala neurons. J Neurophysiol 104:218–229

Jin X, Semenova S, Yang L, Ardecky R, Sheffler DJ, Dahl R, Conn PJ, Cosford ND, Markou A (2010) The mGluR2 positive allosteric modulator BINA decreases cocaine self-administration and cue-induced cocaine-seeking and counteracts cocaine-induced enhancement of brain reward function in rats. Neuropsychopharmacol 35:2021–2036

Jones B (2005) Basic mechanism of sleep-wake states. Principles and practice of sleep medicine. Elsevier Saunders, Philadelphia

Kalinichev M, Rouillier M, Girard F, Royer-Urios I, Bournique B, Finn T et al (2013) ADX71743, a potent and selective negative allosteric modulator of metabotropic glutamate. Receptor 7: in vitro and in vivo characterization. J Pharmacol Exp Ther 344:624–636

Kaschel T, Schubert M, Albrecht D (2004) Long-term depression in horizontal slices of the rat lateral amygdala. Synapse 53:141–150

Katayama T, Jodo E, Suzuki Y, Hoshino KY, Takeuchi S, Kayama Y (2009) Phencyclidine affects firing activity of basolateral amygdala neurons related to social behavior in rats. Neuroscience 159:335–343

Kellner M, Muhtz C, Stark K, Yassouridis A, Arlt J, Wiedemann K (2005) Effects of a metabotropic glutamate(2/3) receptor agonist (LY544344/LY354740) on panic anxiety induced by cholecystokinin tetrapeptide in healthy humans: preliminary results. Psychopharmacol 179:310–315

Kemper TL, Bauman ML (1993) The contribution of neuropathologic studies to the understanding of autism. Neurol Clin 11:175–187

Kim J, Lee S, Park K, Hong I, Song B, Son G et al (2007a) Amygdala depotentiation and fear extinction. Proc Natl Acad Sci U S A 104:20955–20960

Kim J, Lee S, Park H, Song B, Hong I, Geum D, Shin K, Choi S (2007b) Blockade of amygdala metabotropic glutamate receptor subtype 1 impairs fear extinction. Biochem Biophys Res Commun 355:188–193

Kim J, Song B, Hong I, Kim J, Lee J, Park S, Eom JY, Lee CJ, Lee S, Choi S (2010) Reactivation of fear memory renders consolidated amygdala synapses labile. J Neurosci 30:9631–9640

Kim J, An B, Kim J, Park S, Park S, Hong I, et al. (2015) mGlu2/3 in the lateral amygdala is required for fear extinction: cortical input synapses onto lateral amygdala as a target site of the mGlu2/3 action. Neuropsychopharmacol Epub 2015 June 17. Pub Med PMID: 26081171

Kinoshita A, Shigemoto R, Ohishi H, van der Putten H, Mizuno N (1998) Immunohistochemical localization of metabotropic glutamate receptors, mGluR7a and mGluR7b, in the central nervous system of the adult rat and mouse: a light and electron microscopic study. J Comp Neurol 393:332–352

Kling A, Steklis HD (1976) A neural substrate for affiliative behavior in nonhuman primates. Brain Behav Evol 13:216–238

Klodzinska A, Tatarczyńska E, Chojnacka-Wójcik E, Nowak G, Cosford ND, Pilc A (2004) Anxiolytic-like effects of MTEP, a potent and selective mGlu5 receptor agonist does not involve GABA(A) signaling. Neuropharmacology 47:342–350

Kolber BJ, Montana MC, Carrasquillo Y, Xu J, Heinemann SF, Muglia LJ, Gereau RW 4th. (2010) Activation of metabotropic glutamate. Receptor 5 in the amygdala modulates pain-like behavior. J Neurosci 30:8203–8213

Krettek JE, Price JL (1978) Amygdaloid projections to subcortical structures within the basal forebrain and brainstem in the rat and cat. J Comp Neurol 178:225–254

Kumaresan V, Yuan M, Yee J, Famous KR, Anderson SM, Schmidt HD, Pierce RC (2009) Metabotropic glutamate receptor 5 (mGluR5) antagonists attenuate cocaine priming- and cue-induced reinstatement of cocaine seeking. Behav Brain Res 202:238–244

Lee B, Platt DM, Rowlett JK, Adewale AS, Spealman RD (2005) Attenuation of behavioral effects of cocaine by the Metabotropic Glutamate. Receptor 5 Antagonist 2-Methyl-6-(phenylethynyl)-pyridine in squirrel monkeys: comparison with dizocilpine. J Pharmacol Exp Ther 312:1232–1240

Lépine JP (2002) The epidemiology of anxiety disorders: prevalence and societal costs. J Clin Psychiatry 63(Suppl 14):4–8

Li W, Neugebauer V (2004) Differential roles of mGluR1 and mGluR5 in brief and prolonged nociceptive processing in central amygdala neurons. J Neurophysiol 91:13–24
Li W, Neugebauer V (2006) Differential changes of group II and group III mGluR function in central amygdala neurons in a model of arthritic pain. J Neurophysiol 96:1803–1815
Li C, Rainnie DG (2014) Bidirectional regulation of synaptic plasticity in the basolateral amygdala induced by the D1-like family of dopamine receptors and group II metabotropic glutamate receptors. J Physiol 592:4329–4351
Li R, Nishijo H, Wang Q, Uwano T, Tamura R, Ohtani O, Ono T (2001) Light and electron microscopic study of cholinergic and noradrenergic elements in the basolateral nucleus of the rat amygdala: evidence for interactions between the two systems. J Comp Neurol 439:411–425
Lin HC, Wang SJ, Luo MZ, Gean PW (2000) Activation of group II metabotropic glutamate receptors induces long-term depression of synaptic transmission in the rat amygdala. J Neurosci 20:9017–9024
Lin CH, Yeh SH, Leu TH, Chang WC, Wang ST, Gean PW (2003) Identification of calcineurin as a key signal in the extinction of fear memory. J Neurosci 23:1574–1579
Lin CH, Lee CC, Huang YC, Wang SJ, Gean PW (2005) Activation of group II metabotropic glutamate receptors induces depotentiation in amygdala slices and reduces fear-potentiated startle in rats. Learn Mem 12:130–137
Linden AM, Johnson BG, Peters SC, Shannon HE, Tian M, Wang Y, Yu JL, Köster A, Baez M, Schoepp DD (2002) Increased anxiety-related behavior in mice deficient for metabotropic glutamate 8 (mGlu8) receptor. Neuropharmacol 43:251–259
Linden AM, Bergeron M, Schoepp DD (2005) Comparison of c-Fos induction in the brain by the mGlu2/3 receptor antagonist LY341495 and agonist LY354740: evidence for widespread endogenous tone at brain mGlu2/3 receptors in vivo. Neuropharmacol 49(Suppl 1):120–134
Linden AM, Baez M, Bergeron M, Schoepp DD (2006) Effects of mGlu2 or mGlu3 receptor deletions on mGlu2/3 receptor agonist (LY354740)-induced brain c-Fos expression: specific roles for mGlu2 in the amygdala and subcortical nuclei, and mGlu3 in the hippocampus. Neuropharmacol 51:213–228
Lu L, Grimm JW, Hope BT, Shaham Y (2004) Incubation of cocaine craving after withdrawal: a riview of preclinical data. Neuropharmacol 47:214–226
Lu L, Uejima JL, Gray SM, Bossert JM, Shaham Y (2007) Systemic and central amygdala injections of the mGluR(2/3) agonist LY379268 attenuate the expression of incubation of cocaine craving. Biol Psychiatry 61:591–598
Lucas SJ, Bortolotto ZA, Collingridge GL, Lodge D (2013) Selective activation of either mGlu2 or mGlu3 receptors can induce LTD in the amygdala. Neuropharmacol 66:196–201
Mahan AL, Ressler KJ (2012) Fear conditioning, synaptic plasticity and the amygdala: implications for posttraumatic stress disorder. Trends Neurosci 35:24–35
Makoff A, Lelchuk R, Oxer M, Harrington K, Emson P (1996) Molecular characterization and localization of human metabotropic glutamate receptor type 4. Brain Res Mol Brain Res 37:239–248
Malherbe P, Kratzeisen C, Lundstrom K, Richards JG, Faull RL, Mutel V (1999) Cloning and functional expression of alternative spliced variants of the human metabotropic glutamate receptor 8. Brain Res Mol Brain Res 67:201–210
Mansouri M, Kasugai Y, Fukazawa Y, Bertaso F, Raynaud F, Perroy J, Fagni L, Kaufmann WA, Watanabe M, Shigemoto R, Ferraguti F (2015) Distinct subsynaptic localization of type 1 metabotropic glutamate receptors at glutamatergic and GABAergic synapses in the rodent cerebellar cortex. Eur J Neurosci 41:157–167
Mao SC, Chang CH, Wu CC, Orejarena MJ, Manzoni OJ, Gean PW (2013) Inhibition of spontaneous recovery of fear by mGluR5 after prolonged extinction training. PLoS One 8:e59580
Maren S (2001) Neurobiology of Pavlovian fear conditioning. Annu Rev Neurosci 24:897–931
Marsicano G, Wotjak CT, Azad SC, Bisogno T, Rammes G, Cascio MG, Hermann H, Tang J, Hofmann C, Zieglgänsberger W, Di Marzo V, Lutz B (2002) The endogenous cannabinoid system controls extinction of aversive memories. Nature 418:530–534

Masugi M, Yokoi M, Shigemoto R, Muguruma K, Watanabe Y, Sansig G et al (1999) Metabotropic glutamate receptor 7 ablation causes deficit in fear response and conditioned taste aversion. J Neurosci 19:955–963

McKernan MG, Shinnick-Gallagher P (1997) Fear conditioning induces a lasting potentiation of synaptic currents in vitro. Nature 390:607–611

Messenger MJ, Dawson LG, Duty S (2002) Changes in metabotropic glutamate receptor 1-8 gene expression in the rodent basal ganglia motor loop following lesion of the nigrostriatal tract. Neuropharmacol 43:261–271

Milella MS, Marengo L, Larcher K, Fotros A, Dagher A, Rosa-Neto P, Benkelfat C, Leyton M (2014) Limbic system mGluR5 availability in cocaine dependent subjects: a high-resolution PET [(11)C]ABP688 study. NeuroImage 98:195–202

Mitsukawa K, Mombereau C, Lotscher E, Uzunov DP, van der Putten H, Flor PJ et al (2006) Metabotropic glutamate receptor subtype 7 ablation causes dysregulation of the HPA axis and increases hippocampal BDNF protein levels: implications for stress-related psychiatric disorders. Neuropsychopharmacol 31:1112–1122

Miyake N, Skinbjerg M, Easwaramoorthy B, Kumar D, Girgis RR, Xu X, Slifstein M, Abi-Dargham A (2011) Imaging changes in glutamate transmission in vivo with the metabotropic glutamate receptor 5 tracer [11C] ABP688 and N-acetylcysteine challenge. Biol Psychiatry 69:822–824

Monn JA, Valli MJ, Massey SM, Wright RA, Salhoff CR, Johnson BG et al (1997) Design, synthesis, and pharmacological characterization of (+)-2-aminobicyclo[3.1.0]hexane-2, 6-dicarboxylic acid (LY354740): a potent, selective, and orally active group 2 metabotropic glutamate receptor agonist possessing anticonvulsant and anxiolytic properties. J Med Chem 40:528–537

Morrison AR, Sanford LD, Ross RJ (2000) The amygdala: a critical modulator of sensory influence on sleep. Biol Signals Recept 9:283–296

Muly EC, Mania I, Guo JD, Rainnie DG (2007) Group II metabotropic glutamate receptors in anxiety circuitry: correspondence of physiological response and subcellular distribution. J Comp Neurol 505:682–700

Navarro JF, Postigo D, Martin M, Buron E (2006) Antiaggressive effects of MPEP, a selective antagonist of mGlu5 receptors, in agonistic interactions between male mice. Eur J Pharmacol 551:67–70

Navarro JF, de Castro V, Martin-Lopez M (2009) Behavioural profile of selective ligands for mGlu7 and mGlu8 glutamate receptors in agonistic encounters between mice. Psicothema 21:475–479

Neugebauer V, Keele NB, Shinnick-Gallagher P (1997) Epileptogenesis in vivo enhances the sensitivity of inhibitory presynaptic metabotropic glutamate receptors in basolateral amygdala neurons in vitro. J Neurosci 17:983–995

Neugebauer V, Li W, Bird GC, Bhave G, Gereau RW 4th. (2003) Synaptic plasticity in the amygdala in a model of arthritic pain: differential roles of metabotropic glutamate receptors 1 and 5. J Neurosci 23:52–63

Neugebauer V, Li W, Bird GC, Han JS (2004) The amygdala and persistent pain. Neuroscientist 10:221–234

Nicoletti F, Bockaert J, Collingridge GL, Conn PJ, Ferraguti F, Schoepp DD et al (2011) Metabotropic glutamate receptors: from the workbench to the bedside. Neuropharmacol 60:1017–1041

Nordquist RE, Durkin S, Jaeschke G, Spooren W (2007) Stress-induced hyperthermia: effects of acute and repeated dosing of MPEP. Eur J Pharmacol 568:199–202

O'Connor RM, Thakker DR, Schmutz M, van der Putten H, Hoyer D, Flor PJ, Cryan JF (2013) Adult si RNA-induced knockdown of mGlu7 receptors reduces anxiety in the mouse. Neuropharmacol 72:66–73

Ohishi H, Shigemoto R, Nakanishi S, Mizuno N (1993a) Distribution of the messenger RNA for a metabotropic glutamate receptor, mGluR2, in the central nervous system of the rat. Neuroscience 53:1009–1018

Ohishi H, Shigemoto R, Nakanishi S, Mizuno N (1993b) Distribution of the mRNA for a metabotropic glutamate receptor (mGluR3) in the rat brain: an in situ hybridization study. J Comp Neurol 335:252–266

Ohishi H, Akazawa C, Shigemoto R, Nakanishi S, Mizuno N (1995) Distributions of the mRNAs for L-2-amino-4-phosphonobutyrate-sensitive metabotropic glutamate receptors, mGluR4 and mGluR7, in the rat brain. J Comp Neurol 360:555–570

Palazzo E, Fu Y, Ji G, Maione S, Neugebauer V (2008) Group III mGluR7 and mGluR8 in the amygdala differentially modulate nocifensive and affective pain behaviors. Neuropharmacol 55:537–545

Palazzo E, Marabese I, de Novellis V, Rossi F, Maione S (2014) Supraspinal metabotropic glutamate receptors: a target for pain relief and beyond. Eur J Neurosci 39:444–454

Palucha A, Pilc A (2007) Metabotropic glutamate receptor ligands as possible anxiolytic and antidepressant drugs. Pharmacol Ther 115:116–147

Pape HC, Pare D (2010) Plastic synaptic networks of the amygdala for the acquisition, expression, and extinction of conditioned fear. Physiol Rev 90:419–463

Pelkey KA, Yuan X, Lavezzari G, Roche KW, McBain CJ (2007) mGluR7 undergoes rapid internalization in response to activation by the allosteric agonist AMN082. Neuropharmacol 52:108–117

Perroy J, Richard S, Nargeot J, Bockaert J, Fagni L (2002) Permissive effect of voltage on mGlu 7 receptor subtype signaling in neurons. J Biol Chem 277:1223–1228

Petralia RS, Wang YX, Niedzielski AS, Wenthold RJ (1996) The metabotropic glutamate receptors, mGluR2 and mGluR3, show unique postsynaptic, presynaptic and glial localizations. Neuroscience 71:949–976

Phelps EA, LeDoux JE (2005) Contributions of the amygdala to emotion processing: from animal models to human behavior. Neuron 48:175–187

Pickel VM, Segal M, Bloom FE (1974) A radioautographic study of the efferent pathways of the nucleus locus coeruleus. J Comp Neurol 155:15–42

Pieretti M, Zhang FP, Fu YH, Warren ST, Oostra BA, Caskey CT, Nelson DL (1991) Absence of expression of the FMR-1 gene in fragile X syndrome. Cell 66:817–822

Pietraszek M, Sukhanov I, Maciejak P, Szyndler J, Gravius A, Wisłowska A, Płaźnik A, Bespalov AY, Danysz W (2005) Anxiolytic-like effects of mGlu1 and mGlu5 receptor antagonists in rats. Eur J Pharmacol 514:25–34

Pitkänen A (2000) Connectivity of the rat amygdaloid complex. In: Aggleton JP (ed) The amygdala: a functional analysis. Oxford University Press, Oxford, pp 31–115

Pitkänen A, Savander V, LeDoux JE (1997) Organization of intra-amygdaloid circuitries in the rat: an emerging framework for understanding functions of the amygdala. Trends Neurosci 20:517–523

Pomierny-Chamioło L, Rup K, Pomierny B, Niedzielska E, Kalivas PW, Filip M (2014) Metabotropic glutamatergic receptors and their ligands in drug addiction. Pharmacol Ther 142:281–305

Porter RH, Jaeschke G, Spooren W, Ballard TM, Büttelmann B, Kolczewski S, Peters JU, Prinssen E, Wichmann J, Vieira E, Mühlemann A, Gatti S, Mutel V, Malherbe P (2005) Fenobam: a clinically validated nonbenzodiazepine anxiolytic is a potent, selective, and noncompetitive mGlu5 receptor antagonist with inverse agonist activity. J Pharmacol Exp Ther 315:711–721

Rainnie DG, Shinnick-Gallagher P (1992) Trans-ACPD and L-APB presynaptically inhibit excitatory glutamatergic transmission in the basolateral amygdala (BLA). Neurosci Lett 139:87–91

Rainnie DG, Holmes KH, Shinnick-Gallagher P (1994) Activation of postsynaptic metabotropic glutamate receptors by trans-ACPD hyperpolarizes neurons of the basolateral amygdala. J Neurosci 14:7208–7220

Rauch AV, Reker M, Ohrmann P, Pedersen A, Bauer J, Dannlowski U, Harding L, Koelkebeck K, Konrad C, Kugel H, Arolt V, Heindel W, Suslow T (2010) Increased amygdala activation during automatic processing of facial emotion in schizophrenia. Psychiatry Res 182:200–206

Ren W, Neugebauer V (2010) Pain-related increase of excitatory transmission and decrease of inhibitory transmission in the central nucleus of the amygdala are mediated by mGluR1. Mol Pain 6:93

Ren W, Palazzo E, Maione S, Neugebauer V (2011) Differential effects of mGluR7 and mGluR8 activation on pain-related synaptic activity in the amygdala. Neuropharmacol 61:1334–1344

Riedel G, Casabona G, Platt B, Macphail EM, Nicoletti F (2000) Fear conditioning-induced time- and subregion-specific increase in expression of mGlu5 receptor protein in rat hippocampus. Neuropharmacol 39:1943–1951

Robbins MJ, Starr KR, Honey A, Soffin EM, Rourke C, Jones GA, Kelly FM, Strum J, Melarange RA, Harris AJ, Rocheville M, Rupniak T, Murdock PR, Jones DN, Kew JN, Maycox PR (2007) Evaluation of the mGlu8 receptor as a putative therapeutic target in schizophrenia. Brain Res 1152:215–227

Roder S, Ciriello J (1993) Innervation of the amygdaloid complex by catecholaminergic cell groups of the ventrolateral medulla. J Comp Neurol 332:105–122

Rodrigues SM, Bauer EP, Farb CR, Schafe GE, LeDoux JE (2002) The group I metabotropic glutamate receptor mGluR5 is required for fear memory formation and long-term potentiation in the lateral amygdala. J Neurosci 22:5219–5229

Rogan MT, Stäubli UV, LeDoux JE (1997) Fear conditioning induces associative long-term potentiation in the amygdala. Nature 390:604–607

Romano C, Sesma MA, McDonald CT, O'Malley K, Van den Pol AN, Olney JW (1995) Distribution of metabotropic glutamate receptor mGluR5 immunoreactivity in rat brain. J Comp Neurol 355:455–469

Ronesi JA, Huber KM (2008) Metabotropic glutamate receptors and fragile x mental retardation protein: partners in translational regulation at the synapse. Sci Signal 1:6

Rudy JW, Matus-Amat P (2009) DHPG activation of group 1 mGluRs in BLA enhances fear conditioning. Learn Mem 16:421–425

Sah P, Faber ES, De L, Armentia M, Power J (2003) The amygdaloid complex: anatomy and physiology. Physiol Rev 83:803–834

Sanacora G, Zarate CA, Krystal JH, Manji HK (2008) Targeting the glutamatergic system to develop novel, improved therapeutics for mood disorders. Nat Rev Drug Discov 7:426–437

Sanford LD, Parris B, Tang X (2002) GABAergic regulation of the central nucleus of the amygdala: implications for sleep control. Brain Res 956:276–284

Saugstad JA, Kinzie JM, Shinohara MM, Segerson TP, Westbrook GL (1997) Cloning and expression of rat metabotropic glutamate. Receptor 8 reveals a distinct pharmacological profile. Mol Pharmacol 51:119–125

Schmid S, Fendt M (2006) Effects of the mGluR8 agonist (S)-3, 4-DCPG in the lateral amygdala on acquisition/expression of fear-potentiated startle, synaptic transmission, and plasticity. Neuropharmacol 50:154–164

Schneider F, Weiss U, Kessler C, Salloum JB, Posse S, Grodd W, Müller-Gärtner HW (1998) Differential amygdala activation in schizophrenia during sadness. Schizophr Res 34:133–142

Schulz B, Fendt M, Gasparini F, Lingenhöhl K, Kuhn R, Koch M (2001) The metabotropic glutamate receptor antagonist 2-methyl-6-(phenylethynyl)-pyridine (MPEP) blocks fear conditioning in rats. Neuropharmacol 41:1–7

Schumann CM, Bauman MD, Amaral DG (2011) Abnormal structure or function of the amygdala is a common component of neurodevelopmental disorders. Neuropsychologia 49:745–759

See RE, Fuchs RA, Ledford CC, McLaughlin J (2003) Drug addiction, relapse, and the amygdala. Ann N Y Acad Sci 985:294–307

Sharma A, Hoeffer CA, Takayasu Y, Miyawaki T, McBride SM, Klann E, Zukin RS (2010) Dysregulation of mTOR signaling in fragile X syndrome. J Neurosci 30:694–702

Shekhar A, Keim SR (2000) LY354740, a potent group II metabotropic glutamate receptor agonist prevents lactate-induced panic-like response in panic-prone rats. Neuropharmacol 39:1139–1146

Shigemoto R, Nakanishi S, Mizuno N (1992) Distribution of the mRNA for a metabotropic glutamate receptor (mGluR1) in the central nervous system: an in situ hybridization study in adult and developing rat. J Comp Neurol 322:121–135

Shigemoto R, Kinoshita A, Wada E, Nomura S, Ohishi H, Takada M, Flor PJ, Neki A, Abe T, Nakanishi S, Mizuno N (1997) Differential presynaptic localization of metabotropic glutamate receptor subtypes in the rat hippocampus. J Neurosci 17:7503–7522

Siegl S, Flor PJ, Fendt M (2008) Amygdaloid metabotropic glutamate receptor subtype 7 is involved in the acquisition of conditioned fear. Neuroreport 19:1147–1150

Sinclair CM, Cleva RM, Hood LE, Olive MF, Gass JT (2012) mGluR5 receptors in the basolateral amygdala and nucleus accumbens regulate cue-induced reinstatement of ethanol-seeking behavior. Pharmacol Biochem Behav 101:329–335

Sławińska A, Wierońska JM, Stachowicz K, Pałucha-Poniewiera A, Uberti MA, Bacolod MA et al (2013) Anxiolytic- but not antidepressant-like activity of Lu AF21934, a novel, selective positive allosteric modulator of the $mGlu_4$ receptor. Neuropharmacol 66:225–235
Spanka C, Glatthar R, Desrayaud S, Fendt M, Orain D, Troxler T, Vranesic I (2010) Piperidyl amides as novel, potent and orally active mGlu5 receptor antagonists with anxiolytic-like activity. Bioorg Med Chem Lett 20:184–188
Spooren W, Gasparini F (2004) mGlu5 receptor antagonists: a novel class of anxiolytics? Drug News Perspect 17:251–257
Spooren WP, Vassout A, Neijt HC, Kuhn R, Gasparini F, Roux S, Porsolt RD, Gentsch C (2000) Anxiolytic-like effects of the prototypical metabotropic glutamate. Receptor 5 antagonist 2-methyl-6-(phenylethynyl) pyridine in rodents. J Pharmacol Exp Ther 295:1267–1275
Spooren W, Lesage A, Lavreysen H, Gasparini F, Steckler T (2010) Metabotropic glutamate receptors: their therapeutic potential in anxiety. Curr Top Behav Neurosci 2:391–413
Stachowicz K, Kłak K, Kłodzińska A, Chojnacka-Wojcik E, Pilc A (2004) Anxiolytic-like effects of PHCCC, an allosteric modulator of mGlu4 receptors, in rats. Eur J Pharmacol 498:153–156
Stachowicz K, Kłak K, Pilc A, Chojnacka-Wójcik E (2005) Lack of the antianxiety-like effect of (S)-3, 4-DCPG, an mGlu8 receptor agonist, after central administration in rats. Pharmacol Rep 57:856–860
Stachowicz K, Gołembiowska K, Sowa M, Nowak G, Chojnacka-Wójcik E, Pilc A (2007) Anxiolytic-like action of MTEP expressed in the conflict drinking Vogel test in rats is serotonin dependent. Neuropharmacol 53:741–748
Stachowicz K, Branski P, Klak K, van der Putten H, Cryan JF, Flor PJ et al (2008) Selective activation of metabotropic G-protein-coupled glutamate 7 receptor elicits anxiolytic-like effects in mice by modulating GABAergic neurotransmission. Behav Pharmacol 19:597–603
Steckler T, Lavreysen H, Oliveira AM, Aerts N, Van Craenendonck H, Prickaerts J, Megens A, Lesage AS (2005) Effects of mGlu1 receptor blockade on anxiety-related behaviour in the rat lick suppression test. Psychopharmacol 179:198–206
Stein MB, Simmons AN, Feinstein JS, Paulus MP (2007) Increased amygdala and insula activation during emotion processing in anxiety-prone subjects. Am J Psychiatry 164:318–327
Sukoff-Rizzo SJ, Leonard SK, Gilbert A, Dollings P, Smith DL, Zhang MY et al (2011) The metabotropic glutamate receptor 7 allosteric modulator AMN082: a monoaminergic agent in disguise? J Pharmacol Exp Ther 338:345–352
Suvrathan A, Chattarji S (2011) Fragile X syndrome and the amygdala. Curr Opin Neurobiol 21:509–515
Suvrathan A, Hoeffer CA, Wong H, Klann E, Chattarji S (2010) Characterization and reversal of synaptic defects in the amygdala in a mouse model of fragile X syndrome. Proc Natl Acad Sci U S A 107:11591–11596
Swanson LW, Petrovich GD (1998) What is the amygdala? Trends Neurosci 21:323–331
Swanson CJ, Bures M, Johnson MP, Linden AM, Monn JA, Schoepp DD (2005) Metabotropic glutamate receptors as novel targets for anxiety and stress disorders. Nat Rev Drug Discov 4:131–144
Takeuchi Y, McLean JH, Hopkins DA (1982) Reciprocal connections between the amygdala and parabrachial nuclei: ultrastructural demonstration by degeneration and axonal transport of horseradish peroxidase in the cat. Brain Res 239:583–588
Tamaru Y, Nomura S, Mizuno N, Shigemoto R (2001) Distribution of metabotropic glutamate receptor mGluR3 in the mouse CNS: differential location relative to pre- and postsynaptic sites. Neuroscience 106:481–503
Tang X, Yang L, Liu X, Sanford LD (2005) Influence of tetrodotoxin inactivation of the central nucleus of the amygdala on sleep and arousal. Sleep 28:923–930
Tatarczyńska E, Klodzińska A, Chojnacka-Wójcik E, Palucha A, Gasparini F, Kuhn R, Pilc A (2001) Potential anxiolytic- and antidepressant-like effects of MPEP, a potent, selective and systemically active mGlu5 receptor antagonist. Br J Pharmacol 132:1423–1430
Toth I, Dietz M, Peterlik D, Huber SE, Fendt M, Neumann ID, Flor PJ, Slattery DA (2012) Pharmacological interference with metabotropic glutamate receptor subtype 7 but not subtype

5 differentially affects within- and between-session extinction of Pavlovian conditioned fear. Neuropharmacol 62:1619–1626

Uejima JL, Bossert JM, Poles GC, Lu L (2007) Systemic and central amygdala injections of the mGluR2/3 agonist LY379268 attenuate the expression of incubation of sucrose craving in rats. Behav Brain Res 181:292–296

Ugolini A, Large CH, Corsi M (2008) AMN082, an allosteric mGluR7 agonist that inhibits afferent glutamatergic transmission in rat basolateral amygdala. Neuropharmacol 55:532–536

Varty GB, Grilli M, Forlani A, Fredduzzi S, Grzelak ME, Guthrie DH, Hodgson RA, Lu SX, Nicolussi E, Pond AJ, Parker EM, Hunter JC, Higgins GA, Reggiani A, Bertorelli R (2005) The antinociceptive and anxiolytic-like effects of the metabotropic glutamate receptor 5 (mGluR5) antagonists, MPEP and MTEP, and the mGluR1 antagonist, LY456236, in rodents: a comparison of efficacy and side-effect profiles. Psychopharmacol 179:207–217

Walker DL, Davis M (2002) The role of amygdala glutamate receptors in fear learning, fear-potentiated startle, and extinction. Pharmacol Biochem Behav 71:379–392

Wang SJ, Gean PW (1999) Long-term depression of excitatory synaptic transmission in the rat amygdala. J Neurosci 19:10656–10663

Waung MW, Huber KM (2009) Protein translation in synaptic plasticity: mGluR-LTD, Fragile X. Curr Opin Neurobiol 19:319–326

Whiteford HA, Degenhardt L, Rehm J, Baxter AJ, Ferrari AJ, Erskine HE, Charlson FJ, Norman RE, Flaxman AD, Johns N, Burstein R, Murray CJ, Vos T (2013) Global burden of disease attributable to mental and substance use disorders: findings from the global burden of disease study 2010. Lancet 382:1575–1586

Whittle N, Schmuckermair C, Gunduz Cinar O, Hauschild M, Ferraguti F, Holmes A, Singewald N (2013) Deep brain stimulation, histone deacetylase inhibitors and glutamatergic drugs rescue resistance to fear extinction in a genetic mouse model. Neuropharmacol 64:414–423

Wierońska JM, Pilc A (2009) Metabotropic glutamate receptors in the tripartite synapse as a target for new psychotropic drugs. Neurochem Int 55:85–97

Wierońska JM, Pilc A (2013) Glutamate-based anxiolytic ligands in clinical trials. Expert Opin Investig Drugs 22:1007–1022

Wierońska JM, Stachowicz K, Pałucha-Poniewiera A, Acher F, Brański P, Pilc A (2010) Metabotropic glutamate receptor 4 novel agonist LSP1-2111 with anxiolytic, but not antidepressant-like activity, mediated by serotonergic and GABAergic systems. Neuropharmacol 59:627–634

Womble MD, Moises HC (1994) Metabotropic glutamate receptor agonist ACPD inhibits some, but not all, muscarinic-sensitive K+ conductances in basolateral amygdaloid neurons. Synapse 17:69–75

Wright RA, Arnold MB, Wheeler WJ, Ornstein PL, Schoepp DD (2001) [3H]LY341495 binding to group II metabotropic glutamate receptors in rat brain. J Pharmacol Exp Ther 298:453–460

Xu J, Zhu Y, Contractor A, Heinemann SF (2009) mGluR5 has a critical role in inhibitory learning. J Neurosci 29:3676–3684

Yasoshima Y, Morimoto T, Yamamoto T (2000) Different disruptive effects on the acquisition and expression of conditioned taste aversion by blockades of amygdalar ionotropic and metabotropic glutamatergic receptor subtypes in rats. Brain Res 869:15–24

Zhu PJ, Lovinger DM (2005) Retrograde endocannabinoid signaling in a postsynaptic neuron/synaptic bouton preparation from basolateral amygdala. J Neurosci 25:6199–6207

Index

R.T. Ngomba et al. (eds.), *mGLU Receptors*, The Receptors 31,
DOI 10.1007/978-3-319-56170-7

CPSIA information can be obtained
at www.ICGtesting.com
Printed in the USA
LVHW000030100122
653954LV00003B/263